W9-CRB-241

CHEMICAL REACTIONS IN COMPLEX MIXTURES

THE MOBIL WORKSHOP

CHEMICAL REACTIONS IN COMPLEX MIXTURES

THE MOBIL WORKSHOP

Edited by

AJIT V. SAPRE

Mobil Research and Development Corporation
Princeton, New Jersey

FREDERICK J. KRAMBECK

Mobil Research and Development Corporation
Paulsboro, New Jersey

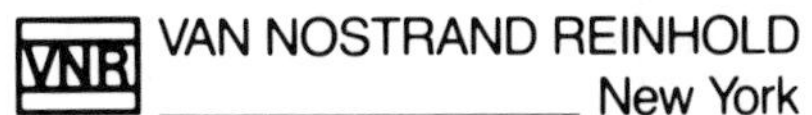

Library of Congress Catalog Card Number 90-23722
ISBN 0-442-00725-6

Printed in the United States of America.

Van Nostrand Reinhold
115 Fifth Avenue
New York, New York 10003

Chapman and Hall
2–6 Boundary Row
London, SE1 8HN, England

Thomas Nelson Australia
102 Dodds Street
South Melbourne 3205
Victoria, Australia

Nelson Canada
1120 Birchmount Road
Scarborough, Ontario MIK 5G4, Canada

16 15 14 13 12 11 10 9 8 7 6 5 4 3 2 1

Library of Congress Cataloging-in-Publication Data

Mobil Workshop on Chemical Reactions in Complex Mixtures (1990: Paulsboro, N.J.)
Chemical reactions in complex mixtures: the Mobil Workshop/ edited by Ajit V. Sapre and Frederick J. Krambeck.
p. cm.
Includes bibliographical references and index.
ISBN 0-442-00725-6
1. Chemical processes—Congresses. 2. Chemical reaction, Conditions and laws of—Congresses. I. Sapre, Ajit V. II. Krambeck, Frederick J. III. Title.
TP155.7.M617 1991
660′.299—dc20 90-23722
CIP

Contents

PART 2 STRUCTURE–ACTIVITY RELATIONSHIPS

PART 3 KINETIC ANALYSIS

PART 4 THERMODYNAMICS

Chapter Summaries

Chemical Reaction Engineering of Multicomponent Mixtures: Open Problems 3

A number of open problems in the chemical reaction engineering of multicomponent mixtures are considered; attention is restricted to the continuous description of such mixtures. The open problems are tentatively classified in the following areas: the description of kinetics; the effect of reactor configuration and mixing conditions; the possible complexity of the topology of reaction networks; selectivity problems in multicomponent mixtures; and reactor stability. A few specific problems in these areas are discussed in some detail. These are problems that still await solution, and only a preliminary sketch of possible ways of attack is presented, as well as the identification of the intrinsic difficulties of each problem.

Multiple Indices, Simple Lumps, and Duplicitous Kinetics 25

This chapter is confined to the mathematical theory of first-order batch or tubular reactions in a complex mixture describable by n indices in which the ultimate components decay by a strictly first-order reaction. It is shown how the general case may be reduced to a single index and hence allow the resources of Laplace transform theory to become available. A continuous mixture (though not necessarily a positive one) can be found to fit an arbitrary kinetic behavior, and the kinetic law can be obtained from observations of the curriculum of the reaction. The case of reversible first-order reactions is also considered.

Continuous Mixtures in Fluid Catalytic Cracking and Extensions 42

The fluid catalytic cracking (FCC) process is a classic example of a system involving a very large number of individual chemical compounds that can be treated by a continuum approach. The early observation that overall conversion can be approximately described by second-order kinetics is elucidated in terms of a continuum analysis and extended to give predictions of the detailed boiling point distribution. The effect of nonideal mixing on selectivity is also analyzed. Interestingly, catalyst deactivation in FCC can be elucidated by a similar approach.

It is then shown how continuous mixture analysis can be extended to systems with more complicated stoichiometry and reversible reactions. The concepts are illustrated with simple examples.

Lumping or Pseudocomponent Identification in Phase Equilibrium Calculations 60

Phase equilibrium calculations for mixtures of many components, such as reservoir fluids or polymer solutions, or even the weathering of oil spills, present special difficulties because the computation time increases with approximately the square of the number of components. The problem is especially severe for intensive calculations such as reservoir simulations. To decrease the computational effort, larger numbers of species are usually lumped into smaller numbers of pseudocomponents. Traditionally, this lumping has been done empirically, by trial-and-error. This chapter will demonstrate the simplicity and success of a lumping (or pseudocomponent identification) method based on well-founded mathematical principles for the calculation of reservoir fluid bubble points and the modeling of enhanced oil recovery by miscible gas injection.

Fundamental Kinetic Modeling of Complex Processes 77

The feedstocks processed in petroleum refining and in many petrochemical operations consist of a large number of components, each reacting according to complicated pathways. Selectivity is often an important feature of these commercial processes. In kinetic modeling the actual reaction network is frequently reduced to a single overall reaction or to a limited number of reactions. Excessive lumping may lead to parameters that depend on the feed composition and even on the type of reactor in which they were determined. As a consequence, extensive experimentation is required for each new feedstock.

The present chapter advocates a fundamental approach in which the detailed reaction network of the feed components is retained to a maximum extent and the kinetics are developed in terms of the elementary steps involved in the reactions. True, for a single component the number of rate parameters is significantly higher than with the usual lumping approach, but their fundamental nature gives them a much wider validity: they do not depend on the chain length and are not affected by a different feed composition. The true benefit of the approach becomes clear when mixtures are dealt with. In that case the lumping approach needs an increasing number of lumps and parameters to retain the significance and accuracy of the modeling. The chapter illustrates to what level of detail the network has to be developed to come to really invariant parameters. Two examples of complete processes will be dealt with. The first is a homogeneous gas phase process: the thermal cracking of naphtha for olefins production. The second is the gas phase hydrocracking of paraffins on a zeolite catalyst loaded with a noble metal.

Structural Models of Catalytic Cracking Chemistry: A Case Study of a Group Contribution Approach to Lumped Kinetic Modeling **101**

This chapter describes a new approach for modeling the kinetics and estimating the properties of petroleum fractions. In this approach, based on chemical structure, several hundred model compounds are used to represent the reactions and properties of the petroleum fraction. The compounds and their relative concentrations are chosen based on chemical and structural data routinely available for fuel fractions. Group contribution methods are used to estimate the reaction rates, reaction pathways, and the properties of both the feed material and the reaction products. Catalytic cracking has been used as a case study for this approach. Based on a series of model compound studies, we established group contribution parameters for cracking rate constants on an amorphous catalyst. These group contributions were successfully used in predicting the reactions of other model compounds and a simulated oil mixture. The output of the kinetic model is used to predict not only fraction yields, but also the properties of each fraction. Sample calculations show the sensitivity of gasoline-octane number yield to the values of cracking rate constants.

Monte Carlo Modeling of Complex Reaction Systems: An Asphaltene Example **126**

This work views the traditional practice of lumping (species aggregation into fewer "pseudo" species) in reaction engineering models in terms of "early" or "late" mixing of information. The concept of late lumping is illustrated via a molecular-structure explicit model for asphaltene thermolysis. This model retains the detailed molecular information throughout the reaction. Lumping only after completion of the reaction simulation provides commercially relevant solubity classes: coke, asphaltene, maltene, and gases. A solution thermodynamics basis for product identification is detailed. Solid–liquid–vapor equilibrium calculations are performed to determine the yields and structural attributes of the solubility classes.

Light olefins in the propene to hexene range react and convert over the ZSM-5 zeolite catalyst to higher molecular weight olefins through a sequence of oligomerization, isomerization, and cracking reactions. These reactions form the basis of the Mobil Olefins to Gasoline and Distillate (MOGD) process and occur readily in the temperature range of 150 to 450 °C. The resulting olefin product carbon number distribution ranges from 2 to over 50 and is very dependent on process conditions and configurations. A kinetic process model incorporating the complex sequence of olefin reactions was developed to assist in the design development of the MOGD process.

Even though the basic concepts of lumping in monomolecular systems have been understood for some years, including continuous mixtures, for other systems several issues remain. One is how to choose the lumps, in practice normally done by trial-and-error requiring a large amount of data and/or chemical knowledge. Another issue is the effect of data error on lumping. Next, there is the question of Wei–Kuo Exact Lumping versus other schemes. The translation of lumping schemes from one reactor type to another has been little explored. Finally, the definition and behavior of "Approximate Wei–Kuo Exact Lumping" needs to be investigated for application to real systems. Each of these topics will be discussed, with the current status and possible future direction emphasized.

Chemical engineers are required to confront an enormous variety of complex chemical systems, each with its own set of chemical species and its own network of chemical reactions. The governing differential equations change markedly from one reactor to the next, but they derive in a rather precise way from the underlying reaction network. The aim of *chemical reaction network theory* is to draw connections between reaction network structure and qualitative properties of the corresponding differential equations. Recent results make possible quick but penetrating inferences about very complicated chemical systems, even by chemists and engineers without much mathematical training. Some of these results are surveyed, in particular those that bear on mechanism discrimination in heterogeneous catalysis.

Development of chemical reaction models involves a number of steps: compilation of a specific reaction network, collection and estimation of reaction rate parameters, analysis and verification of the composed reaction mechanism, and reduction of the model to the form and size suitable for practical applications. The first two parts rely on chemical insights and intuition of the modeler. The other two require rigorous mathematical methods. The current needs in various fields of science and engineering require development of realistic models with a detailed coupling of gas dynamics and chemical kinetics, optimization of complex models, and reduction in size and computational speed of these reactive-flow models so they can fit a design-engineer work station. The main focus of the chapter is on several methods developed for application to model development and mechanism reduction. Examples from the areas of hydrocarbon pyrolysis and combustion, chemical vapor deposition and aerosol reactors are included.

Kinetic Modeling at Mobil: An Historical Perspective 222

In the midst of the pursuit of progress in a given research field, it is sometimes fruitful to pause occasionally and take an historical stock of what has been accomplished so far. Although the focus in this chapter is restricted mainly to the kinetic modeling efforts at Mobil, it gives a flavor of what we believe to be a realistic picture of the industry as a whole. We will give a brief account of Mobil's FCC and reforming modeling technology developed over the last three decades and highlight key milestones. Also, practical aspects of integration of the kinetic modeling technology with the overall process model for commercial plant monitoring and optimization will be discussed. With advances in analytical chemistry and computer technology, the future of developing accurate kinetic models for complex feedstocks is very bright. Some of the challenges facing us will be highlighted.

Uses and Needs of Thermodynamics in the Oil Industry 257

Uses and needs of thermodynamics in the oil industry are discussed. The driving forces behind this key activity are analyzed. The objectives of "lumping" of mixtures are defined. The areas of interest in developing thermophysical data bases of pure compounds are explored. The importance of "isomer lumping" is emphasized as well as the construction of data bases of thermophysical properties for "isomer lumps." Concentration of "lumps" and their thermophysical data base is one of the most important issues confronting many of today's industrial researchers. Depending on the objectives of the lumping and on how much information one has accumulated on the lumped mixtures, several lumping approaches are suggested and discussed. A strong relationship between some thermophysical properties and isomer molecular structures, particularly those with high degree of molecular symmetry, is emphasized. Special attention is also given to the lumping of mixtures related to certain isomerization reactions. With great progress in both analytical and computational capabilities taking place today and expected in the future, one should be able to construct a thermophysical data base of any isomer lump starting with isomer enumeration and molecular structure elucidation using expert systems. Five important but nonlumping areas concerning complex mixtures are also discussed.

In calculating equilibrium compositions of chemically reacting complex organic systems, the number of species to be considered can be greatly reduced by use of isomer groups. This also makes it possible to extrapolate to higher carbon numbers where data are lacking. Another way to simplify calculations is to use pressures of reactants as independent variables. For example, in considering the benzene series of polycyclic aromatic hydrocarbons in the presence of acetylene and molecular hydrogen, the Gibbs energies of the successive isomer groups can be adjusted to the specified partial pressures of acetylene and hydrogen to obtain $\Delta_f G_n^*$ and can be used to calculate the equilibrium distribution of isomer groups within the homologous series. Because the standard Gibbs energy of formation of an isomer group in the benzene series is linear in carbon number n, $\Delta_f G_n^\circ = A + Bn$, the equilibrium distribution of isomer groups at specified partial pressures of acetylene and hydrogen is given by a geometric distribution that depends only on B. On the other hand, the equilibrium distribution of isomer groups at a specified partial pressure of molecular hydrogen is given by a geometric distribution that depends only on A. To illustrate various choices of independent variables and the effect that a mechanism can have on the equilibrium composition, the benzene series of polycyclic aromatic hydrocarbons in the presence of acetylene and molecular hydrogen is discussed for six choices of independent variables. In three of the cases, the equilibrium distributions can also be calculated with geometric distributions because the Gibbs energy of formation in the homologous series is linear in carbon number.

Recent developments related to the calculation of chemical equilibrium in complex mixtures are reviewed, and current and future research directions are discussed. Topics considered include alternative mathematical formulations of the problem, the structure of numerical algorithms, sensitivity analysis, stoichiometric restrictions, and equilibrium constraints. Some commercially available microcomputer-based software packages for performing equilibrium calculations are also discussed.

Foreword: Welcoming Remarks

I would like to welcome you all. We at Mobil Research and Development Corporation are honored to sponsor this Workshop on Chemical Reactions in Complex Mixtures. Mobil, based on the work of Dwight Prater, Paul Weisz, Jimmy Wei, and a number of others through the years, has developed a substantial history and culture at our laboratories in reaction kinetics, and kinetic models for our processes. Although some of the systems are old and a lot of work has been done in the past, a lot more is yet to be explored. The technology in this area is advancing very rapidly.

The only commercial message that you will hear in this Workshop is from me. From the business side, we see a tremendous need for improved understanding of complex reaction mixtures. Therefore, technology in this area is critical. You may be aware that Mobil, with thirteen other oil companies, has joined the three major U.S. automobile manufacturers to go into what is called the auto/oil program for air quality improvement. Phase I of the program will cost about $15 million and has just started. We hope it will be completed this year. Basically it is driven by the demands to improve our air quality and is directed to develop data on effects of fuel composition on emissions from automobiles. The reason this program and this Workshop are important to Mobil is that we sell complex mixtures, fuels, based on their physical properties. Rapidly that is changing. We will continue to sell complex mixtures, but they will be mixtures of chemicals with known composition driven by emission regulations. Petroleum refiners of today will become petrochemical manufacturers of tomorrow. We need the technology that you will be discussing in kinetic modeling, coupled with the analytical techniques that are developing very rapidly and process optimization and control technologies to produce these new fuels. We will have to do this with limited capital and with environmental constraints. We will not be able to build as many new refineries, especially in the United States. We will have to do better with what we already have. We hope the technology we will discuss in the next two days will provide us guidance to meet tomorrow's challenges in an economical and efficient fashion.

I am very pleased to host this Workshop. I certainly appreciate the presence of so many eminent representatives of the academic profession. We look forward to this exchange between academia and industry. This event will certainly strengthen the relationship between the universities and Mobil. I once again appreciate your being here, and look foward to a very stimulating and productive Workshop.

H. A. McVeigh
Manager
Paulsboro Research Laboratory

LIST OF PARTICIPANTS

Invited Speakers

Bob Alberty, MIT
David Allen, UCLA
Gianni Astarita, University of Delaware/Naples
Rutherford Aris, University of Minnesota
Ken Bischoff, University of Delaware
Martin Feinberg, University of Rochester
Michael Frenklach, Pennsylvania State University
Gilbert Froment, University of Ghent
Michael Klein, University of Delaware
Stan Sandler, University of Delaware
Bill Smith, University of Guelph

Panel Discussion Coordinators

Dan Luss, University of Houston
Dwight Prater, California Institute of Technology and University of Nevada
Reuel Shinnar, City College of New York
James Wei, Massachusetts Institute of Technology

Mobil Attendees

PRL
P. Adornato
D.H. Anderson
A.A. Avidan
N.A. Bhore
B. Choi
N. Collins
P.P. Durand
K.R. Graziani
R. Hu
S.M. Jacob
S.B. Jaffe
D.V. Jorgensen
C.R. Kennedy
F.J. Krambeck
J.C.W. Kuo
J.S. Liou
H.A. McVeigh
T. Mo
J.F. Quanci
R.J. Quann
A.V. Sapre
S. Sankhe
S. Shih
C.M. Sorensen
T.R. Stein

CRL
N.Y. Chen
H.J. Schoennagel
P.B. Venuto
V. W. Weekman. Jr.

Engineering
P.H. Schipper

DRL
J.A. Pita

Preface

In recent years there has been a convergence of trends in chemical reaction engineering and chemistry which have set the stage for significant advances in kinetic and thermodynamic modeling of processes. New analytical chemistry methods, new mathematical methods, and new computational tools facilitate a more fundamental approach and a deeper understanding of chemical reactions in complex mixtures with very large numbers of compounds, such as petroleum fractions. This fortunate state of affairs has stimulated important new work both in academia and industrial research labs. The purpose of the workshop that led to this book was to bring together researchers at the forefront of this field to review the state of the art, stimulate communication and cooperation between industry and academia, and develop a cohesive picture of research trends and future directions.

The chapters of the book have been organized into four main areas:

- *Continuous mixtures*, where the very large numbers of discrete compounds present are regarded as making up a continuum,
- *Structure–activity relationships*, where the nature and rates of the reactions that a particular molecule undergoes are correlated with its chemical structure, thus allowing the kinetics of very large numbers of compounds to be described by a few parameters,
- *Kinetic analysis*, where mathematical techniques are applied to analyze the behavior of kinetic networks, and
- *Thermodynamics*, emphasizing the practical and computational aspects of chemical equilibrium in complex mixtures.

Taking an overall perspective on these areas makes it very clear that they are complementary views on a single unified subject and all have a contribution to make in progressing our understanding of it.

Through the discussions between the authors and others at the workshop it became very clear that there is a sense of excitement about this field of study, a feeling that a new direction in research is just beginning and that it will be both theoretically fruitful and industrially valuable.

Contributors

Robert Alberty, Department of Chemistry, Massachusetts Institute of Technology, Cambridge, MA 02139

David T. Allen, Department of Chemical Engineering, University of California, Los Angeles, CA 90024

Rutherford Aris, Department of Chemical Engineering and Materials Science, University of Minnesota, Minneapolis, MN 55455

Gianni Astarita, Dipartimento di Ingegneria Chimica, University di Napoli, Plazzale Tecchio, 80125 Napoli, Italy

Kenneth B. Bischoff, Center for Catalytic Science and Technology, Department of Chemical Engineering, University of Delaware, Newark, DE 19716

Martin R. Feinberg, Department of Chemical Engineering, University of Rochester, Rochester, NY 14627

Michael Frenklach, Department of Materials Science and Engineering, The Pennsylvania State University, University Park, PA 16802

G. F. Froment, Laboratorium voor Petrochemische Techniek, University of Ghent, Rijksuniversiteit, Krijgslaan 281, B9000 Gent, Belgium

Michael T. Klein, Center for Catalytic Science and Technology, Department of Chemical Engineering, University of Delaware, Newark, DE 19716

F. J. Krambeck, Mobil Research and Development Corporation, Paulsboro, NJ 08006-0480

James C. W. Kuo, Mobil Research and Development Corporation, Paulsboro, NJ 08066-0480

Cristian Libanati, Center for Catalytic Science and Technology, Department of Chemical Engineering, University of Delaware, Newark, DE 19716

Dimitris Liguras, Department of Chemical Engineering, University of Delaware, Newark, DE 19716

Matthew Neurock, Center for Catalytic Science and Technology, Department of Chemical Engineering, University of Delaware, Newark, DE 19716

Abhash Nigam, Center for Catalytic Science and Technology, Department of Chemical Engineering, University of Delaware, Newark, DE 19716

Richard J. Quann, Mobil Research and Development Corporation, Paulsboro, NJ 08066-0480

Stanley I. Sandler, Department of Chemical Engineering, University of Delaware, Newark, DE 19716

Ajit V. Sapre, Mobil Research and Development Corporation, Princeton, NJ 08543

William R. Smith, Department of Mathematics and Statistics, University of Guelph, Guelph, Ontario, Canada N1G 2W1

PART 1

CONTINUOUS MIXTURES

1

Chemical Reaction Engineering of Multicomponent Mixtures: Open Problems

GIANNI ASTARITA

The continuous description of multicomponent mixtures where all reactants undergo irreversible first-order reactions (such as is expected to be the case in thermal cracking) has been presented in the classic paper by Aris (1968). Extension of this type of analysis to the case where the individual chemical reactions are not governed by first-order kinetics turned out to reserve subtle difficulties, which were elucidated only 20 years later by Ho and Aris (1987). After the publication of the latter article, there has been widespread renewed interest in the subject, and a number of articles addressing the issue have been published in the last 2 years (Aris 1989; Aris and Astarita 1989, 1990; Astarita 1989, 1990; Astarita and Nigam 1989; Astarita and Ocone 1988, 1989a, 1989b, 1990; Chou and Ho 1988, 1989; Ho 1990; Hu and Ho 1990; Li and Ho 1990; Li and Rabitz 1989; White and Ho 1990).

What understanding is available today on the continuous description of mixtures with intrinsically nonlinear kinetics may perhaps be compared to that in the early 1950s on single reactant systems. In somewhat simplified terms, one may say that, at least under appropriate qualifying conditions, *we understand how to describe the kinetics, but we still need to develop the chemical reaction engineering of such systems*. This chapter—in keeping with the spirit of this volume—is meant to identify areas of ignorance (and hence of possibly fruitful research). I do not in any way try to be exhaustive; completely covering some area of knowledge is a difficult task, but completely covering some area of ignorance is a hopeless one. I therefore restrict attention to those open problems for which I have at least been able to identify an initial way of attack—sometimes in a very preliminary form.

The chapter is organized as follows. First, a very brief review of available results is given, which tries to stress the essential conceptual issues rather than concentrating on details. Next, some classes of open problems are identified. The body of the chapter is then dedicated to the discussion of preliminary ideas on how to approach some specific problems within such classes. Finally, some issues are distilled and presented for consideration.

BACKGROUND AND NOTATION

This section is not meant to be a review of the continuous description of nonlinear kinetics, for which the reader is referred to the review by Aris and Astarita (1989). I simply set down the background and notation used in subsequent sections, and try to isolate a few definitions and concepts that play a crucial role in the following.

As for notation, all concentrations are normalized with respect to the total lumped concentration in the feed and are dimensionless. Individual species are identified by a label x, ranging in some appropriate range, and integrals are intended over that range unless otherwise indicated. When the label is to be regarded as a dummy one, it will be indicated with y. The concentration distribution in the feed is $G(x)$, in the sense that $G(x)\,dx$ is the (dimensionless) concentration in the feed of species with labels between x and $x + dx$. The label is always rescaled, and the distribution $G(x)$ normalized, so that

$$\int G(y)\,dy = \int yG(y)\,dy = 1. \tag{1–1}$$

Except when otherwise noted, $G(x)$ will be taken to be a gamma distribution:

$$G(x) = \frac{\alpha^{\alpha} x^{\alpha-1} e^{-\alpha x}}{\Gamma(\alpha)}, \quad \alpha > 0. \tag{1–2}$$

Given any property $Q(x)$ defined for every species in the mixture, its initial average value is indicated with Q^*:

$$Q^* = \int Q(y)G(y)\,dy. \tag{1–3}$$

Given a frequency factor $k(x)$ [which can be identified under mild conditions of well-behavedness near equilibrium (Astarita and Ocone 1988)], x

should have been rescaled so that $k(x) = k^*x$. Hence all time variables are made dimensionless multiplying them by k^*; in particular, the residence time in a batch reactor is indicated with $\tau = k^*t$. Whenever a time-type variable is to be regarded as a dummy one, it will be indicated with u.

In a batch reactor, which is the one considered when developing the description of the kinetics, the concentration at time τ of species with labels between x and $x + dx$ is $g(x, \tau)\, dx$, with $g(x, 0) = G(x)$. The total lumped concentration at time τ, $C(\tau)$, is

$$C(\tau) = \int g(y, \tau)\, dy, \quad C(0) = 1. \tag{1–4}$$

The rate of reaction of component x at time τ, $r(x, \tau)$, is defined as the rate of change of $g(x, \tau)$ in a batch reactor:

$$r(x, \tau) = -\frac{\partial g(x, \tau)}{\partial \tau} \tag{1–5}$$

and the question that has been addressed in the recent literature is the problem of how to express $r(x, \tau)$ when it is not linear in $g(x, \tau)$. I now identify two main lines of thought: independent kinetics and interactive kinetics.

Independent Kinetics

The assumption is that $r(x, \tau)$ depends *only* on $g(x, \tau)$—each reactant goes its own separate way, independently of what other reactants may be present and in what concentration. Apart from the mathematical difficulties of describing such a situation, I have difficulties envisioning any realistic chemical mechanism where this is likely to be true—except in the case of intrinsically first-order kinetics. The difficulty can be concisely presented in the following way. The *specific* rate of reaction, $r(x, \tau)/g(x, \tau)$, represents the probability that any given molecule of species x may undergo the reaction. Now if that probability is a constant, then one has first-order kinetics. But if it is not a constant, it means that the presence of other molecules affects it. If that is the case, and if one has very many reactants of similar chemical structure and reactivity, it is hard to justify the assumption that species x reacts at a specific rate that is independent of the concentration of any species except x itself. This is, admittedly, a heuristic argument (somebody might call it a handwaving one). A possibly stronger argument has been advanced by Froment (1991, this volume), since his detailed analysis of elementary steps in naphtha cracking clearly shows the fact that even elementary steps take place at rates that are affected by the environment. In any event, because I am not capable of analyzing a situation of which I do not understand the underlying physics, I will not discuss independent kinetics. The reader is referred to several articles by Ho and co-workers on this point; these articles deal with the issue by the "coordinate transformation" method introduced by Chou and Ho (1988). An alternate approach to the problem of describing independent nonlinear kinetics can be found in the two-labels formalism introduced by Aris (1989).

Interactive Kinetics

Here the assumption is that $r(x, \tau)$ may very well depend on the entire spectrum of concentrations of reactants $g(y, \tau)$. Given some weighting factor distribution $K(x, y)$ that measures the interaction of the generic species y with the specific species x considered, a very general description of interactive kinetics is (Astarita and Ocone 1988)

$$r(x, \tau) = xg(x, \tau)\mathfrak{f}[K(x, y)g(y, \tau)], \tag{1–6}$$

where $\mathfrak{f}[\ ]$ is a functional of the weighted concentration distribution $K(x, y)g(y, \tau)$. This is almost hopelessly complex for attacking the solution, and a very powerful assumption (but not necessarily a physically realistic one) is that the kinetics are *uniform*, namely, that $K(x, y)$ in fact depends only on y—so that the value of $\mathfrak{f}$ is the same for all components. In particular, the following special case has been often considered:

$$r(x, \tau) = xg(x, \tau)F\left[\int K(y)g(y, \tau)\, dy\right]. \tag{1–7}$$

with $F[\ ] = 1$ for the first-order case. The weighting factor distribution needs to be assigned, and the following one has been proposed (Astarita 1989):

$$K(x) = K^*(\alpha x)^\beta \Gamma(\alpha)/\Gamma(\alpha + \beta). \tag{1–8}$$

The distributions in Eqs. (1–2) and (1–8) have been called the α, β, Γ ones (Aris 1989).

Substitution of Eqs. (1–5) and (1–7) into Eq. (1–6) yields a quasilinear differential equation that has the following solution:

$$g(x, \tau) = G(x) \exp[-xw(\tau)], \quad w(0) = 0, \tag{1–9}$$

where the warped time w is the solution of

$$\frac{dw}{d\tau} = F\left[\int G(y)e^{-yw(\tau)}\, dy\right]. \tag{1–10}$$

With the α, β, Γ distributions, w can always be eliminated from the equations to obtain an explicit kinetic equation for the lumped concentration: given the function $F[\ \ ]$ and the values of α and β, one can calculate the *lumped kinetic function* $R(\ \ )$ such that:

$$-dC/d\tau = R(C). \tag{1–11}$$

Conversely, given $R(C)$—as possibly obtained from experiments—and the values of α and β, it is always possible to find the function $F[\ \ ]$ that will generate $R(\ \ )$.

IDENTIFICATION OF OPEN PROBLEMS

In this brief section, I try to present my views on what are the important problems in the area that still await solution. Five classes of problems are tentatively identified; these are problems in *analysis*, where the necessary information is regarded as being available. A more philosophical kind of problem is discussed at the end of this section, which has to do with the actual availability of the information needed for analysis.

The first class is that of *kinetics* itself. As very concisely discussed in the previous section, under appropriate simplifying assumptions we know how to deal with kinetic lumping in nonlinear systems. However, the simplifying assumptions are strong ones, particularly that of uniformity, and some analysis of cases where it is untenable is in order. I also believe (somewhat tentatively) that the description of *independent* kinetics, to within its being a significant problem for some systems, still needs some improvement over what has been obtained so far. A preliminary approach to a specific problem of nonuniform nonlinear kinetics is discussed in the next section.

The second class of problems is that of *reactor configurations and mixing conditions*. Having obtained a description of the lumped kinetics really means that we know how to deal with a batch reactor, or, equivalently, with a plug flow one. However, we are still at a loss concerning other reactor configurations. The problem one faces is the following one. Consider the very simple case of a continuous stirred tank reactor (CSTR) with a dimensionless residence time T. Should Eq. (1–11) be a *true* kinetic equation, the lumped concentration in the exit stream would be given by the solution of $TR(C) + C = 1$. However, Eq. (1–11) is not a true kinetic equation, but only the lumped kinetic equation applying to a batch reactor, and one cannot use it to solve the CSTR case. The solution of the CSTR has been discussed by Astarita and Nigam (1989), and it turns out that one has to work on a case-by-case basis—a lumped kinetic equation applying to the CSTR cannot be found in general. Of course the CSTR case is the simplest possible one, and more complex reactor configurations can be envisaged. Some progress toward the analysis of maximum mixedness reactors with arbitrary residence time distributions has been made (Astarita and Ocone 1990), and this will be briefly reviewed in the next section.

The third class of problems is that of the *topology of reaction networks*. In a system containing very many reactants, reaction networks of remarkable topological complexity are possible. First steps in this direction have been taken only for the case where the intrinsic kinetics are linear: Aris (1989) has considered the sequential case $A \rightarrow B \rightarrow C$, and Astarita and Ocone (1989a) the parallel case $C \leftarrow A \rightarrow B$. A tentative attack to a partly nonlinear sequential case is discussed in the next section.

The consideration of reaction networks leads to the consideration of the fourth class of problems, i.e., problems of *selectivity*—some of the intermediate products in the network may well be the desired ones. Selectivity in multicomponent mixtures may be quite tricky: the linear parallel case $C \leftarrow A \rightarrow B$ has no selectivity aspect for the single-component case (the ratio of the concentrations of the two products is constant in time), but in the mixture case the ratio of the lumped concentrations of B-type and C-type products may go through a minimum or a maximum as time proceeds (Astarita and Ocone 1989a). This is concisely discussed in the next section, where I also discuss an initial approach to the case where both (parallel) steps are governed by nonlinear kinetics, as well as another problem of selectivity in the sequential case.

Finally, the fifth class of problems is that of the *stability* of reactors fed with multicomponent reacting mixtures. Again a simple example clarifies the type of problem. For a Langmuir isotherm type of kinetics, with appropriate values of α and β, the lumped kinetic function has a negative apparent order at large values of C (Astarita 1989). Should the lumped kinetic function represent a true kinetics, this would imply that a CSTR could, for some residence times and feed concentrations, exhibit multiple steady states. However, that is not the case when the mixture is considered (Nigam, personal communication, 1989). This conclusion shows that our understanding of stability of reactors for single reactants cannot easily be extended to the case where the feed is a multicomponent mixture. Some minor progress in this area has been made (Astarita 1990), and this will be discussed concisely in the next section.

The philosophical problem that needs serious consideration is the following one. From the viewpoint of analysis, it is convenient to rescale the label x so that it is proportional to the frequency factor of the component considered. However, in actual fact a concentration distribution is more likely to be known in terms of such measurable "labels" as boiling point or carbon number. The question that arises is of course that one would need to know the relationship between the frequency factors and the "measurable" label—or, in other words, one needs to have some understanding of the relationship between kinetic reactivity and molecular structure. The analogous type of information required in thermodynamics is generally available, at least in tentative form; our understanding in the case of kinetics is unfortunately much less developed.

SOME EXAMPLES OF APPROACH TO OPEN PROBLEMS

In this section, preliminary ideas about the approach to some open problems of analysis of the type sketched in the previous section are presented. The aim is *not* to give solutions, but to submit for discussion rather vague and possibly largely wrong viewpoints about such problems. The problems discussed below may look like a list of possible PhD theses, and this is simply due to the fact that they are indeed such a list. To the extent that any results are presented for such problems, these are mostly original results.

Kinetics: Nonuniform Interactive Kinetics

The assumption of uniformity for interactive kinetics is a very strong one, and some initial analysis of cases where it is untenable seems to be in order. A first step in this direction has recently been taken by Li and Ho (1990), who make three very strong assumptions:

1. They restrict attention to bimolecular reactions.
2. They assume that all bimolecular kinetic constants have a leading term that is the same for all couples of reactants.
3. They analyze only discrete cases, with a finite number of components, without considering the problem of a continuous description.

It is perhaps possible to attack the problem by the same essential technique of Li and Ho (a perturbation expansion), maintaining the first one of their assumptions but relaxing the other two. Specifically, I consider a continuous mixture where components undergo bimolecular reactions: a concrete example could be that of oligomerization of olefins on phosphoric acid. The rate of the reaction between the species considered (label x) and the generic species (label y) is assumed to be expressible as follows:

$$r(x, y; \tau) = [xy + b(x, y)]g(x, \tau)g(y, \tau). \tag{1–12}$$

The quantity in square brackets in Eq. (1–12) is the sum of a leading term xy and a "correction" $b(x, y)$. Function $b(x, y)$ needs of course to be symmetric, $b(x, y) = b(y, x)$. The label has been rescaled in the usual way, and $b(x, y)$ itself can be normalized so that:

$$\iint b(x, y)G(x)G(y)\,dx\,dy = 0. \tag{1–13}$$

The form of the leading term (xy) is partially justified for some systems. For instance, in the oligomerization of olefins, the reactions proceed via the formation of carbocations, which are very reactive; as a crude first-order approximation, one could say that the reactivity of a carbocation depends only weakly on the molecular structure of the olefin it reacts with, and that therefore the bimolecular kinetic constant k_{IJ} may be expressed as the product of the intrinsic reactivities of the two olefins—perhaps as $\sqrt{(k_{II}k_{JJ})}$. Indeed, Quann and Krambeck (1991, this volume) use the carbon number as the label I (or J), and they assume that k_{IJ} is proportional to some power of the product IJ—again, as the product of two intrinsic reactivities. From then on, obtaining the form xy is simply a question of rescaling and normalization.

Be that as it may, the kinetic equation for the continuous mixture becomes, under the assumptions stated:

$$-\frac{\partial g(x, \tau)}{\partial \tau} = g(x, \tau) \int [xy + b(x, y)]g(y, \tau)\,dy. \tag{1–14}$$

The condition that xy should be the leading term can perhaps be expressed as follows (but there are subtleties involved):

$$\varepsilon = \max|b(x, y)| \ll 1. \tag{1–15}$$

It is now useful to define $f(x, y)$ as follows:

$$f(x, y) = b(x, y)/\varepsilon, \quad |f(x, y)| \le 1 \tag{1–16}$$

so that the differential equation becomes

$$-\frac{\partial g(x, \tau)}{\partial \tau} = xg(x, \tau) \int yg(y, \tau)\,dy + \varepsilon g(x, \tau) \int f(x, y)g(y, \tau)\,dy. \tag{1–17}$$

It is now possible to attack a perturbation expansion by assuming that $g(x, \tau)$ is expressible as:

$$g(x, \tau) = g_0(x, \tau) + \varepsilon g_1(x, \tau) + \cdots \tag{1–18}$$

The zero-order term is delivered by Eq. (1–17) with the second term on the right-hand side set to zero; this shows that at the zero-order level the kinetics are indeed uniform, and the solution is given by:

$$g_0(x, \tau) = G(x) \exp[-xw(\tau)], \quad w(0) = 0. \tag{1–19}$$

where $w[\tau]$ is the solution of

$$\frac{dw}{d\tau} = \int yG(y) \exp[-yw(\tau)]\, dy = \left[\frac{\alpha}{\alpha + w}\right]^{\alpha+1}. \tag{1–20}$$

This yields

$$w = [\alpha^{\alpha+2} + \alpha^{\alpha+1}(\alpha + 2)\tau]^{1/(\alpha+2)} - \alpha \tag{1–21}$$

$$\alpha = \infty, \quad w = \ln(1 + \tau) \tag{1–22}$$

The total lumped concentration is, at the zero-order level,

$$C_0(\tau) = \left[\frac{\alpha}{\alpha + w}\right]^{\alpha} \tag{1–23}$$

which yields, of course, $C_0 = 1/(1 + \tau)$ for the single reactant case $\alpha = \infty$. The lumped kinetic function is

$$-\frac{dC_0}{d\tau} = R(C_0) = C_0^{(2\alpha+2)/\alpha} \tag{1–24}$$

So far, so good—at the zero-order level we are still within the realm of uniform kinetics. At the first-order level, the differential equation for $g_1(x, \tau)$, subject now to the initial condition $g_1(x, 0) = 0$, is

$$\begin{aligned} -\frac{\partial g_1(x, \tau)}{\partial \tau} &= xG(x)e^{-xw(\tau)} \int yg_1(y, \tau)\, dy \\ &\quad + xg_1(x, \tau)\left[\frac{\alpha}{\alpha + w}\right]^{\alpha+1} \\ &\quad + G(x)e^{-xw(\tau)} \int f(x, y)G(y)e^{-yw(\tau)}\, dy \end{aligned} \tag{1–25}$$

In order to proceed, one needs, of course, to assign the function $f(x, y)$. However, before doing so, it is useful to discuss the nature of the mathematical problem one is facing. The very last term on the right-hand side of Eq.

(1–25) is a known term, which makes the equation nonhomogeneous but does not give any additional problem. The second term on the right-hand side is linear in g_1, and by itself could easily be dealt with. It is the first term on the right-hand side that is the culprit: it makes the equation an integrodifferential one, and it is quite obvious that the same would happen at the higher-order levels. Hence one concludes (perhaps) that even a perturbation expansion approach around a uniform kinetics approximation leads to partial integrodifferential equations at any level of nontrivial perturbation.

A possible form for the function $f(x, y)$, which satisfies all the constraints, is (D being a parameter):

$$f(x, y) = \frac{e^{-D(x+y)} - [\alpha/(\alpha + D)]^{2\alpha}}{1 - [\alpha/(\alpha + D)]^{2\alpha}}. \tag{1–26}$$

The parameter D is subject to

$$D \geq \alpha[2^{1/2\alpha} - 1] = D_{\text{MIN}}. \tag{1–27}$$

Restriction (1–27) is a mild one: for $\alpha = 1$, $D_{\text{MIN}} = \sqrt{2} - 1 = 0.414$, and its value decreases steadily with α toward the asymptotic value corresponding to $\alpha = \infty$, which is $\ln 2/2 = 0.347$.

The maximum positive value of f is 1, while its minimum value exceeds -1 unless $D = D_{\text{MIN}}$, when it is -1. The f distribution in Eq. (1–26) means that the actual reactivity of components with a high leading term is lower than the leading term, and vice versa.

That is as far as I have gotten on this problem. Of course, numerical solutions could (perhaps) be easily obtained for assigned forcing functions, though the presence of an integral term in the differential equation may make even the numerical approach rather cumbersome.

Reactor Configuration: Maximum Segregation and Maximum Mixedness

If one focuses attention on reactors with arbitrary residence time distributions (RTD), one immediately realizes that, since we are dealing with nonlinear kinetics (and, in the case of mixtures, even intrinsically linear kinetics may be nonlinear at the lumped level), the RTD by itself is not enough to characterize the reactor. The two extreme cases of maximum segregation and maximum mixedness seem obvious candidates for a preliminary analysis.

The maximum segregation case is in a sense trivial. Equation (1–11) can in general be integrated to yield the lumped concentration as an explicit function of the dimensionless reaction time, $C = U(\tau)$. Now given a RTD

$m(\tau)$—whose zeroth moment will in general not be unity, since times have been normalized by multiplication with k^*—the residual concentration in the exit stream, C_E, will be given by

$$C_E = \int_0^\infty m(u)U(u)\, du. \tag{1–28}$$

This is easily read off the two-label formalism of Aris (1989), if one interprets the second label as distributing over residence times. The interesting case is the one of maximum mixedness, where one needs to work with the life expectancy distribution $m(\mu)$, where μ is the dimensionless life expectancy. A preliminary approach to this case has been discussed by Astarita and Ocone (1990), and this is concisely presented here. The Zwietering (1959) mass balance equation for a maximum mixedness reactor is trivially extended to the case of a mixture with uniform interactive nonlinear kinetics, since the reactant label x simply plays the role of a parameter:

$$\frac{\partial g(x, \mu)}{\partial \mu} = xg(x, \mu)F\left[\int K(y)g(y, \mu)\, dy\right] + \frac{m(\mu)}{M(\mu)}[g(x, \mu) - G(x)], \tag{1–29}$$

where $M(\mu)$ is the complementary cumulative RTD:

$$M(\mu) = \int_\mu^\infty m(u)\, du, \quad M(0) = 1. \tag{1–30}$$

The interesting point is that a formal solution of the Zwietering equation can be found, in spite of the fact that, as far as has been able to ascertain, such a solution has not been available in the literature even for the single-reactant case. Let the *warped life expectancy* $w(\mu)$, and the *warped RTD* $H(\mu)$ be defined from

$$\frac{dw}{d\mu} = F\left[\int K(y)g(y, \mu)\, dy\right], \quad w(0) = 0, \tag{1–31}$$

$$H(\mu) = m(\mu)W'(w), \tag{1–32}$$

where W' is the ordinary derivative of the inverse of the $w(\ \ )$ function, $W' = d\mu/dw$. One obtains, after some algebra,

$$g(x, \mu) = \frac{G(x)e^{xw(\mu)}}{M(w(\mu))} \int_{w(\mu)}^\infty H(u)e^{-xu}\, du. \tag{1–33}$$

The lumped exit concentration is thus given by

$$C_E = \int_0^\infty H(u)\left[\int G(y)e^{-uy}\,dy\right]du. \tag{1–34}$$

Notice that the integral within square brackets represents the lumpe concentration at time u which one would obtain should the kinetics be linea and hence for the α, β, Γ distributions it is given by

$$\int G(y)e^{-uy}\,dy = \left[\frac{\alpha}{\alpha + u}\right]^{1/\alpha} \tag{1–35}$$

and that therefore all the nonlinearities of the problem have been con centrated in the warped residence time distribution $H(\mu)$: the lumped exi concentration is the one that would be observed, should the kinetics be linea in a maximum segregation reactor endowed with a RTD given by $H(\mu$ However, the function $H(\mu)$ is delivered by the solution of the rathe formidable integrodifferential equation which is obtained by substitutin Eqs. (1–33) and (1–31) into Eq. (1–32).

Details on the analysis of maximum mixedness reactors are given b Astarita and Ocone (1990). Here it suffices to say that the problem is tractable one, at least in the sense that it is as tractable as the single-reactan case; however, difficulties arise, some of which are discussed in the subsectio on stability.

Topology: Sequential Reactions with Nonlinear Kinetics

Consider a mixture of very many reactants where the sequential reactio scheme A → B → C may take place. The mixture consists initially of onl reactants of type A. The desired products are components of type B, so tha a selectivity problem arises. Let x be the label identifying A-type reactant and suppose the A → B step is governed by uniform nonlinear kinetics:

$$-\partial g(x, \tau)/\partial \tau = xg(x, \tau)F\left[\int K(y)g(y, \tau)\,dy\right]. \tag{1–36}$$

I will only consider the case where the second step B → C has first-orde kinetics. This, of course, makes the mathematics somewhat simpler, and ha some partial justification if one considers a problem where selectivity fo B-type components is sought, and the latter may decompose to undesire products only thermally.

Since the kinetics of disappearance of A-type reactants is not affected b the fact that the B-type components may in turn be consumed by the secon tep of the reaction, all the results available for the A → B reaction topology pply to the case at hand as far as $g(x, \tau)$ is concerned. In particular, $g(x, \tau)$ s given by Eqs. (1–9) and (1–10).

In the following, $w(\tau)$ will be regarded as a known function, since in rinciple it can be obtained by integration of Eq. (1–10) for any assigned rm of the kinetic function $F[\ \ ]$. In the analysis of only the first step, one ften does not need to actually integrate Eq. (1–10) (Astarita 1989; Aris 1989), ince often the function $w(\tau)$ can be eliminated from the equations. This is nfortunately not the case for the problem at hand.

Now consider B-type components, and let x be the label identifying the component resulting from conversion of the corresponding A-type compo- ent. Let $v(x, \tau)\, dx$ be the (dimensionless) concentration of B-type compo- ents resulting from conversion of A-type components identified by labels etween x and $x + dx$. For the problem considered, $v(x, 0) = 0$, since one tarts with a mixture of only A-type reactants. Components of type B are roduced by conversion of the corresponding A-type reactants and are onsumed by the second step of the reaction. It follows that the rate of roduction of B-type components is given by

$$\partial v(x, \tau)/\partial \tau = -\partial g/\partial t - h(x)v(x, \tau). \tag{1–37}$$

The quantity $h(x)$ in Eq. (1–37) is the kinetic constant of B-type component , divided by the initial average frequency factor of A-type reactants, k^*. ince x has already been normalized so that $k(x) = k^*x$, no additional ormalization can be performed here, and $h(x)$ has to be left as some arbitrary unction, having some "average" initial value h^*.

The problem to be addressed is the solution of Eq. (1–37), and the stimation of the time at which some appropriate lumped value of $v(x, \tau)$ eaches a maximum. Notice that once $v(x, \tau)$ has been determined, the istribution of concentrations of C-type products, $z(x, \tau)$, is obtained directly y an overall mass balance:

$$z(x, \tau) = G(x) - g(x, \tau) - v(x, \tau). \tag{1–38}$$

Function $q(x, \tau) = v(x, \tau)/G(x)$ is subject to the following differential equa- ion and boundary condition:

$$\partial q/\partial \tau = xw'(\tau)\exp(-xw) - h(x)q, \quad q(x, 0) = 0, \tag{1–39}$$

here w' is the ordinary derivative of $w(\tau)$. Equation (1–39) can be solved ormally almost by inspection:

$$q = e^{-h(x)\tau}x \int_0^\tau e^{h(x)u}e^{-xw(u)}w'(u)\, du. \tag{1–40}$$

For the linear case where $w(\tau) = \tau$, Eq. (1–40) reduces to the Aris (1989) result:

$$q = \frac{x}{h(x) - x}\,[e^{-x\tau} - e^{-h(x)\tau}]. \tag{1–41}$$

Things are made easier by working with the warped time $w(\tau)$, rather than with time itself, as the independent variable. Since the value of $F[\ \ ]$ is bound to be positive, Eq. (1–10) guarantees that $w(\tau)$ is invertible, say $\tau = W(w)$. Thus Eq. (1–39) can be integrated to yield

$$q(x, w) = e^{-h(x)W(w)}x \int_0^w e^{h(x)W(u)}e^{-xu}\,du. \tag{1–42}$$

In fact, function $W(w)$ can be obtained formally from Eq. (1–10). Let $K(w)$ be the argument of the $F[\ \ ]$ function in Eq. (1–10), i.e., the average value of K when the warped time is w:

$$K(w) = \int K(y)g(y, 0)e^{-yw(\tau)}\,dy. \tag{1–43}$$

Equation (1–10) thus reduces to $dW/dw = 1/F[K(w)]$, and $W(w)$ is obtained by quadrature:

$$W(w) = \int_0^w \frac{du}{F[K(u)]}. \tag{1–44}$$

Choosing the α-β-Γ distributions, one obtains

$$K(w) = K^*\left[\frac{\alpha}{\alpha + w}\right]^{\alpha + \beta}. \tag{1–45}$$

The problem considered is one where numerical solution should not present any significant difficulty once the relevant kinetic information is available.

Selectivity: The Linear Parallel Case, an Approach to the Nonlinear Parallel One, and a Problem for the Sequential Case

For the linear parallel case $C \leftarrow A \rightarrow B$, let x be scaled so that it is proportional to the sum of the kinetic constants of the two steps, and let $g(x, \tau)$, $v(x, \tau)$, and $z(x, \tau)$ be the concentration distributions of components f type A, B, and C, respectively, with C, V, and Z their lumped values. As ar as components of type A are concerned, the classic linear analysis of Aris 1968) applies, and $g(x, \tau) = G(x)e^{-x\tau}$. If $\theta(x)$ is the kinetic constant of he A → B step, normalized with respect to the average overall kinetic onstant [notice that $\theta(x) \leq x$], the differential equation for $v(x, \tau)$ is

$$\partial v/\partial \tau = \theta(x)G(x)e^{-x\tau}, \quad v(x, 0) = 0, \tag{1–46}$$

vhich has the solution

$$v(x, \tau) = \theta(x)G(x)(1 - e^{-x\tau})/x. \tag{1–47}$$

In order to proceed, one needs to assign $\theta(x)/x$, and the obvious symmetry f the problem suggests that one consider only two typical cases. For he first case, $\theta(x)/x$ steadily decreases with x, $\theta(x)/x = e^{-x}$. We define $\Omega = \alpha/(\alpha + 1)$ and the selectivity $S(\tau) = V(\tau)/Z(\tau)$—noticing that for the ingle-component case S is independent of τ. One obtains

$$V(\tau) = \Omega^{\alpha} - [\Omega/(1 + \tau/\Omega)]^{\alpha}, \tag{1–48}$$

$$S(0) = \Omega^{\alpha+1}/(1 - \Omega^{\alpha+1}), \tag{1–49}$$

$$S(\infty) = \Omega^{\alpha}/(1 - \Omega^{\alpha}) \geq S(0), \tag{1–50}$$

$$dS/d\tau \geq 0. \tag{1–51}$$

The equal sign applies only for the single reactant case. S steadily increases n time, and it reaches the single reactant value only at $\tau = \infty$. For $\alpha = 1$, he initial selectivity is one-third of what it would be for the equivalent ingle-reactant case.

For the second case $\theta(x)/x$ goes through a maximum, $\theta(x)/x = xe^{-x}$; its verage initial value is $\Omega^{\alpha+1}$. One obtains

$$V(\tau) = \Omega^{\alpha+1} - [\Omega/(1 + \tau/\Omega)]^{\alpha+1}, \tag{1–52}$$

$$S(0) = S(\infty) = \Omega^{\alpha+1}/(1 - \Omega^{\alpha+1}). \tag{1–53}$$

The selectivity has the same value it would have for an equivalent single eactant at both time zero and time infinity, but it goes through a maximum, o that in general the mixture will have a higher selectivity than the single eactant (and, of course, conversely if C-type products are the desired ones). problem of optimal residence time thus arises for the mixture, whereas one exists in the equivalent single-reactant case.

I now move to an initial approach to the case where both parallel steps nay be governed by nonlinear uniform kinetics. Let the label x be rescaled so that it is proportional to the frequency factor of the $A(x) \rightarrow B(x)$ step, an let the dimensionless time τ be defined as the product of the actual time times the initial average value of the same frequency factor, $\tau = k^*t$. Th frequency factor for the $A(x) \rightarrow C(x)$ step can be written as $k^*l(x)$, wit function $l(x)$ having some average initial value l^* which cannot be normalize to be unity. Let unprimed quantities refer to the $A(x) \rightarrow B(x)$ step (the desire one), and let primed quantities refer to the undesired step $A(x) \rightarrow C(x)$: sc e.g., $F[\ \ ]$ and $F'[\ \ ]$ are the kinetic functions for the two steps. Finally, le $g(x, \tau)$, with $g(x, 0) = G(x)$, be the concentration distribution at time τ c A-type reactants. The kinetic equation for $g(x, \tau)$ is

$$-\frac{\partial g(x, \tau)}{\partial \tau} = xg(x, \tau)F\left[\int K(y)g(y)\,dy\right] + l(x)g(x, \tau)F'\left[\int K'(y)g(y)\,dy\right]. \quad (1\text{–}5$$

The solution is, as can easily be checked by substitution

$$g(x, \tau) = G(x)e^{-xw(\tau) - l(x)p(\tau)}, \quad (1\text{–}5$$

where the functions $w(\ \ )$ and $p(\ \ )$, which are subject to $w(0) = p(0) =$ (are the solutions of the following two ordinary differential equations:

$$\frac{dw}{d\tau} = F\left[\int K(y)g(y, \tau)\,dy\right], \quad w(0) = 0, \quad (1\text{–}56$$

$$\frac{dp}{d\tau} = F'\left[\int K'(y)g(y, \tau)\,dy\right], \quad p(0) = 0. \quad (1\text{–}5$$

These are two coupled nonlinear ordinary differential equations for $w($ and $p(\ \ )$, which can always be solved numerically. Once $g(x, \tau)$ has bee obtained, it is of course easy to calculate the concentration distributions c B- and C-type products and hence to calculate the selectivity as a functio of time. However, before doing so for any special case of the definin functions, it is useful to consider a somewhat special case where the governin equations can be solved analytically.

The special case is the one where the α, β, Γ distributions hold for bot steps, with of course β in general different from β'. Furthermore, conside the case where $l(x) = l^*x$, i.e., the frequency factors for the two steps ar proportional to each other (this, again, has some handwaving physic justification if one regards the frequency factors as yardsticks of som intrinsic reactivity). With this assumption, the governing equation for $g(x,$ becomes quasilinear, and it is useful to define the quantity $a(\tau) =$ $w(\tau) + l^*p(\tau)$, so that Eqs. (1–56) and (1–57) can be collapsed into a singl differential equation for $a(\tau)$. One obtains after some algebra

$$g(x, \tau) = G(x) \exp(-xa), \tag{1–58}$$

where $a(\tau)$, subject to $a(0) = 0$, is the solution of

$$\frac{da}{d\tau} = F\left[K^*\left[\frac{\alpha}{\alpha + a}\right]^{\alpha+\beta}\right] + l^*F'\left[K^{*\prime}\left[\frac{\alpha}{\alpha + a}\right]^{\alpha+\beta'}\right]. \tag{1–59}$$

One can now obtain the lumped concentration of A-type reactants as a function of time:

$$C = \left[\frac{\alpha}{\alpha + a}\right]^{\alpha}. \tag{1–60}$$

By substitution, one can eliminate $a(\ )$ between the equations to obtain a lumped kinetic function $R(C)$ which gives directly the rate of change of the lumped concentration of reactants:

$$-\frac{dC}{d\tau} = C^{(\alpha+1)/\alpha}[F[K^*C^{(\alpha+\beta)/\alpha}] + l^*F'[K^{*\prime}C^{(\alpha+\beta')/\alpha}]] = R(C). \tag{1–61}$$

Now let $v(x, \tau)$ be the concentration distribution of B-type components, which is subject to $v(x, 0) = 0$. The differential equation for $v(x, \tau)$ is

$$\frac{\partial v(x, \tau)}{\partial \tau} = xG(x)e^{-xa}F\left[K^*\left[\frac{\alpha}{\alpha+a}\right]^{\alpha+\beta}\right]. \tag{1–62}$$

This is an explicit differential equation which can be solved in general (at least numerically). Furthermore, one may directly calculate [by elimination of $a(\ )$ between the relevant equations] the rate of change of the lumped concentration of B-type products, $V(\tau)$, which, after some algebra, is seen to be

$$\frac{dV}{d\tau} = F[KC^{(\alpha+\beta)/\alpha}]c^{(\alpha+1)/\alpha}. \tag{1–63}$$

Equation (1–63), subject to $V(0) = 0$, can, in principle, be integrated to yield $V(\tau)$ in terms of the (known) function $C(\tau)$. The selectivity $S(\tau)$, defined as the ratio of the lumped concentrations of B-type and C-type products (the latter is, of course, $1 - C(\tau) - V(\tau)$, as calculated from an overall mass balance], is thus obtained as

$$S(\tau) = \frac{V(\tau)}{1 - C(\tau) - V(\tau)}. \qquad (1\text{–}64$$

The sequential case which is worth considering is the following one $A \leftrightarrow B \rightarrow C$. For this case, the single-reactant problem has been widel studied, and two singular asymptotic cases are well established: the on where the $A \leftrightarrow B$ step may be regarded as being always at equilibrium (excep in the appropriate boundary layer), and the case where the quasi-steady-stat approximation $dB/d\tau \approx 0$ can be used (Bowen *et al.* 1963). It seems worth while to investigate whether analogous singular asymptotic solutions ma be established for the mixture case, whether the singularities are of the sam well-behaved form, and whether the answers to such questions do or do no depend on the details of the initial distributions.

Stability: Maximum Mixedness Reactors

Maximum segregation reactors are known never to exhibit multiple stead states, so that the question of stability does not arise. One may therefor begin the consideration of stability problems for multicomponent mixture by considering the case of maximum mixedness reactors. One first has t deal with the Shinnar paradox, which, in essence, has to do with the fac that the classic Zwietering definition of a maximum mixedness reactor is on where there is no backmixing, and therefore it would seem that multipl steady states are impossible—contrary to the known fact that multiple stead states are possible in a CSTR which is a special case of maximum mixednes reactor.

The Shinnar paradox was analyzed, and essentially solved, by Glasser *e al.* (1986); the problem has recently been reconsidered from a somewha different viewpoint (Astarita 1990). The problem may be concisely describe as follows. Equation (1–33) is an exact solution of the Zwietering equation however, the numerical procedure for integrating the Zwietering differentia equation requires using a boundary condition for $g(x, \infty)$. Now Eq. (1–33 can be used to ask oneself what the value of $g(x, \infty)$ might be, and the answe is that there are two possible cases, as concisely discussed below.

The first case is the one where the limit that defines the quantity T^* belov exists:

$$T^* = \lim_{\mu = \infty} \frac{M(\mu)}{m(\mu)} \qquad (1\text{–}65$$

In this first case, $g(x, \infty)$ is the concentration distribution one woul calculate in a CSTR with dimensionless residence time T^*. Hence if such reactor exhibits multiple steady states, so will the solution of the Zwietering equation, and vice versa. Thus, the stability analysis for this first case reduces to that of a CSTR.

The second case is the one where the limit does not exist. A plug flow reactor with recycle is an example of this second case. In this second case the Zwietering procedure cannot be applied at all, for lack of a boundary condition, and hence one is left with only two possibilities: either one solves the integrodifferential equation for $H(\mu)$ and then uses Eq. (1–33), or one needs to have a complete description of the micromixing function (Glasser and Jackson 1984; Jackson and Glasser 1986). It is interesting to note that whether one falls in the first or the second case only depends on the RTD, independently of whether one is dealing with a single reactant or a mixture, and independently of what the kinetic functions might be.

CONCLUSION

What has been presented in the body of this chapter is, and purposely so, very tentative. When trying to extract some conclusions of a general character from such material, it is not surprising that the conclusions should be equally tentative, and that in fact recommendations for discussion are limited.

A first point that comes to mind is that the analysis of what appear to be significant problems in the area requires mathematical tools not usually at the disposal of chemical engineers, the most important one of those being the ability to deal with integrodifferential equations. Major a difficulty as this might be, it is not one that necessarily scares an academic, who is accustomed to the leisurely pace of research in universities, which leaves time to study some mathematics when one convinces oneself it is needed for the problem at hand.

The second point is that progress could be made by well-planned numerical work. The area we are considering is one where the stage of *modeling* is under scrutiny, and the best way of scrutinizing a tentative model is to find out what it predicts—and how do such predictions compare with experimental data. But, before embarking on a possibly meaningless program of numerical work on different problems, it would be useful to know where such a comparison is both significant and possible. Here is an area where cooperation with industrial practitioners is, I believe, of crucial importance, in two respects. First, academicians may try—as I have done—to identify open problems, and dedicate their attention to them; but it is only industrial practitioners who can set some meaningful scale of priority to such problems and help the academician to concentrate his or her efforts on the important ones. Second, modeling is a somewhat hollow exercise if one cannot compare theory and experiments—and experiments in this area are likely to be expensive and difficult, possibly to the point of being outside of what can meaningfully be done in academia. Industry could do well, in my opinion, in making experimental data, even of proprietary nature, available to academics for comparison with theoretical models.

The third point has to do with the philosophical problem discussed at the end of the "open problems" section of this chapter. Kinetic data are available, but they are not easily retrieved, categorized, and analyzed. Perhaps this volume may be a focal point for deciding the best way industry and academia may cooperate to make the best use of whatever information is available and to plan in a rational way the collection of what useful information is not available.

A fourth point is as follows: The problems that have been illustrated as potential of possible analysis may sound as a collection of "academic" problems, with little—if any—relationship to reality. This is only partly true: by identifying which problems are possibly potential of a solution, one also identifies which assumptions (strong as they may be) make the analysis possible, and which ones do not play a crucial role in the analysis and may therefore be relaxed. However, whether the *desirable* (from the viewpoint of analysis) assumptions are, at least in some cases, realistic ones, or whether they are not, is the type of information that can emerge only from a well planned and significant cooperation of industry and academia. This volume could be the beginning of such a cooperation.

NOMENCLATURE

$a(\,)$	warped time for quasilinear parallel case
$b(\,)$	correction term in bimolecular kinetics
C	total lumped concentration of initial reactants
C_E	lumped concentration in exit stream
C_N	N^{th} order term for C
D	parameter for $f(\,)$ distribution
D_{MIN}	minimum value for D
$f(\,)$	$= b(x, y)/\varepsilon$
$F[\,]$	kinetic function
$F'[\,]$	kinetic function for undesired step
$g(\,)$	concentration distribution of A-type reactants
$g_N(\,)$	N^{th} order term for $g(\,)$
$G(\,)$	concentration distribution in feed
$H(\,)$	warped RTD
$h(\,)$	kinetic constant distribution for $B \rightarrow C$ step
$k(\,)$	frequency factor distribution, sec^{-1}
k^*	average frequency factor, sec^{-1}
$K(\,)$	weighting factor distribution

$K'(\,)$ weighted factor distribution for undesired step

K^* average value of K in feed

$K^{*\prime}$ average value of K' in feed

K average value of K

$l(\,)$ frequency factor distribution for undesired parallel step

l^* initial average value for $l(\,)$

$m(\,)$ RTD

$M(\,)$ complementary cumulative RTD

$p(\,)$ secondary warped time

$Q(\,)$ distribution of generic property

Q^* average value of Q in feed

$q(\,)$ $= v(x, \tau)/G(x)$

$r(\,)$ reaction rate of individual reactant

$R(\,)$ lumped reaction rate

$S(\,)$ selectivity

t time, sec

T residence time in CSTR

T^* defined in Eq. (1–54)

u dummy time variable

$U(\,)$ function assigning C

$v(\,)$ concentration distribution of B-type components

V lumped value of $v(\,)$

w warped time or life expectancy

w' $dw/d\tau$

W inverse $w(\,)$ function

W' derivative of $W(\,)$, $= d\mu/dw$

x component label

Y dummy component label

$z(\,)$ concentration distribution of C-type products

Z lumped value of $z(\,)$

α parameter in $G(x)$ distribution

β parameter in $K(x)$ distribution

β' parameter in $K'(x)$ distribution

ε perturbation parameter

$\theta(\,)$ kinetic constant distribution for $A \rightarrow C$ step

μ life expectancy

τ dimensionless time

Ω $= \alpha/(\alpha + 1)$

REFERENCES

Aris, R. (1968) Prolegomena to the rational analysis of systems of chemical reactions. II: Some addenda. *Arch Ratl. Mech. Anal.* **27**:35.

Aris, R. (1989) Reactions in continuous mixtures. *A.I.Ch.E.J.* **35**:539.

Aris, R. and G. Astarita (1989) Continuous lumping of nonlinear chemical kinetics. *Chem. Engng. Proc.* **26**:63.

Aris, R. and G. Astarita (in press) On aliases of differential equations. *Rend. Acc. Lincei, Matematica,* **LXXXIII**.

Astarita, G. (1989) Lumping nonlinear kinetics: apparent overall order of reaction. *A.I.Ch.E.J.* **35**:529.

Astarita, G. (in press) Multiple steady states in maximum mixedness reactors. *Chem. Engng. Commun.*

Astarita, G. and A. Nigam (1989) Lumping nonlinear kinetics in a CSTR. *A.I.Ch.E.J.*, **35**:1927.

Astarita, G. and R. Ocone (1988) *Lumping nonlinear kinetics.* **34**:1299.

Astarita, G. and R. Ocone (1989a) Accorpamento continuo di miscele complesse. *Chim. Ind. (Milan)* **71**:78.

Astarita, G. and R. Ocone (1989b) Heterogeneous chemical equilibria in multicomponent mixtures. *Chem. Engng. Sci.* **44**:2323.

Astarita, G. and R. Ocone (in press) Continuous lumping in a maximum mixedness reactor. *Chem. Engng. Sci.*

Bowen, J. R., A. Acrivos, and A. K. Oppenheim (1963) Singular perturbation refinement to quasi steady state approximation in chemical kinetics. *Chem. Engng. Sci.* **18**:177.

Chou, M. Y. and T. C. Ho (1988) Continuum theory for lumping nonlinear reaction mixtures. *A.I.Ch.E.J.* **34**:1519.

Chou, M. Y. and T. C. Ho (1989) Lumping coupled nonlinear reactions in continuous mixtures. *A.I.Ch.E.J.* **35**:533.

Glasser, D., C. M. Crowe and R. Jackson (1986) Zwietering's maximum mixed model and the existence of multiple steady states. *Chem. Engng. Commun.* **40**:41.

Glasser, D. and R. Jackson (1984) A generalized residence time distribution model for a chemical reactor. *A.I.Ch.E. Symp. Series, N.* 87, 535.

Jackson, R. and D. Glasser (1986) A general mixing model for steady flow chemical reactors. *Chem. Engng. Commun.* **42**:17.

Ho, T. C. A simple expression for the overall behavior of a continuum of mixtures.

Ho, T. C. and R. Aris (1987) On apparent second order kinetics. *A.I.Ch.E.J.* **33**:1050.

Hu, R. and T. C. Ho A numerical study of lumping nth-order reactions.

Li, B. Z. and T. C. Ho Lumping weakly nonuniform bimolecular reactions. *Chem. Engng. Sci.* (submitted).

Li, G. and H. Rabitz (1989) A general analysis of exact lumping in chemical kinetics. *Chem. Engng. Sci.* **44**:1413.

Quann, R. J. and F. J. Krambeck (1990) Olefin oligomerization kinetics over ZSM-5. (this volume).

White, B. S. and T. C. Ho Lumped kinetics for a continuum of parallel reactions (submitted).

Zwietering, Th. N. (1959) The degree of mixing in continuous flow systems. *Chem. Engng. Sci.* **11**:1.

2

Multiple Indices, Simple Lumps, and Duplicitous Kinetics

RUTHERFORD ARIS

PRELIMINARIES

We shall be concerned with the mathematics of continuous mixtures and will not attempt to survey the engineering literature at this time. Suffice it to say that the concept goes back at least to DeDonder (1931) and that it has found application in several branches of engineering as well as in the natural philosophy of chemical reaction. A convenient entry point to the literature can be found in a group of three articles by Astarita, Chou and Ho, and Aris in the AIChE Journal for April 1989 (pp. 529–548). Some priority issues in the question of multiple indices are mentioned in a letter later in the same journal (p. 695).

By an n-index mixture we mean a distribution of reactive matter such that the concentration of material with i^{th} index in the interval $(x_i, x_i + dx_i)$ at time t is

$$u(x_1 \ldots x_n; t)\, dx_1 \ldots dx_n. \tag{2–1}$$

The function $u \geq 0$ for a "real" mixture, but we shall allow u to be negative in the general case. For convenience and without loss of generality, we may choose the units so that

$$\int_\Omega u(x_1, \ldots, x_n; 0)\, dx_1 \ldots dx_n = 1, \tag{2–2}$$

where Ω is the index domain to be specified later. Discrete species in the mixture are specified by Dirac delta functions. Thus a single-index mixture with a finite or denumerable number of species of indices $x_1 = \{k_1, \ldots, k_n, \ldots\}$, $x_2 = x_3 \ldots = 0$ is

$$u(x_1, t) = \sum_0^\infty u_n \delta(x_1 - k_n). \tag{2–3}$$

For example, we might have

$$u_n = (1 - \rho)\rho^n, \quad 0 < \rho < 1, \tag{2–4}$$

since this would satisfy the normalization (2–2).

The total concentration of the mixture at time t is

$$U(t) = \int_\Omega u(x_1, \ldots, x_n; t)\, dx_1 \ldots dx_n. \tag{2–5}$$

Hence, by Eq. (2–2), $U(0) = 1$. Sometimes it may be desirable to take a non-negative weighting function and write

$$U_w(t) = \int_\Omega w(x, \ldots, x_n) u(x_1, \ldots, x_n; t) dx_1 \ldots dx_n. \tag{2–6}$$

Only strictly linear kinetics will be considered here. Astarita's "uniform kinetics," the most interesting recent development in the subject, has been, and will be, treated elsewhere by its inventor and others. See, for example, Astarita and Ocone (1988).

By first-order reaction we understand that the concentration of each "component" at time t is related to its concentration at time 0 by

$$u(x_1, \ldots, x_n; t) = u(x_1, \ldots x_n; 0)e^{-kt}. \tag{2–7}$$

The rate constant $k = k(x_1, \ldots, x_n)$ is not negative and the domain of the index space Ω is defined by

$$k(x_1, \ldots, x_n) \geq 0. \tag{2–8}$$

In this domain let the hypersurfaces of constant k be denoted by

$$\Omega_k: (x_1, \ldots, x_n) \ni k(x_1, \ldots, x_n) = k. \tag{2–9}$$

We lose no generality by supposing k goes from zero to infinity, since any lesser range can be accommodated by setting $u = 0$ outwith it.

It would be surprising if $k(x_1, \ldots, x_n)$ were not monotonic with respect to at least one of its variables, and we shall assume that these have been arranged so that

$$\frac{\partial k}{\partial x_n} > 0. \tag{2–10}$$

Then k can be inverted and x_n written as a function of $x_1, \ldots, x_{n-1}$ and k

$$x_n = \xi(x_1, \ldots, x_{n-1}, k). \tag{2–11}$$

It follows that

$$\begin{aligned}\int_\Omega F(x_1, \ldots, x_n)\, dx_1 \ldots dx_n &= \int_0^\infty \left[\int_{\Omega_k} F(x_1, \ldots, x_{n-1}, \xi)\left(\frac{\partial \xi}{\partial k}\right) dx_1 \ldots dx_{n-1} \right] dk \\ &= \int_0^\infty \Phi(k)\, dk. \end{aligned} \tag{2–12}$$

REDUCTION TO ONE INDEX

Combining Eq. (2–7) and Eq. (2–5) and using Eq. (2–12) we have

$$\begin{aligned} U(t) &= \int_\Omega u(x_1, \ldots, x_n; 0) e^{-k(x_1, \ldots, x_n)t}\, dx_1 \ldots dx_n \\ &= \int_0^\infty e^{-kt} \Gamma(k)\, dk, \end{aligned} \tag{2–13}$$

where

$$\Gamma(k) = \int_{\Omega_k} u(x_1, \ldots, \xi; 0)(\partial \xi / \partial k)\, dx_1 \ldots dx_{n-1}. \tag{2–14}$$

Thus $U(t)$ is the Laplace transform of $\Gamma(k)$ with k playing the role of the variable usually denoted by t and t playing the role of the Laplace transform variable usually denoted by p or s.

As an example, consider the case

$$k = \kappa x_1, \ldots, x_n = \kappa \prod_1^n x_m$$

so that

$$\xi = (k/\kappa)\left(\prod_{1}^{n-1} x_m\right)^{-1}$$

and

$$\Gamma(k) = \frac{1}{\kappa}\int_{\Omega_k} u\left(x_1, \ldots, x_{n-1}, k/\kappa \prod_{1}^{n-1} x_m; 0\right)\prod_{1}^{n-1}\left(\frac{dx_m}{x_m}\right) \tag{2–15}$$

In particular, let

$$x_1 = x, \quad x_2 = y, \quad \kappa = 1, \quad \sigma = \kappa t/\alpha\beta, \quad \alpha > 0, \beta > 0,$$

and

$$u(x_1, x_2; 0) = \frac{\alpha^\alpha x^{\alpha-1}}{\Gamma(\alpha)}\frac{\beta^\beta y^{\beta-1}}{\Gamma(\beta)}e^{-\alpha x-\beta y} \tag{2–16}$$

Then

$$\Gamma(k) = \frac{2(\alpha\beta)^{(\alpha+\beta)/2}}{\Gamma(\alpha)\Gamma(\beta)}k^{(\alpha+\beta)/2-1}K_{\alpha-\beta}(2\sqrt{\alpha\beta}) \tag{2–17}$$

and

$$U(t) = e^{1/2\sigma}\sigma^\lambda W_{\lambda,\mu}(1/\sigma) \tag{2–18}$$

with

$$\lambda = (1-\alpha-\beta)/2, \quad \mu = (\alpha-\beta)/2. \tag{2–19}$$

or

$$U = \sigma^{-\alpha}\psi(\alpha; 1+\alpha-\beta; 1/\sigma). \tag{2–20}$$

Here $K_\nu(z)$ is the modified Bessel function of the second kind and $W_{\lambda,\mu}(z)$ and $\psi(a; c; z)$ are Whittaker's and Tricomi's forms of the confluent-hypergeometric function, respectively. (These integrals are to be found in Erdelyi, Magnus, Oberhettinger, and Tricomi's, *Tables of Integral Transforms*, Eqs. 4.5.2 and 4.16.37. See also Spanier and Oldham's *An Atlas of Functions*, Chaps. 47–48.)

APPARENT KINETICS

By the curriculum of a reaction in a reactor is meant the history of the concentration as a function of the reaction time. The ordinary time t is this reaction time for a batch reactor; the length divided by the velocity is the reaction time for the tubular; and e^{-kt} is the curriculum of a reaction of the first order. In general, the curriculum of a reaction depends on kinetic parameters, and two reactions that display the same curriculum, save for possibly different values of the kinetic parameters, are said to have the same apparent kinetics. Thus, the apparent kinetics of the lump given by

$$x_1 = x, \quad u(x, 0) = \alpha^\alpha x^{\alpha-1} e^{-\alpha x}/\Gamma(\alpha), \quad k = \kappa x \tag{2–21}$$

is

$$U(t) = \{1 + (n-1)\kappa t\}^{-1/(n-1)}, \quad n = (1+\alpha)/\alpha$$

which is of the nth order, since

$$U = -kU^n. \tag{2–22}$$

In general, the apparent kinetic law of the total concentration of the mixture is obtained by eliminating t between

$$U(t) = \int_0^\infty \Gamma(k) e^{-kt}\, dk \quad \text{and} \quad \dot{U}(t) = -\int_0^\infty k\Gamma(k) e^{-kt}\, dk \tag{2–23}$$

to give

$$\dot{U} = -\kappa R(U), \tag{2–24}$$

where κ is a kinetic parameter of the dimensions of reciprocal time.

The curriculum can be obtained from the kinetic equation by quadrature

$$\kappa t = \int_{U(t)}^1 du/R(u), \tag{2–25}$$

but in many cases this has to be left as an implicit formula. For example,

$$R(U) = \kappa U/(1 + \lambda U) \tag{2–26}$$

and

$$\kappa t = \lambda(1 - U) + \ln(1/U). \tag{2–27}$$

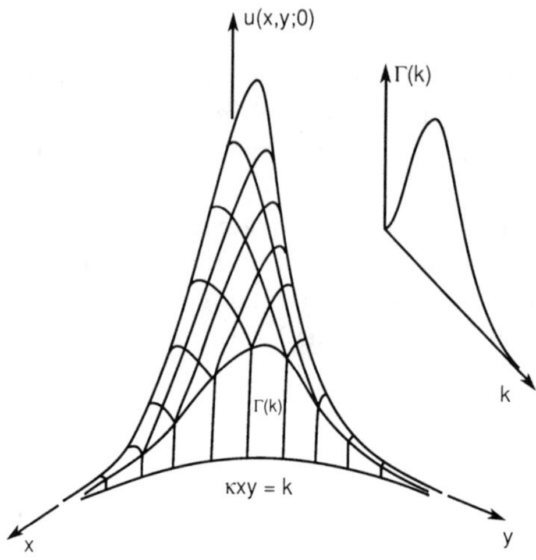

Figure 2–1 The surface $u(x, y, 0)$ and its section by constant $k(x, y)$.

For any distribution $\Gamma(k)$, a graph of $R(U)$ can be obtained by plotting $-\dot{U}(t)$ versus $U(t)$ (often on log log scale) using t as a parameter. Let κ be chosen to make $-\dot{U}(0) = 1$; i.e., $R(1) = 1$ and a dimensionless time

$$\tau = kt \tag{2–28}$$

is defined, e.g., $\kappa = 1 + \lambda$ in Eq. (2–25). For Figure 2–1, $R(U)$ should be obtained by eliminating σ between Eq. (2–18) and

$$-\dot{U}(t) = \kappa e^{1/2\sigma}\sigma^{\lambda-1}W_{\lambda-1,\mu}(1/\sigma) = \kappa\sigma^{-(\alpha+1)}\psi(\alpha+1;\alpha-\beta+1;1/\sigma), \tag{2–29}$$

where λ and μ are given by Eq. (2–19). Clearly this would be difficult to do, and the parametric plot is the answer. For small t,

$$-\dot{U}(t) = \kappa\left\{1 - \frac{(\alpha+1)(\beta+1)}{\alpha\beta}\kappa t + \frac{(\alpha+1)(\alpha+2)(\beta+1)(\beta+2)}{\alpha^2\beta^2}\frac{\kappa^2 t^2}{2}\cdots\right.$$

So $\kappa = 1$ normalizes $\dot{U}(0)$. Then

$$U(t) = 1 - t + \frac{(\alpha+1)(\beta+1)}{\alpha\beta}\frac{t^2}{2}\cdots,$$

so

$$R = 1 - \frac{(\alpha+1)(\beta+1)}{\alpha\beta}(1-U) + \frac{\alpha+\beta+3}{2}\frac{(\alpha+1)(\beta+1)}{\alpha^2\beta^2}(1-U)^2\cdots.$$

THE INVERSE PROBLEM

Given the curriculum of reaction $U(t)$, the distribution $\Gamma(k)$ that yields this behavior is

$$\Gamma(k) = L^{-1}[U] = \frac{1}{2\pi i}\int_{\gamma-i\infty}^{\gamma+i\infty} e^{kt}U(t)\,dt. \tag{2–30}$$

Since

$$U(0) = \int_0^\infty \Gamma(k)\,dk = 1, \tag{2–31}$$

γ can be taken as small as we please. If $t = i\omega$ and $U(i\omega) = V(\omega) + iW(\omega)$

$$\Gamma(k) = \frac{1}{\pi}\int_0^\infty \{V(\omega)\cos\omega k + W(\omega)\sin\omega k\}\,d\omega. \tag{2–32}$$

An alternative formula available if the nth derivative of $U(t)$ can be calculated is

$$\Gamma(k) = \lim_{n\to\infty} \frac{(-)^k}{n!}\left(\frac{n}{k}\right)^{n+1} U^{(n)}\left(\frac{n}{k}\right). \tag{2–33}$$

For example, an apparent second-order reaction for which

$$U(t) = (1 + k_2 t)^{-1} \tag{2–34}$$

gives

$$v(\omega) = \frac{1}{1 + k_2^2\omega^2}, \quad w(\omega) = -\frac{k_2\omega}{1 + k_2^2 w^2}$$

and from tables of Fourier sine and cosine transforms

$$\Gamma(k) = \frac{1}{\pi}\int_0^\infty \frac{\cos\omega + k_2\omega\sin k\omega}{1 + k_2^2\omega^2}\,d\omega = \frac{1}{k_2}e^{-k/k_2}. \tag{2–35}$$

Similarly,

$$U(t) = \left(1 + \frac{\tau}{\alpha}\right)^{-\alpha} \text{ gives } U^{(n)}(\tau) = (-)^n \frac{(\alpha)_n}{\alpha^n}\left(1 + \frac{\tau}{\alpha}\right)^{-n-\alpha} \tag{2–36}$$

and

$$\Gamma(k) = Lt\,\frac{\alpha(\alpha+1)\ldots(\alpha+n-1)}{n!\alpha^n}\left(\frac{n}{k}\right)^{n+1}\left(1+\frac{n}{\alpha k}\right)^{-\alpha-n}$$

$$= Lt\,\frac{\Gamma(\alpha+n)}{\Gamma(\alpha)n!}\,\alpha\left(1+\frac{\alpha k}{n}\right)^{-\alpha-n}\left(\frac{n}{\alpha k}\right)^{1-\alpha}$$

$$= Lt\,\frac{\alpha}{\Gamma(\alpha)}\sqrt{\frac{n+1}{n+\alpha}}\,e^{-\alpha-n+n+1}\,\frac{(\alpha+n)^{\alpha+n}}{(1+n)^{n+1}}\left(1+\frac{\alpha k}{n}\right)^{-\alpha-n}\left(\frac{n}{\alpha k}\right)^{1-\alpha}$$

$$= \frac{\alpha^\alpha k^{\alpha-1}e^{-\alpha k}}{\Gamma(\alpha)}. \tag{2–37}$$

These are well-known results, serving to confirm this method approach.

There is a numerical method of getting $\Gamma(k)$ from a discrete series of values of $U(t)$. In the integral

$$U(t) = \int_0^\infty \Gamma(k)e^{-kt}\,dk$$

let

$$e^{-\sigma k} = \cos\theta, \qquad \Gamma\left(\frac{1}{\sigma}\csc\theta\right) = \Phi(\theta) \tag{2–38}$$

so that

$$\sigma U(t) = \int_0^{\pi/2} (\cos\theta)^{(t/\sigma)-1}\sin\theta\Phi(\theta)\,d\theta. \tag{2–39}$$

Then

$$\sigma U((2n+1)\sigma) = \int_0^{\pi/2} \cos^{2n}\theta\sin\theta\Phi(\theta)\,d\theta. \tag{2–40}$$

Now think of $\Phi(\theta)$ as a Fourier series,

$$\Phi(\theta) = \sum_{m=0}^{\infty} c_m \sin(2m+1)\theta \tag{2–41}$$

and determine the coefficients as follows. Since

$$\cos^{2n}\theta\sin\theta = 2^{2n}\sum_{k=0}^{n}\left\{\binom{2n}{k}-\binom{2n}{k-1}\right\}\sin(2(n-k)+1)\theta, \quad \text{where } \binom{2n}{-1}=0. \tag{2–42}$$

As we can use the orthogonality of the sine functions,

$$\int_0^{\pi/2}\sin(2k+1)\theta\sin(2m+1)\theta\,d\theta = \frac{\pi}{4}\delta_{km}, \tag{2–43}$$

to give

$$\left[\binom{2n}{n}-\binom{2n}{n-1}\right]c_0+\cdots+\left[\binom{2n}{k}-\binom{2n}{k-1}\right]c_{n-k} + \cdots + c_n = \frac{4^{n+1}}{\pi}\sigma U((2n+1)\sigma), \tag{2–44}$$

this gives recursively

$$\begin{aligned} c_0 &= \frac{4}{\pi}\sigma U(\sigma),\\ c_1 &= \frac{4^2}{\pi}\sigma U(3\sigma)-\frac{4}{\pi}\sigma U(\sigma),\\ c_2 &= \frac{4^3}{\pi}\sigma U(5\sigma)-3\frac{4^2}{\pi}\sigma U(3\sigma)+\frac{4}{\pi}\sigma U(\sigma). \end{aligned} \tag{2–45}$$

For small k, σ should be large and conversely, since

$$\Gamma(k) = \sum_{0}^{\infty} c_m \sin[(2m+1)\cos^{-1}e^{-\sigma k}]. \tag{2–46}$$

Figure 2–1 summarizes this method. When the behavior of $U(t)$ is known for large t we can make some statements about $\Gamma(k)$ for small k. For example,

$$\Gamma(0) = \underset{t\to\infty}{Lt}\, tU(t) \tag{2–47}$$

when this limit exists. Also if the asymptotic behavior of $U(t)$ is dominantly exponential, with

$$U(t) \sim e^{-\lambda t} V(t) \tag{2–48}$$

where V is bounded as $t \to \infty$, then

$$\Gamma(k) = 0, \quad 0 \le k < \lambda. \tag{2–49}$$

This follows immediately from the shift operation. We can also use the convolution theorem to see that if $U(t)$ can be decomposed into the product

$$U(t) = U_1(t) U_2(t) \tag{2–50}$$

with two factors being derived from distributions $\Gamma_2(k)$ and $\Gamma_2(k)$s, then

$$\Gamma(k) = \int_0^k \Gamma_1(k - l)\Gamma_2(l)\, dl. \tag{2–51}$$

It is sometimes helpful to let U_1 account for the asymptotic behavior so that $U_2 \sim 1$.

If $U(t)$ can be expanded in an absolutely convergence series of the form $U(t) = \sum_1^\infty a_n t^{-\lambda_n}$, $0 < \lambda_1 < \lambda_2 <, \ldots,$ then $\Gamma(k) = \sum_1^\infty a_n k^{\lambda_n - 1}/\Gamma(\lambda_n)$ is convergent for all k. This includes the special case of $\lambda_n = n$. If an expansion

$$U(t) = \sum_1^\infty a_n e^{-knt}(t + \mu_n)^{-\lambda_n} \tag{2–52}$$

is possible, then

$$\Gamma(k) = \sum_1^\infty \frac{a_n}{\Gamma(\lambda_n)} (k - v)^{\lambda_n - 1} e^{-\mu_n(k - v_n)} H(k - v_n). \tag{2–53}$$

Of course, the normalization condition requires that

$$\sum_1^\infty a_n \mu_n^{-\lambda_n} = 1. \tag{2–54}$$

If the expansion contains isolated exponentials, i.e., terms such as $a_n e^{-v_n}$ with $\lambda_n = 0$, then $\Gamma(k)$ contains discrete components $a_n \delta(k - v_n)$. We shall see that most reversible reactions fall within this category.

An irreversible autocatalysis, normalized to $R(1) = 1$, is given by

$$-\dot{U} = \frac{1}{\varepsilon} U(1 + \varepsilon - U) \tag{2–55}$$

and its solution is

$$U(t) = \frac{1 + \varepsilon}{1 + \varepsilon e^{(1+\varepsilon)t/\varepsilon}} = \frac{1 + \varepsilon}{\varepsilon} e^{-(1+\varepsilon)t/\varepsilon}\left\{1 + \frac{1}{\varepsilon} e^{-(1+\varepsilon)t/\varepsilon}\right\}^{-1}$$

$$= (1 + \varepsilon) \sum_{1}^{\infty} (-)^{n-1} \varepsilon^{-n} e^{-n(1+\varepsilon)t/\varepsilon}. \tag{2–56}$$

This expansion converges for $t > \varepsilon \ln \varepsilon/(1 + \varepsilon)$, which vanishes in the limit $\varepsilon \to 0$. The spectrum of the mixture is

$$\Gamma(k) = (1 + \varepsilon) \sum_{1}^{\infty} (-)^{n+1} \varepsilon^{-n} \delta\left(k - n \frac{1 + \varepsilon}{\varepsilon}\right), \tag{2–57}$$

and the discrete components alternate in sign, thus raising the eyebrows of the physically minded.

It is interesting to enquire what conditions must be imposed on $U(t)$ to ensure that $\Gamma(k)$ is positive. We take a theorem from Doetsch (*Guide for the Applications of the Laplace Transform*, 1961, p. 204). If $\Gamma(k)$ is real and has n changes of sign in $k > 0$, then the derivatives $U^{(m)}(t)$ have exactly n real zeros for all sufficiently large m. If $\Gamma(k)$ changes sign at k_0, the corresponding zero of $U^{(m)}(t_m)$ satisfies $\lim(m/t_m) = k_0$. Note that a zero does not necessarily imply a change of sign. Thus, $\Gamma(k)$ is positive if $U^{(m)}(t)$ has no zeros for large m. This is a difficult test unless there is a formula for the nth derivative. Even so simple a curriculum as

$$U(t) = A + B/(e^t + C)$$

seems to have no such formula. In the Appendix we show that if U satisfies $-\dot{U} = R(U)$, then $(-)^{n+1} U^{(n+1)}(t)$ can be written as

$$\sum B_n \prod R^{(a_1)} R^{(a_x)} \ldots R^{(a_n)}$$

where

$$a_1 + a_2 + \cdots a_n = n, \quad a_1 \geq a_2 \geq \cdots \geq a_n \geq 0$$

a partition of n into n parts (using zero). The numbers B are each associated with a partition and are calculated by means of a recurrence relation. Certainly, $\Gamma(k)$ will be positive if none of the derivatives of R has a zero for $0 < U \leq 1$ (apart from the possibility of its being identically zero), i.e., but this is probably a wastefully sufficient condition. If the derivatives alternate in sign, then, since $\sum (a_n - 1) = 0$, each term in the sum of products is positive and again we can ensure the positivity of $\Gamma(k)$.

REVERSIBLE KINETICS

The curriculum of a first-order reversible reaction is

$$u(x, t) = u(x, 0)\frac{1 + K(x)e^{-xt}}{1 + K(x)} = u_e(x) + u(x, 0)\frac{K(x)}{1 + K(x)} \tag{2–58}$$

giving, for the lump,

$$U(t) = U_e + \int_0^\infty u(x, 0)\frac{K(x)}{1 + K(x)}e^{-xt}\,dx, \tag{2–59}$$

where

$$U_e = \int_0^\infty u_e(x)\,dx = \int_0^\infty u(x, 0)\frac{dx}{1 + K(x)}. \tag{2–60}$$

If this has to imitate a second-order reaction $2A_1 \rightleftharpoons 2A_2$ with $U(0) = K_2^2 = k_1/k_2$, for which the curriculum is

$$U(t) = U_e + \frac{2K_2}{(1 + K_2)^2}\frac{e^{-k_2t}}{1 - [(K_2 - 1)/(K_2 + 1)]e^{-2k_2t}} \tag{2–61}$$

for some k_2, K_2 by U_e: note $U_e = 1/(1 + K_2)$ since $U(0) = 1$ if

$$u(x, 0) = \frac{1 + K(x)}{K(x)}\frac{2K_2}{(1 + K_2)^2}\sum_{n=0}^{\infty}\left\{\frac{K_2 - 1}{K_2 + 1}\right\}^n\delta(x - 2(n + 1)k_2), \tag{2–62}$$

then

$$\begin{aligned} U(t) &= U_e + \frac{2K_2}{(1 + K_2)^2}\sum_0^\infty\left\{\frac{K_2 - 1}{K_2 + 1}\right\}^n e^{-2k_2t(n+1)} \\ &= U_e + 2U_e(1 - U_e)\sum_0^\infty (1 - 2U_e)^n e^{-2k_2(n+1)t}. \end{aligned} \tag{2–63}$$

We confirm $U(0) = 1$ by observing that

$$U_e = \int_0^\infty \frac{u(x, 0)}{1 + K(x)}\, dx = 1 - \int_0^\infty \frac{K(x)}{1 + K(x)}\, u(x, 0)\, dx = 1 - \frac{2U_e(1 - U_e)}{(1 - 1 + 2U_e)}.$$

Thus, the mixture has a denumerable number of components, $x = 2(n + 1)k_2$, $n = 0, 1, \ldots,$ the nth being present in concentration

$$2\{1 + 1/K(2k_2(n + 1))\}U_e(1 + U_e)(1 - 2U_e)^n.$$

If $U_e > 1/2$, alternate concentrations are negative and the kinetics cannot be simulated by non-negative distributions.

In general, a reversible reaction can be written as

$$R(U) = (U - U_e)f(U)/(1 - U_e)f(1),$$

where $f(U)$ has no zeros in $1 \geq U \geq U_e$ and

$$t = \int_U^1 \frac{du}{R(u)} = \int_{U_\varepsilon}^1 \frac{du}{R(u)} + \frac{f(1)(1 - U_e)}{f(U_\varepsilon)} \int_U^{U_e} \frac{f(U_\varepsilon)}{f(u)} \frac{du}{u - U_e}.$$

If U_ε is chosen so that $1 - \varepsilon \leq |f(U_\varepsilon)/f(u)| \leq 1 + \varepsilon$ for k in (U_e, U_ε), and l_ε denotes the integral from U_ε to 1, then

$$(1 - \varepsilon) \ln \frac{U_\varepsilon - U_e}{U - U_e} < \frac{(kt - l_\varepsilon)f(U_e)}{(1 - U_e)f(1)} < (1 + \varepsilon) \ln \frac{U_\varepsilon - U_e}{U - U_e}$$

or

$$(U_\varepsilon - U_e)e^{-(\hat{k}t - \hat{l})/(1 + \varepsilon)} \leq U(t) - U_e \leq (U_\varepsilon - U_e)e^{-(\hat{k}t - \hat{l})/(1 - \varepsilon)}$$

where

$$\hat{k} = f(U_e)/(1 - U_e)f(1).$$

Thus, we should expect the most general reversible reaction to have a $\Gamma(k)$ that is zero for

$$k < \operatorname*{Lt}_{U \to U_e} R(U)/(U - U_e).$$

This limit can be zero if, for example,

$$R(U) = (U - U_e)^p f(U)/(1 - U_e)^p f(1).$$

THE INTERPRETATION OF DISTRIBUTIONS

Though we have shown that the excursion into many indices was unnecessary in the sense that the interpretation of the kind is clearer in the multi-index case. Thus, Eq. (2–16) has been shown to prepresent a Γ-distribution of nth-order reactants where α is the parameter of the distribution and $\beta = 1/(n - 1)$ (Aris 1989). Alternatively, it is a completely segregated n^{th} order reactant in a sequence of β stirred tanks, $n = (\alpha + 1)/\alpha$.

The great difficulty of the present approach which on the face of it claims that any monotone lumped kinetics can be represented by a simple spectrum is that it involves the inversion of a Laplace transform in an inherently ill-conditioned process. Moreover, two functions that have the same Laplace transforms are equal almost everywhere and certainly at points of continuity. Yet we are constantly seeing delta-functions arising naturally out of asymptotic behavior. The absence of a workable test for the positivity of the spectrum is also a difficulty. The greater flexibility of Astarita's uniform kinetics seems to get around some of these difficulties.

ACKNOWLEDGMENTS

I am greatly indebted to correspondents T. C. Ho and Charles Wittman for keeping me in touch with present and past work and particularly to a stimulating collaboration with G. Astarita.

APPENDIX

A partition of n into m parts is a set of m integers $\{a_1, \ldots a_m\} = \varpi^n_m$, such that

$$a_1 \geq a_2 \geq \cdots \geq a_m \geq 1 \quad \text{and} \quad a_1 + a_2 + \cdots + a_m = n. \tag{A–1}$$

$P(n, m)$ is the number of distinct ways in which this division of n into m parts may be effected, and

$$P(n) = P(n, 1) + P(n, 2) + \cdots + P(n, n). \tag{A–2}$$

An elementary theorem in the theory of partitions gives

$$P(n) = P(2n, n) \tag{A–3}$$

and a beautiful, but recondite, theorem of Ramanujan and Hardy gives

$$P(n) \sim \exp(\pi\sqrt{2n/3})/4n\sqrt{3}, \quad n \to \infty. \tag{A–4}$$

Clearly,

$$P(n, 1) = P(n, n) = 1. \tag{A–5}$$

Let the partitions of n be arranged in descending order; i.e., $\varpi_p^n = \{a_{1_p}, a_{2_p}, \ldots a_{m_p}\}$ precedes $\varpi_q^n = \{a_{1_q}, \ldots a_{m_q}\}$ if the first a_{r_p} that is not equal to a_{r_q} is greater that it. Thus,

n	$P(n)$								
1	1	ϖ_1^1	1	0	1				
2	2	ϖ_1^2	2	0	0	1			
		ϖ_2^2	1	1	0	1			
3	3	ϖ_1^3	3	0	0	0	1		
		ϖ_2^3	2	1	0	0	4		
		ϖ_3^3	1	1	1	0	1		
4	5	ϖ_1^4	4	0	0	0	0	1	
		ϖ_2^4	3	1	0	0	0	7	
		ϖ_3^4	2	2	0	0	0	4	
		ϖ_4^4	2	1	1	0	0	11	
		ϖ_5^4	1	1	1	1	0	1	
5	7	ϖ_1^5	5	0	0	0	0	0	1
		ϖ_2^5	4	1	0	0	0	0	11
		ϖ_3^5	3	2	0	0	0	0	15
		ϖ_4^5	3	1	1	0	0	0	32
		ϖ_5^5	2	2	1	0	0	0	34
		ϖ_6^5	2	1	1	1	0	0	26
		ϖ_7^5	1	1	1	1	1	0	1

where array of $P(n)$ rows for n has been filled out to $(n + 1)$ columns by added zeros. Let the entries of such an $P(n) \times (n + 1)$ array be denoted by A_{ij}^n, so that $A_{ij}^n = a_{ji}$ of ϖ_i^n if this exist and zero otherwise.

If $DU = \dot{U} = -R(U)$ and $D^n(U)$ denotes the nth derivative of $U(D^0U = U)$, then

$$-DU = R,$$

$$D^2U = DR.R,$$

$$-D^3U = D^2R.R^2 + (DR)^2R,$$

$$D^4U = D^3R.R^3 + 4D^2R.DR.R^2 + (DR)^3R,$$

$$-D^5U = (D^4R)R^4 + 7(D^2R)(DR)R^3 + 4(D^2R)R^3 + 11(D^2R)(DR)^2R^2 + (DR)^4R.$$

We see that in the expression for $D^{(n+1)}U$, each term has $(n + 1)Rs$ while the powers of D involved are a partition of n, i.e.,

$$(-)^{n+1}D^{(n+1)}U = \sum_{i=1}^{P(n)} B_i^n \prod_{j=1}^{n+1} D^{A_{ij}^n}R,$$

where B_i^n are integer coefficients we would like to generate recursively. Clearly,

$$B_1^n = B_{P(n)}^n = 1.$$

Let a partition ϖ_p^n be said to be an offspring of ϖ_q^{n-1} if its first n columns A_{pj}^n differ from the n columns of ϖ_q^{n-1} by 1 in one, and only one, column. However, two offspring are indistinguishable members of an r-plet if, when ordered by Eq. (A–1) they give the same offspring. Thus, ϖ_1^3 has two distinguishable offspring ϖ_1^4 and ϖ_2^4 but ϖ_2^4 is really a set of indistinguishable triplets since (3, 1, 0, 0) (3, 0, 1, 0) and (3, 0, 0, 1) are all equivalent to (3, 1, 0, 0), which is properly ordered. If ϖ_p^n is an offspring ϖ_q^{n-1}, then ϖ_q^{n-1} is a progenitor of ϖ_p^n. Let r_{pq}^n be the number of indistinguishable ϖ_p^n offspring of ϖ_q^{n-1}, e.g., $r_{21}^4 = 3$, $r_{33}^5 = 2$, $r_{53}^5 = 3$. Then

$$B_p^n = \sum_q r_{pq}^n B_q^{n-1},$$

where the summation is taken over all progenitors of ϖ_p^n.

The final column is B_p^n and to take an example, we note that ϖ_5^5 has progenitors ϖ_4^4 and ϖ_3^4. Now $r_{54}^5 = 2$, $r_{53}^5 = 3$, $B_4^4 = 11$, $B_3^4 = 4$ so $B_5^5 = 2.11 + 4.3 = 34$. The sequence $B_1^6, \ldots, B_{11}^6$ is

1, 11, 26, 76, 15, 252, 96, 136, 180, 57, 1.

They are not, so far as I know, any of the standard combinatorial sequences.

NOMENCLATURE

k	First-order rate constant
$R(U)$	kinetic law for lump
t	time
U	total concentration of lump
U_e	total concentration of equilibrium
u	concentration density in mixture
$w_{k,m}$	Whittaker's hypergeometric function (2–18)
x_i	i^{th} index
w	weighting function
α, β	parameters in Γ distinction
$\Gamma(k)$	reduced distribution
δ	delta function
τ	dimensionless function
φ	Tricomi's function (2–20)
Ω	index domain
Ω_k	hypersurface of constant k in Ω.

REFERENCES

Aris, R. (1989) Reactions in continuous mixtures. *A.I.Ch.E.J.* **35**:539, 1398.

Astarita, G. (1989) Lumping nonlinear kinetics: apparent overall order of reaction. *A.I.Ch.E.J.* **35**:529.

Astarita, G. and R. Ocone (1988) Lumping nonlinear kinetics. *A.I.Ch.E.J.* **34**:1299.

Chou, M. Y. and T. C. Ho (1989) Lumping continuous nonlinear reactions in continuous mixtures. *A.I.Ch.E.J.* **35**:533.

deDonder, Th. (1931) *L'affinite.* 2^e partie. Gauthier-Villars, Paris.

Erdelyi, A., W. Magnus, F. Oberhettinger, and F. G. Tricomi. (1954) *Tables of integral transforms.* McGraw-Hill, New York.

3

Continuous Mixtures in Fluid Catalytic Cracking and Extensions

F. J. KRAMBECK

THE FCC PROCESS

The cracking of heavy petroleum fractions into lighter ones is the oldest and best example of a chemical process involving so many individual compounds that a continuum approach seems very natural. Fluid catalytic cracking (FCC) is the currently dominant version of this technology. In this process the heavy feed, generally produced by vacuum distillation of the bottoms from atmospheric distillation of crude oil, is contacted with hot (about 750°C) catalyst particles in the 20–100-μm size range, thus vaporizing and heating the feed to reaction temperature of about 550°C. The weight ratio of catalyst to feed is generally about 6:1. The mixture then flows upward in an entrained bed reactor for several seconds until the catalyst and products are separated in cyclones. Many complex reactions occur but the ones of primary interest are those that crack large molecules into smaller ones and thus reduce their boiling point to the more useful range of gasoline and light fuel oil. At the same time a small portion of the feed, typically about 5%, is transformed into a very high boiling material that adheres to the catalyst and is called coke. After separation of the catalyst from the products, the catalyst is regenerated by burning the coke deposits, thus raising the catalyst temperature for another cycle. Typical distillation curves, showing the percent of material vaporized versus temperature, of the feed and products at different

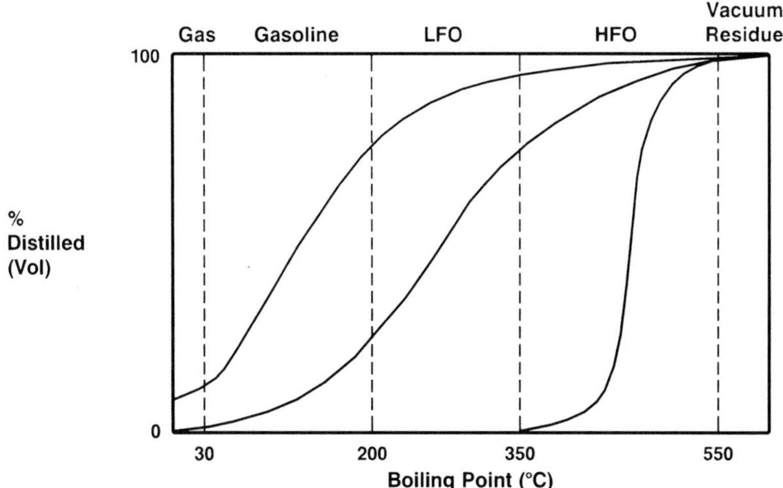

Figure 3–1 Typical distillation curves.

extents of reaction are shown in Figure 3–1. The figure also shows typical boiling point boundaries for the various petroleum products.

OVERALL CONVERSION KINETICS

Because gasoline is the most valuable product resulting from the process the extent of reaction is generally expressed as the conversion of the feed to materials in the gasoline boiling range or lighter, typically about 200°C as shown. This is just the ordinate of Figure 3–1 at 200°C. Early studies of the variation of conversion with space velocity showed that the unconverted material (that boiling above 200°C) decays according to a second-order rate law (Blanding 1953). This is in spite of the fact that individual hydrocarbons are known to crack according to first-order kinetics. The reason that the apparent kinetics are higher than first order is that there are a great many compounds with different reaction rates, so that the fastest reacting species are depleted fastest. Thus the reaction rate slows down faster with conversion than that of a single compound would.

The particular significance of second-order kinetics was pointed out earlier (Krambeck 1984) and has been more recently discussed by Ho and Aris (1987) and by Krambeck (1988). We assume that a particular fraction contains a very large number of species whose first-order reaction rate coefficients k can be described by a probability distribution $f(k) = f_0(k) \exp(-kt)$, where t is the reaction time and $f_0(k)$ is the distribution at time $t = 0$. Then the total amount of the lump remaining at time t, $w(t)$, is given by the Laplace transform of $f_0(k)$

$$w(t) = \hat{f}_0(t) = \int_0^\infty f_0(k) \exp(-kt)\, dk. \tag{3–1}$$

Because of the limit properties of Laplace transforms (see, e.g., Feller 1966) the asymptotic behavior of $w(t)$ at large times depends only on the behavior of $f_0(k)$ at small values of k. In this context it is because those species with larger values of k disappear quickly. More specifically,

$$\lim_{t \to \infty} w(t) \sim t^{-q} \qquad \text{iff} \quad \lim_{k \to 0} f_0(k) \sim k^{q-1}. \tag{3–2}$$

This is valid for any q greater than zero, not, as some have implied, only for $q > 1$. The behavior at large times given by Eq. (3–2) is equivalent to that of power law kinetics with an apparent reaction order of $1 + 1/q$. Thus whenever $f_0(k)$ has a finite intercept at $k = 0$, $q = 1$ and apparent second-order kinetics will result at large times. In practice it turns out that not very much time is required to reach that asymptote.

The second-order approximation has been very useful in FCC as shown below. However it should be clearly understood that this is only an empirical observation for this particular system and not every system of parallel first-order reactions with large numbers of species will have distributions $f_0(k)$ with intercepts greater than zero and less than infinity. One very significant exception is, in fact, the steam deactivation of FCC catalyst, which is also discussed below. This follows apparent kinetics significantly higher than second order, implying a distribution $f_0(k)$ with an infinite intercept at $k = 0$.

CATALYST DEACTIVATION BY COKE

The activity of FCC catalyst at typical operating conditions is extremely high, with only a few seconds being required for reaction. However the coke that deposits on the catalyst during this time, on the order of 0.5%–1% by weight of the catalyst, dramatically decreases its activity. It was shown by Voorhies (1945) that the rate of coke buildup follows an equation of the form

$$\frac{dc}{dt} = \phi(c) = kc^{1-1/b} \tag{3–3}$$

or

$$c = (kt/b)^b. \tag{3–4}$$

Here t is the time that the catalyst is exposed to the reaction mixture and c is the coke content of the catalyst in wt% of the clean catalyst. The value of b in Eq. (3–3) or (3–4) was about one-half for the old nonzeolitic catalyst in use in the 1940s and 1950s but modern cracking catalysts containing Y-type zeolite have values of b of approximately 1/3.

The deactivation of the catalyst for cracking reactions essentially parallels its deactivation for coke production, so the same function, $\phi(c)$, can be used to describe both. The conversion and coke kinetics can then be combined together. Here the coke production is expressed in terms of its yield as a percent of feed oil, u

$$\frac{dw}{d\tau} = -k\phi(c)w^2, \tag{3–5}$$

$$\frac{du}{d\tau} = \phi(c). \tag{3–6}$$

Here w is the unconverted percentage of the feed oil and τ is the space time (W_{cat}/F_{oil}) or $(t \cdot F_{cat}/F_{oil})$. Note that by dividing Eq. (3–5) by Eq. (3–6) we obtain a relationship just between conversion and coke yield with no explicit dependence on space velocity or residence time

$$\frac{dw}{du} = -kw^2. \tag{3–7}$$

Integrating gives

$$\frac{100 - w}{w} = ku, \tag{3–8}$$

or, in terms of conversion, x,

$$\frac{x}{100 - x} = ku. \tag{3–9}$$

Figure 3–2 shows a comparison of this equation with some actual cracking data on a number of different feedstocks. The figure illustrates how well the equation matches the data and, also, how very different is the coke selectivity of various feedstocks.

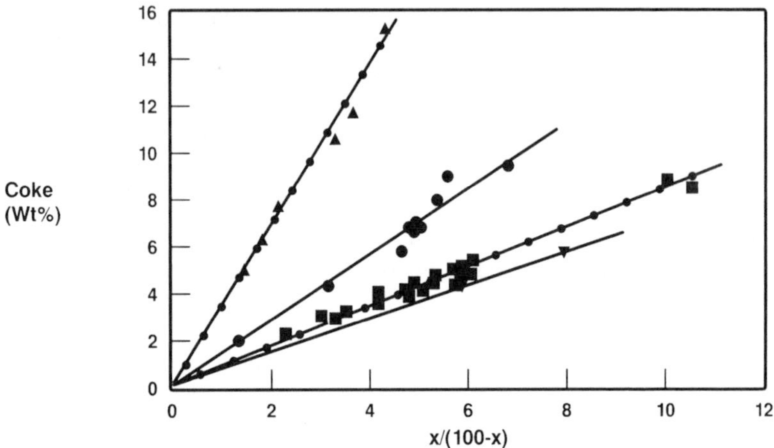

Figure 3–2 Coke yield versus conversion.

GASOLINE SELECTIVITY

Of the material converted in FCC most is converted to gasoline, but a significant fraction is also converted to lighter compounds with from one to four carbon atoms. The split between these is very significant economically with the gasoline having the higher value. Thus the next question to be addressed by FCC kinetics is the yield of gasoline. This was approached by Weekman and Nace (1970) by assuming that the gasoline formed in the above second-order reaction itself cracks by a first-order reaction. This is fairly successful over the range of conditions considered. However a more general approach, which can be applied to predict the entire boiling point distribution of FCC products, is to assume that the lump of material boiling heavier than any particular cutpoint temperature follows second-order kinetics with a rate coefficient that depends on the cutpoint. In particular this applies to the initial boiling point of the gasoline fraction (about 40°C). This gives, for the gasoline yield g,

$$g = \frac{(1 - k_g)(100 - x)}{100 - (1 - k_g)x}. \tag{3–10}$$

This expression is compared with some typical FCC data for a number of different feedstocks in Figure 3–3 and for the same feedstocks for two different catalysts in Figure 3–4. The ordinate in these tables is the gasoline

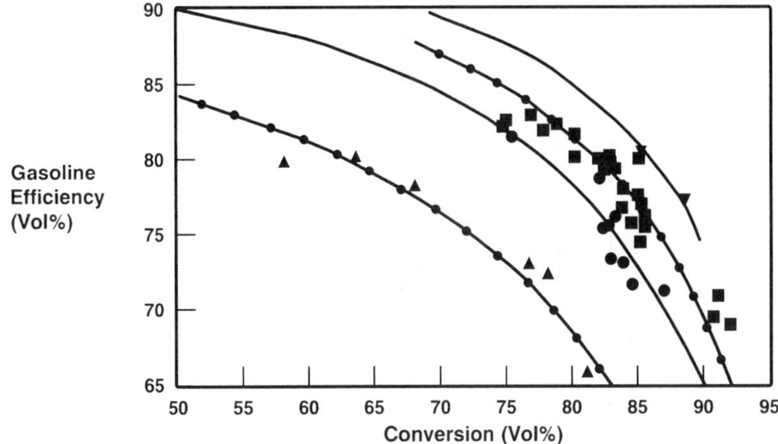

Figure 3–3 Gasoline selectivity; feedstock effect.

efficiency, defined as g/x. It can be seen that this equation does quite a good job of predicting the effect of conversion on gasoline yield.

Since Eq. (3–10) contains only the one parameter k_g it can be used to predict an entire gasoline versus conversion curve from one point or to determine if several experimental points lie on the same curve. Figure 3–5 shows a family of such curves and is very useful for analyzing experimental data, interpolating, and extrapolating.

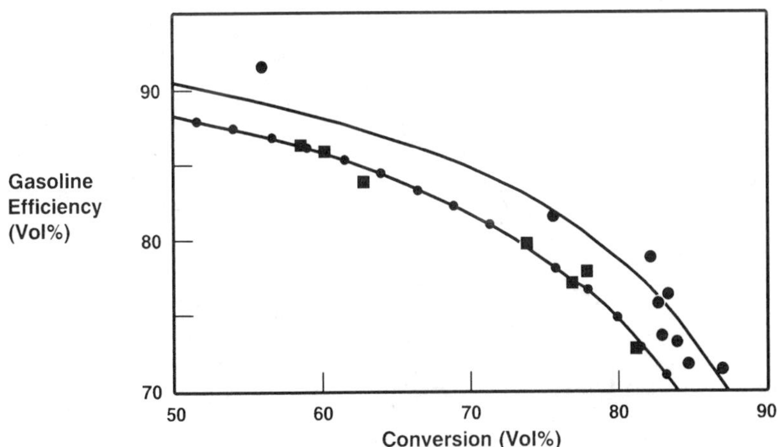

Figure 3–4 Gasoline selectivity; catalyst effect.

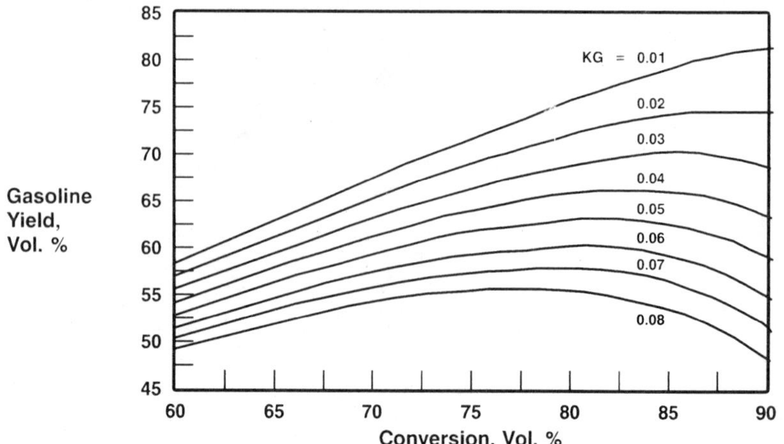

Figure 3–5 Second-order gasoline selectivity curves.

BOILING POINT DISTRIBUTION

The above equations can be generalized to predict the entire boiling point distribution

$$w_0(T)/w(T) - 1 = k(T)x/(100 - x). \tag{3–11}$$

Here $w_0(T)$ is the percent of the feedstock boiling above temperature T and $w(T)$ is the percent of the total reaction products boiling above temperature T. In particular $k(200°C) = 1$. It is assumed here that all the feedstock

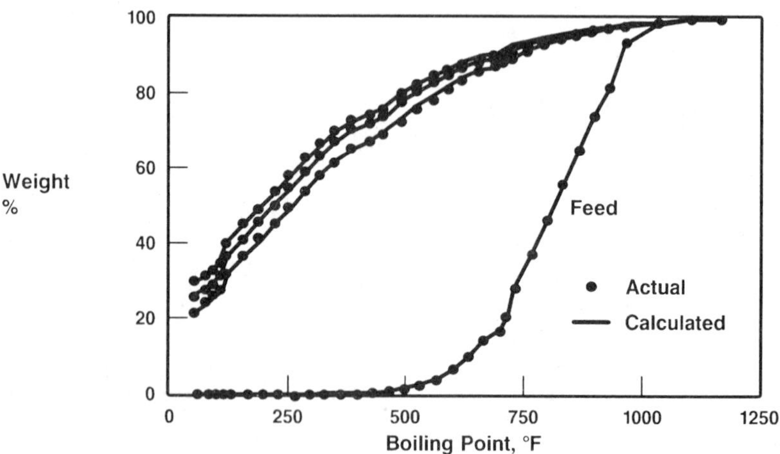

Figure 3–6 Boiling point distribution.

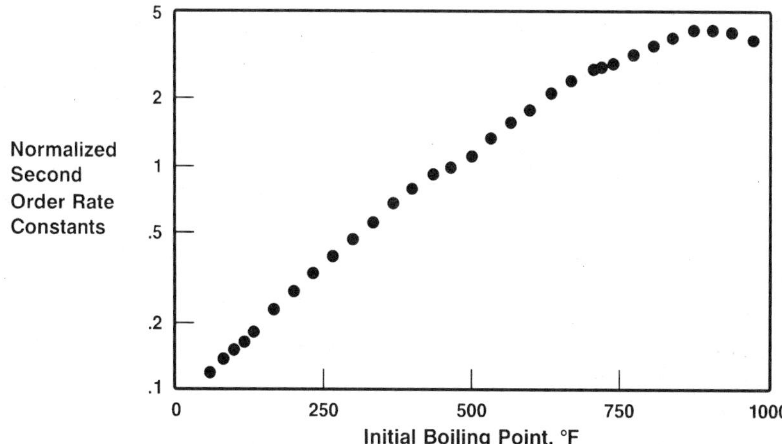

Figure 3–7 Rate coefficient versus cut temperature. Courtesy of Raymond Hu, Mobil R & D.

boils above 200°C [$w_0(200°\text{C}) = 100$]. Equation (3–11) is compared with some typical data in Figure 3–6. The ordinate in that figure is the percent boiling *below* T [i.e., $100 - w(T)$] versus T. Again, good predictions can be made of the effect of conversion. The corresponding rate coefficients $k(T)$ are shown in Figure 3–7.

MIXING EFFECTS

One rather subtle point about Eq. (3–11) is that it is the *ratio* $w(T)/w_0(T)$ that disappears by second-order kinetics rather than $w(T)$ itself. This is because the underlying kinetics of cracking are first order as explained above. For the same reason, the effect of mixing is also much different than it would be for true second-order kinetics. If $r(t)$ is a residence time distribution function,

$$\frac{w}{w_0} = \int_0^\infty r(t)w(t)\,dt = \int_0^\infty \int_0^\infty r(t)f_0(k)\exp(-kt)\,dk\,dt, \qquad (3\text{–}12)$$

where w/w_0 is the outlet unconverted fraction.

As an example, suppose $r(t) = (1/\tau)\exp(-t/\tau)$, the residence time distribution for a well-mixed reactor with overall space–time τ. The unconverted fraction, $w(\tau)/w_0$, would then be given by

$$\frac{w(\tau)}{w_0} = \int_0^\infty \frac{(1/\tau)\exp(-t/\tau)}{1 + kt}\,dt = \frac{e^{1/k\tau}}{k\tau}\,E_1\!\left(\frac{1}{k\tau}\right). \qquad (3\text{–}13)$$

Here $E_n(x)$ is the exponential integral as defined, for example, by Abramowitz (1964). This is quite different from a true second-order reaction in a well-mixed reactor

$$\frac{w(\tau)}{w_0} = \frac{1}{2k\tau w_0}\left(\sqrt{1 + 4k\tau w_0} - 1\right). \tag{3–14}$$

Of course it is difficult to imagine carrying on the FCC reactions in a well-mixed reactor but the true mixing situation in the bottom zone of an FCC riser reactor is far from plug flow. Thus if one can model the mixing situation in this zone, Eq. (3–12) can be used to evaluate the impact of imperfect mixing on the cracking unit performance.

CATALYST DEACTIVATION BY STEAM

The rapid catalyst deactivation that occurs by coke in the FCC reactor is reversible by burning the coke deposits off the catalyst in the regenerator. However there is also a slow irreversible deactivation process that occurs as the catalyst is exposed in the regenerator to high temperatures (700–750°C) in the presence of water vapor that results from the hydrogen content of the burning coke (H/C $\sim$ 1). This deactivation occurs over a period of months and requires the addition of a fresh catalyst to the unit to maintain catalyst activity. It has been found that this type of deactivation also follows a power-law type of kinetics with an order of 2–5, depending on the type of catalyst. This is undoubtedly due to the presence in each catalyst particle of a large number of active sites that deactivate at different rates. The orders higher than 2 correspond to values of $q < 1$ in Eq. (3–2) and imply a pole at the origin for the rate distribution $f_0(k)$. On the time scale of the catalyst makeup rate, about 1% of the inventory per day, the catalyst very closely approximates perfect mixing. Since each catalyst particle deactivates independently of the others the integration of Eq. (3–12) is certainly valid. For apparent reaction order n, Eq. (3–13) then becomes

$$\frac{w(\tau)}{w_0} = \frac{e^{1/(n-1)k\tau}}{(n-1)k\tau} E_{[1/(n-1)]}[1/(n-1)k\tau]. \tag{3–15}$$

LIMITATIONS OF THE CONTINUOUS APPROACH

The approach described above is very useful for predicting how an operating unit will change from a given initial operation in response to various adjustments to the controllable variables. It is a relatively simple matter to add the effects of temperature and pressure through activation energies (about 56 kJ/mol for cracking, about 0 for coking) and partial pressure exponents. However, the performance with feedstocks of differing chemical composition can vary dramatically as shown in Figures 3–2 and 3–3 and there is no straightforward way of incorporating these effects except through various empirical correlations. Thus once one gets beyond the simpler types of operating condition variations toward real refinery optimization, a more fundamental treatment in terms of chemical composition is required.

Another problem arises when one examines the chemistry in more detail. The reactions are not strictly irreversible, in that condensation reactions as well as cracking reactions occur. Indeed coke formation is such a process. Thus it is interesting to consider how this type of model might be extended to begin to incorporate more chemical information and allow for reversibility.

MORE GENERAL CONTINUOUS MIXTURE MODELS

One particular way of analyzing the general properties of chemical kinetic equations for discrete mixtures was discussed in some detail previously (Krambeck 1970). It will now be shown how this approach can be extended to continuous mixtures.

The stoichiometry of a chemical system can be expressed in terms of a set of balance equations

$$\sum_{j=1}^{N} a_{ij}n_j = b_i, \quad i = 1, \ldots, M. \tag{3–16}$$

Here the n_j is the number of moles of the jth chemical compound. The coefficients a_{ij} express the conservation of atoms and whatever other restrictions there are on the allowable chemical transformations. Thus for a particular system of hydrocarbons the a_{ij} could include the number of carbon atoms, the number of hydrogen atoms, and also the number of aromatic rings in each molecule if the particular catalyst did not catalyze ring saturation. The number of independent constraints, M, must be at least one less than the number of compounds, N, in order for any reactions to be possible. The parameters b_i are invariants of the reactions and are calculated from the system's initial condition.

Extension of Eq. (3–16) to the continuous case is straightforward. We define a probability density function $n(x)$ to represent the mixture composition. Here x is a continuous variable that is assumed to completely define the properties of a particular compound in the mixture. The stoichiometric constraint equation then becomes

$$\int a_i(x)n(x)\,dx = b_i, \quad i = 1, \ldots, M. \tag{3–17}$$

Note that by formulating the reaction stoichiometry in terms of the constraints rather than attempting to list a set of independent reactions we require only a finite set of conditions. In general the number of constraints, M, will be no greater for a continuous system, or one with a very large number of components, than for a discrete system.

The thermodynamics for a discrete system can be expressed in terms of its Gibbs free energy as a function of temperature, pressure, and composition. It is conveniently represented in terms of chemical potentials μ_j

$$G = \sum_{j=1}^{N} \mu_j n_j. \tag{3–18}$$

The criterion for chemical equilibrium is easily seen to be

$$\mu_j = \sum_{i=1}^{M} \lambda_i a_{ij} \tag{3–19}$$

For any choice of the λ_i the μ_j from Eq. (3–19) will be orthogonal to any direction dn that leaves the b_i invariant, i.e., for any allowable reaction. Thus $dG = 0$. The λ_i are actually the Lagrange multipliers for the minimization of G subject to the constraints of Eq. (3–16). The equilibrium problem can be solved once the composition dependence of the μ_j are specified. For example if the system is an ideal solution,

$$\mu_j = \mu_j^0(T, P) + RT \ln\left(n_j \middle/ \sum_p n_p\right) \tag{3–20}$$

where μ_j^0 is the chemical potential of pure compound j. Equations (3–16), (3–19), and (3–20) may be solved by the following procedure. We first solve Eqs. (3–19) and (3–20) for the mole fractions $y_j = n_j/\Sigma\, n_p$ in terms of the Lagrange multipliers λ_i

$$y_j = \exp\left(\sum_{i=1}^{M} \lambda_i a_{ij} - \mu_j^0/RT\right). \tag{3–21}$$

Equation (3–16) may be rewritten in terms of the mole fractions by subtracting the first equation multiplied by appropriate constants from all the others.

Substituting Eq. (3–21) into it then gives M equations for the M unknowns λ_i which are then solved numerically, for example by Newton's method. Note that only M equations are involved no matter how many compounds are in the system. A derivation and computer program for this are given in the Appendix. This method was published in a preprint some time ago (Krambeck 1978) but has not appeared in a more accessible form.

We now consider the equilibrium problem for continuous mixtures. The Gibbs free energy, Eq. (3–18) becomes

$$G = \int \mu(x)n(x)\, dx \tag{3–22}$$

and the equilibrium criterion, Eq. (3–19), becomes

$$\mu(x) = \sum_{i=1}^{M} \lambda_i a_i(x). \tag{3–23}$$

The next task is recasting Eq. (3–20), the definition of an ideal solution, for a continuous mixture. Because we are assuming a very large number of compounds parameterized by the variable x, it would seem unavoidable that there would be essentially identical compounds in the immediate neighborhood of any point x. When identical compounds are lumped together for equilibrium calculations (enantiomers, for example) the pure component chemical potential must be corrected by the quantity $-RT\ln(N)$, where N is the number of compounds lumped together. This suggests the form

$$\mu(x) = \mu^0(T, P, x) - RT\ln[N(x)] + RT\ln[y(x)], \tag{3–24}$$

where $y(x)$ is a "mole fraction density" defined by $y(x) = n(x)/\int n(s)\, ds$ and $N(x)$ is a "number density" of compounds. Of course one could lump the term $-RT\ln[N(x)]$ into the function μ^0 but this would create problems in predicting μ^0 from group contributions and causes poor behavior when the scale of x is changed. Since both $N(x)$ and $y(x)$ are density functions and appear as a quotient $(y(x)/N(x))$ Eq. (3–24) is invariant with respect to change of the variable x.

One problem with Eq. (3–24) arises when a mixture contains both discrete and continuous components. In this case the function $y(x)$ will contain delta functions and $\mu(x)$ will not be defined. Since the function $N(x)$ is a property of the system and not the particular composition, it cannot be expected to have a compensating delta function at just the right place. Actually this situation could not occur at an equilibrium point unless one of the $a_i(x)$

functions contained a delta function. However, a possible way to allow for this situation would be to define a mole fraction function $y_\varepsilon(x)$ by

$$y_\varepsilon(x) = \int_{x-\varepsilon/2}^{x+\varepsilon/2} y(s)\, ds. \tag{3–25}$$

We then let $\varepsilon = 1/N(x)$ and replace Eq. (3–24) by

$$\mu(x) = \mu^0(T, P, x) + RT \ln[y_\varepsilon(x)]. \tag{3–26}$$

If $y(x)$ is smooth and $N(x)$ is fairly large, it can be seen that Eq. (3–26) approaches (3–24), and if $y(x)$ contains delta functions, Eq. (3–26) gives the discrete case Eq. (3–20).

To illustrate these concepts let us consider the example of a continuous approximation to olefin oligomerization. In this system there is only one independent stoichiometric constraint in Eq. (3–17) which expresses the conservation of carbon atoms. The hydrogen to carbon ratio of all the olefins is 2, so hydrogen balance is not independent. In this case no additional kinetic constraints arise. We take the continuous variable x to be equal to the degree of oligomerization, that is, the number of carbon atoms added to ethylene. Thus $a(x)$ in Eq. (3–17) becomes

$$a(x) = 2 + x, \quad 0 \le x < \infty. \tag{3–27}$$

In this system the function $N(x)$ increases quite sharply with carbon number. However, it has been shown by Alberty (1986) that the isomer group free energies are linear with carbon number. This is equivalent to

$$\frac{\mu^0(x)}{RT} - \ln[N(x)] = \alpha + \beta x. \tag{3–28}$$

We thus obtain from Eqs. (3–23) and (3–24)

$$y(x) = e^{2\lambda - \alpha} e^{-(\beta - \lambda)x}. \tag{3–29}$$

The constraint Eq. (3–17) in this case reduces to the requirement that $\int y(x)\, dx = 1$ (see Appendix). Defining γ as $\beta - \lambda$ gives

$$y(x) = \gamma e^{-\gamma x} \tag{3–30}$$

where γ satisfies

$$\gamma e^{\gamma} = e^{2\beta - \alpha}. \tag{3–31}$$

For situations in which γ is small, so that the distribution (3–30) extends up to very high carbon numbers, this gives a reasonable approximation to the discrete solution to this problem.

One further constraint on chemical kinetics is the principle of detailed balance which is actually equivalent to the Onsager reciprocity condition (Krambeck 1970). This can also be extended from the discrete mixture case to the continuum case. Thus we represent the reaction rate function for the discrete mixture case as $r_j(T, P, y)$, which is the rate of production of compound j per unit volume of reactor. The detailed balance condition then becomes

$$\frac{\partial r_i}{\partial \mu_j} = \frac{\partial r_j}{\partial \mu_i}. \tag{3–32}$$

at chemical equilibrium. For the ideal solutions we are considering here this can be written in terms of mole fractions by making use of Eq. (3–20)

$$\frac{\partial r_i}{\partial y_j} y_j = \frac{\partial r_j}{\partial y_i} y_i. \tag{3–33}$$

To extend this idea to continuous mixtures we need to use functional derivatives of the reaction rate expression. Thus if the reaction rate as a function of composition $y(x)$ is expressed as $r(x, y(\cdot))$, the functional derivative, $r_y(x, y(\cdot), z)$ satisfies

$$r(x, y(\cdot) + \Delta y(\cdot)) = r(x, y(\cdot)) + \int r_y(x, y)(\cdot), z)\, \Delta y(z)\, dz + 0(\Delta y^2). \tag{3–34}$$

In these terms the detailed balance condition becomes

$$r_y(x, y(\cdot), z)y(z) = r_y(z, y(\cdot), x)y(x). \tag{3–35}$$

Condition (3–35) will be satisfied at every $y(x)$ that is an equilibrium composition.

An example of such a rate expression can be constructed for the olefin oligomerization system discussed above

$$r(x, y(\cdot)) = k \int_0^\infty [e^{-\gamma x} y(z) - e^{-\gamma z} y(x)]\, dz. \tag{3–36}$$

This expression is identically zero at the equilibrium composition of Eq. (3–30), $y(x) = \gamma \exp[-\gamma x]$, and satisfies the detailed balance condition (3–35) at that composition

$$r_y(x, y(\cdot), z) = ke^{-\gamma x}, \quad x \neq z. \tag{3–37}$$

Clearly Eq. (3–37) satisfies Eq. (3–35) at equilibrium. For a plug-flow or batch reactor the reaction rate given by Eq. (3–36) may easily be integrated to give

$$y(\tau, x) = e^{-k\tau/\gamma}y(0, x) + (1 - e^{-k\tau/\gamma})\gamma e^{-\gamma x}. \tag{3–38}$$

CONCLUSION

Thus we see how to extend the continuum approach to chemical systems with more complex stoichiometry and thermodynamics than the parallel irreversible cases treated previously. One method of handling combined discrete and continuous mixtures through delta functions in the mole fraction distribution was shown, although it would probably be more convenient just to handle the discrete and continuous components in a mixture separately. The particular example treated above is rather idealized to allow simple analytical solutions but there is no basic reason why more realistic assumptions could not be incorporated, such as quadratic rather than linear kinetics. I suspect, however, that real progress in complex kinetics will only come with more detailed chemical understanding of the system through the conventional discrete component approach. Once this is achieved, however, the continuum approach may very well provide useful approximations for efficient computer simulation.

APPENDIX: CHEMICAL EQUILIBRIUM CALCULATIONS (KRAMBECK 1978)

The chemical equilibrium problem is to solve Eqs. (3–16), (3–19), and (3–20) for the unknowns n_i, λ_i, and μ_i. This can be reduced to just M equations in the λ_i by first solving for the mole fractions, y_i, in terms of the λ_i as shown in Eq. (3–21). We then rewrite Eq. (3–16) in terms of mole fractions

$$\sum_{j=1}^{N} \left[a_{ij} - \frac{b_i}{b_1} a_{ij} \right] y_j = 0, \quad i = 2, \ldots, M. \tag{A–1}$$

Note that the first equation, for $i = 1$, has been nullified. We then add the condition

$$\sum_{j=1}^{N} y_j = 1 \tag{A–2}$$

to give a complete set of M equations. Substitution of Eq. (3–21) into (A–1) and (A–2) thus gives M equations in the M unknown λ_i. These are solved using the Newton–Raphson method. Substitution into (3–21) then gives the equilibrium mole fractions and renormalization, using any row of (3–16), gives the mole numbers n_j.

The following computer program in APL accomplishes this task. To use this the global variable AS (a matrix) and LNK (a vector) are set up to describe the system

$$\text{AS} = \|a_{ij}\|, \quad \begin{matrix} i = 1, M \\ j = 1, N \end{matrix} \tag{A–3}$$

$$\text{LNK} = -\mu_j^0(T, P)/RT, \quad j = 1, N. \tag{A–4}$$

```
∇ N←EQUCALC NO;S;L;X;E;ITER;B;AC
  ⍝ SETUP
  S←AS+.×NO
  AC←AS-(S÷S[1])∘.×AS[1;]
  AC[1;]←1
  B←(⍴S)↑1
  ITER←1
  ⍝ INITIAL GUESS
  L←-(LNK+⍟⍴LNK)⌹⍉AS
  ⍝ CONVERGE ON SOLUTION
  LOOP:E←B-AC+.×X←*LNK+L+.×AS
  →((10×⎕CT)>⌈/|E)/FINISH
  L←L+E⌹AC+.×⍉AS×(⍴AS)⍴X
  →(100≥ITER←ITER+1)/LOOP
  'ALGORITHM FAILED'
  →0
  ⍝ RESCALE X TO PRODUCE N
  FINISH:N←X×1↑S÷AS+.×X
∇
```

Note that μ_j^0 is the free energy of formation, ΔG_f^0. Thus LNK is the natural log of the equilibrium constant of formation. The function is then executed by

$$\text{N} \leftarrow \text{EQUCALC NO} \tag{A–5}$$

where NO is the vector of starting mole numbers and N is the vector of equilibrium mole numbers.

NOMENCLATURE

$a_i(x)$	stoichiometric constraint function
a_{ij}	stoichiometric constraint matrix
b_i	invariant
c	coke content of catalyst, wt%
$f(k)$	distribution of reactant as a function of rate coefficient
$f_0(k)$	initial distribution as a function of rate coefficient
$f_0(t)$	Laplace transform of $f_0(k)$
G	Gibbs free energy
g	gasoline yield, % of feed
k	rate coefficient
$k(x, y)$	reaction rate coefficient function
k_g	gasoline cracking factor
M	number of stoichiometric constraints
n	reaction order
$n(x)$	mole number density function
n_j	mole number of species j
R	ideal gas constant
$r(t)$	residence time distribution
$r(x, y(\cdot))$	continuous reaction rate density function
$r_j(T, P, y)$	discrete reaction rate function
$r_y(x, y(\cdot), z)$	kernel of functional derivative of $r(x, y(\cdot))$
T	temperature
t	time
u	coke yield, wt% of feed
w	amount of lump
w_0	initial amount of lump
$y(x)$	mole fraction density function
y_i	mole fraction of species i
$y_\varepsilon(x)$	defined by eq. (3–25)
$\mu(x)$	chemical potential function
μj	chemical potential of species j
μ_j^0	chemical potential of pure species j
τ	space–time (weight basis)
λ_j	Lagrange multiplier
$\phi(c)$	deactivation function

REFERENCES

Alberty, R. A. (1986) Extrapolation of standard chemical thermodynamic properties of alkene isomer groups to higher carbon numbers. *J. Phys. Chem.* **87**:4999–5002.

Blanding, F. H. (1953) Reaction rates in the catalytic cracking of petroleum. *I/EC* **45**:1186–1197.

Feller, W. (1966) *An Introduction to Probability Theory and its Applications, Vol II*, 418–422. John Wiley & Sons, New York.

Ho, T. C. and Aris, R. (1987) On apparent second-order kinetics. *A.I.Ch.E.J.* **33**:1050.

Krambeck, F. J. (1970) The mathematical structure of chemical kinetics in homogeneous, single-phase systems. *Arch. Ration. Mech. Anal.* **38**:317–347.

Krambeck, F. J. (1978) APL in an industrial environment. In: *71st A.I.Ch.E. Annual Meeting*, Miami Beach, FL (November 16, 1978).

Krambeck, F. J. (1984) Computers and modern analysis in reactor design. *Inst. Chem. Eng. Symp. Ser.* **87**:733–754.

Krambeck, F. J. (1988) Letter to editor: *A.I.Ch.E.J.* **34**:877.

Voorhies, A. (1945) Carbon formation in catalytic cracking. *I/EC* **37**:318.

Weekman, V. W., Jr. and D. M. Nace (1970) Kinetics of catalytic cracking selectivity in fixed, moving, and fluid-bed reactors. *A.I.Ch.E.J.* **16**:397.

4

Lumping or Pseudocomponent Identification in Phase Equilibrium Calculations

STANLEY I. SANDLER

There are a number of situations that involve phase equilibrium calculations with very many components. The classic example is reservoir fluids. While the light ends of such fluids contain identifiable components, the complete speciation and quantification of all components with carbon numbers greater than about 7 is not possible at present. The problem is complicated further by both the many isomeric species and the different families of species, such as the paraffins, the aromatics, the olefins, and the naphthenes, which may be present. Though any one of these high-molecular-weight species will be present in very small concentrations, the totality of them can have an important effect on the dew point and retrograde behavior of the fluid, and on the location of the critical point. Therefore, some method of representing a large number of components is required for the thermodynamic and phase equilibrium descriptions to be sufficiently accurate for reservoir simulation or downstream processing.

There are other situations in which mixtures of very many components need to be considered. A less pleasant example is to model the weathering of an accidental oil discharge, such as that at Valdez. The environmental fate of each component in a spill will be different due to differences in volatility and solubility in water. Another example of a system with very many components is any polymer solution. As a result of the method of manufacture, a polymer is a large collection of chemically similar molecules with a molecular weight distribution about a nominal mean value. This dispersion in molecular weight can have an important effect on the liquid–liquid equilibrium behavior of polymer solutions, such as the cloud point, the binodal curve, and the plait point. Finally, aerosols and particulates are another example of systems with a dispersion in particle size which will effect their properties.

Here we will consider only vapor–liquid equilibrium in reservoir fluids which is of interest in reservoir simulation and in modeling enhanced oil recovery by miscible gas injection. Other applications include liquid–liquid equilibrium in polymer solutions which is of current interest in the aqueous two-phase extraction of biomolecules, viruses, etc. (Kang and Sandler 1988).

A characteristic of phase equilibrium calculations for mixtures is that the computer time required increases with approximately the square of the number of components. Consequently, calculations for systems with a large number of components may be very time consuming. This problem is especially serious if the phase equilibrium calculation is embedded in what is already a complex calculation, such as reservoir simulation. This chapter is concerned with the mathematical approximation of lumping, that is, representing the many components in a mixture with just a few in the context of phase equilibrium calculations.

DESCRIPTION OF RESERVOIR FLUIDS

Vogel et al. (1983) provide several from among the many published examples of the extended compositional analysis of a crude oil. The main conclusion that can be drawn from such data is that above approximately C_7 the variation of mole fraction with carbon number (or molecular weight or boiling point) becomes very regular. In particular, the mole fraction of the heavy components decreases approximately exponentially with carbon number. Here we will explore the advantage of considering a mixture of very many components to be a semicontinuous mixture, that is, a mixture of light, identifiable components and a continuous spectrum of heavy components. Thus we will represent such a mixture as consisting of a series of light components with specified properties and mole fractions, and a continuous spectrum of heavy components characterized by the index I whose composition is given by a distribution function $F(I)$ so defined that

$$\int_J^{J+dJ} F(I)\,dI = \text{mole fraction of species with the index } I \text{ between } J \text{ and } J + dJ. \quad (4\text{–}1)$$

The idea of using a distribution function to represent the composition of a continuous or semicontinuous distribution of species in a crude oil or reservoir fluid is not new. Bowman (1949) and Edmister (Bowman and Edmister 1951; Edmister and Buchanan 1953) were perhaps the first to do this, though largely in the context of Raoult's law. Only recently was this concept used again, first with the simple van der Waals equation by Gualtieri et al. (1981, 1982) and more recently with more realistic equations (Cotterman and Prausnitz 1985; Cotterman et al. 1985, 1986).

Based on the compositional data of Vogel et al. (1983) and others, we will use only the simple exponential distribution

$$F(I) = qe^{-mI} \tag{4–2}$$

to represent the heavy ends of reservoir fluids, though other distributions, such as the Gaussian and Schultz–Flory distributions, could be used and will be considered shortly. Also we will take the index I to be the carbon number, though boiling point, molecular weight, etc., could have been used. Because a number of defined components are also present, the normalization condition for this distribution is

$$\int_{C_N+0.5}^{\infty} F(I)\, dI = 1 - \sum x_i. \tag{4–3}$$

Here the summation is over all defined components (carbon number C_N and below, and the inorganic gases), and we have set the lower limit of the integral at one-half a carbon number above that of the last identified light component (C_N). The upper limit of the integral deserves some discussion. Even though only finite carbon numbers have been observed analytically in reservoir fluids, we have set the upper limit for the distribution at infinity because that makes the treatment that follows a little simpler. Behrens et al. (Behrens and Sandler 1988; Shibata et al. 1987) consider the case of a finite upper bound on the carbon number; the difference between a finite and infinite upper bound is mainly of importance for very heavy crude oils. This generalization is discussed briefly in a later section.

Because of the computationally intensive nature of reservoir engineering calculations, relatively simple equations of state, for example cubic equations of state, are used. For illustration we consider the Peng–Robinson (1976) equation

$$P = \frac{RT}{\mathbf{v} - b} - \frac{a(T)}{\mathbf{v}(\mathbf{v} + b) + b(\mathbf{v} - b)} \tag{4–4}$$

with the generalized parameters

$$a = 0.45724\,\frac{(RT_c)^2}{P_c}\left[1 + \kappa\left(1 - \sqrt{\frac{T}{T_c}}\right)^2\right], \tag{4–5}$$

$$b = 0.07780\,\frac{RT_c}{P_c},$$

and

$$\kappa = 0.37464 + 1.54226\omega - 0.26992\omega^2,$$

where P is pressure, T is temperature, and $\mathbf{v}$ is molar volume; the subscript c denotes a critical property, R is the gas constant, and ω is the acentric factor.

For a mixture of the only identified components, cubic equations of state are usually extended to mixtures using the van der Waals one-fluid mixing rules

$$a = \sum \sum x_i x_j a_{ij} \tag{4–6}$$

and

$$b = \sum x_i b_i \tag{4–7}$$

and the combining rule

$$a_{ij} = \sqrt{a_{ii} a_{jj}}(1 - k_{ij}). \tag{4–8}$$

The binary interaction parameter k_{ij} is adjusted to give the best fit of vapor–liquid equilibrium data for each of the constituent binary pairs in the mixture. Empirically, this parameter is found to be small or zero for a pair of similar species, and nonzero if the components are dissimilar. In reservoir simulation it is generally assumed that

$$k_{ij} = \begin{cases} \text{same nonzero value for an inorganic gas with all hydrocarbons} \\ \text{same nonzero value for } C_1 \text{ with all } C_{7+} \text{ hydrocarbons} \\ \text{very small, same nonzero value for } C_2 \text{ with all } C_{7+} \text{ hydrocarbons} \\ 0 \text{ otherwise} \end{cases}$$

Typically the values of these binary parameters are adjusted to match experimental data as part of the tuning process in reservoir simulation. An important empirical observation is that the value of the binary interaction parameter depends on the inorganic gas or small hydrocarbon species, but is the same for all large hydrocarbon molecules.

For semicontinuous mixtures the mixing rules become

$$b = \sum x_i b_i + \int_{\eta}^{\infty} F(I) b(I)\, dI \tag{4–9}$$

and

$$\begin{aligned} a = {} & \sum \sum x_i x_j \sqrt{a_{ii} a_{jj}} (1 - k_{ij}) \\ & + 2 \sum x_j \sqrt{a_{jj}} \int_{\eta}^{\infty} F(I) \sqrt{a(I, I)} (1 - k_{Ij})\, dI \\ & + \int_{\eta}^{\infty} \int_{\eta}^{\infty} F(I) F(J) \sqrt{a(I, I) a(J, J)} (1 - k_{IJ})\, dI\, dJ \end{aligned} \tag{4–10a}$$

where the lowercase indices designate defined components, and the uppercase indices continuous components. From the discussion above, we have that $k_{IJ} = 0$ and that k_{Ij} is independent of I. Therefore, the mixing rule for the equation of state a parameter reduces to

$$\begin{aligned} a = {} & \sum \sum x_i x_j \sqrt{a_{ii} a_{jj}} (1 - k_{ij}) \\ & + 2 \sum x_0 \sqrt{a_{jj}} (1 - k_{ij}) \left(\int_{\eta}^{\infty} F(I) \sqrt{a(I, I)}\, dI \right) \\ & + \left(\int_{\eta}^{\infty} F(I) \sqrt{a(I, I)}\, dI \right)^2 . \end{aligned} \tag{4–10b}$$

Consequently, to use an equation of state for a mixture of continuously distributed components we must be able to evaluate the integrals

$$\int_{\eta}^{\infty} F(I) \sqrt{a(I, I)}\, dI \tag{4–11a}$$

and

$$\int_{\eta}^{\infty} F(I) b(I)\, dI, \tag{4–11b}$$

where $F(I)$ is some distribution, such as the exponential distribution considered earlier, and $a(I, I)$ function of temperature and the index I, while $b(I)$ depends only on the index.

Proceeding to vapor–liquid equilibrium calculations one can show that by starting from the criterion of equality of species fugacities (or their logarithms) in each phase, one obtains

$$\ln \frac{y_i}{x_i} = (K_1^L - K_1^V) + (K_2^L - K_2^V)b_i + (K_3^L - K_3^V)\sqrt{a(i, i)} \quad \text{(identified components)} \quad (4\text{–}12a)$$

and

$$\ln \frac{F^V(I)}{F^L(I)} = (K_1^L - K_1^V) + (K_2^L - K_2^V)b(I) + (K_3^L - K_3^V)\sqrt{a(I, I)} \quad \text{(continuous components).} \quad (4\text{–}12b)$$

An important observation is that K_1, K_2, and K_3 are functionals of the sums and integrals discussed earlier, but not explicit functions of the compositions x_i or the distribution function $F(I)$. These equations must be solved with the mass balances

$$x_i^F = x_i L + y_i V \quad \text{(identified components),} \quad (4\text{–}13a)$$

$$F^F(I) = F^L(I)L + F^V(I)V \quad \text{(continuous components),} \quad (4\text{–}13b)$$

and

$$1 = L + V \quad (4\text{–}13c)$$

where L and V are the fractions of the feed which, after the phase equilibrium calculation, are liquid and vapor, respectively.

LUMPING IN PHASE EQUILIBRIUM CALCULATIONS

There are now two ways to proceed. One is to continue with the continuous or semicontinuous approach in which one evaluates these integrals analytically, and then proceeds to the phase equilibrium calculation. The second approach is to evaluate these integrals numerically. In this case we would write

$$\int_{\eta}^{\infty} F(I)b(I)\, dI = \sum w_j b(z_j), \quad (4\text{–}14)$$

where w_j is weight factor at the jth quadrature point, and z_j is the location of that point. Comparing Eqs. (4–7) and (4–14) it is clear that the numerical quadrature method is functionally equivalent to choosing a set of pseudocomponents or lumps, each of which represents a collection of components. The jth pseudocomponent is represented by the value of the index I equal to z_j and its mole fraction is equal to the weight factor w_j. The advantage of using the quadrature method is that current phase equilibrium computational methods used in reservoir simulation, oil processing, etc., can be retained even though the fluid is composed of an essentially continuous distribution of species.

The questions that arise are: (1) How many quadrature points should one choose? and (2) How should one choose the location and weights of these quadrature points in a mathematically optimal manner? The first of these questions can be answered only by trial and error, and will depend on the application. For example, few pseudocomponents may be satisfactory in reservoir simulation, whereas at least one or two may be needed to represent each product stream in downstream processing. The second question is answered in a rigorous way with the family of Gauss quadrature methods (Stroud and Secrest 1966); as will be described shortly, the specific member of the family used depends on the form of the distribution function used.

For the exponential distribution of Eq. (4–2), letting $\theta(I)$ be either $\sqrt{a(I, I)}$ or $b(I)$, we have

$$\int_{\eta}^{\infty} qe^{-mI}\theta(I)\,dI = \int_{\eta}^{\infty} e^{-J}\Phi(J)\,dJ,$$

where

$$J = m(I - \eta) \qquad \text{and} \qquad \Phi(J) = qe^{-m\eta}\theta\left(\frac{J}{m} + \eta\right), \tag{4–15}$$

we have that the numerical quadrature formula is

$$\int_{\eta}^{\infty} e^{-J}\Phi(J)\,dJ = \sum_{j=1}^{N} w_j \theta\left(\frac{z_i}{m} + \eta\right). \tag{4–16}$$

Here, as shown in Figure 4–1, z_j in the jth zero of the Laguerre polynomial of order n, $L_n(x)$, and

$$w_j = \frac{(n!)^2 z_j}{[L_{n+1}(z_j)]^2}. \tag{4–17}$$

The two-point Gauss–Laguerre evaluation of

$$\int_0^\infty e^{-x} f(x)\,dx = w_1 f(x) + w_2 f(x_2)$$

can be made exact for polynomial of $f(x)$ of order 3 or less

$$f(x) = 1 \Rightarrow \int_0^\infty e^{-x}\,dx = 1 = w_1 + w_2,$$

$$f(x) = x \Rightarrow \int_0^\infty xe^{-x}\,dx = 1 = w_1 x_1 + w_2 x_2,$$

$$f(x) = x^2 \Rightarrow \int_0^\infty x^2 e^{-x}\,dx = 2 = w_1 x_1^2 + w_2 x_2^2,$$

$$f(x) = x^3 \Rightarrow \int_0^\infty x^3 e^{-x}\,dx = 6 = w_1 x_1^3 + w_2 x_2^3.$$

The solution to this set of four equations and four unknowns is

$$w_1 = (\sqrt{2} + 1)/2\sqrt{2}, \qquad w_2 = (\sqrt{2} - 1)/2\sqrt{2}, \qquad x_1 = 2 - \sqrt{2},$$
$$\text{and } x_2 = 2 + \sqrt{2}$$

More generally for an n-point integration scheme we have

$$x_k = \text{zeros of Laguerre polynomial } L_n(x)$$

and

$$w_k = \frac{(n!)^2 x_k}{[L_{n+1}(x_k)]^2}.$$

Figure 4–1 Gauss–Laguerre integration scheme.

The values of z_j and w_j for the Gauss–Laguerre quadrature procedure appear in numerous reference books (Abramowitz and Stegun, 1972; Katz and Firoozabadi, 1978). Consequently, once one has decided upon the number of pseudocomponents (quadrature points) to be used, the Gauss–Laguerre procedure provides the optimal method for picking that number of pseudocomponents.

Figure 4–2 presents an example of identifying lumped pseudocomponents based on the Gauss–Laguerre integration scheme follows. The data for this

$$\text{MW}(I) \cong 14I - 4,$$

$$\overline{\text{MW}} = 170 = \frac{\int_{6.5}^{\infty} qe^{-mI}\text{MW}(I)\,dI}{\int_{6.5}^{\infty} qe^{-mI}\,dI} = 14\left(\frac{1}{m} + 6.5\right) - 4$$

so that $m = 0.16867$ or $1/m = 5.9286$.

The two-point Gauss–Laguerre quadrature (see Figure 4–1)

$$z_1 = 0.5858, \qquad w_1 = 0.8536,$$

$$z_2 = 3.4142, \qquad w_2 = 0.1464,$$

with

$$\text{C}_k = \frac{z_k}{m} + 6.5 \quad \text{and} \quad x_k = 0.0913\ (w_k)$$

gives

$$\begin{matrix} \text{C}_1 = 9.973, & & x_1 = 0.0779 \\ & \text{and} & \\ \text{C}_2 = 26.741, & & x_2 = 0.0134 \end{matrix}$$

or

lumped cut 1 (mole fraction $= 0.0779) \Rightarrow \text{C}_7\text{–C}_{17}$ $(\text{C}_{9.973})$

lumped cut 2 (mole fraction $= 0.0134) \Rightarrow \text{C}_{18+}$ $(\text{C}_{26.741})$

Vogel *et al.* (1983), after extensive trial and error found Cut I $\text{C}_7\text{–C}_{16}(\text{C}_{9.9})$ and Cut II C_{17+} $(\text{C}_{26.3})$.

Figure 4–2 Example based on data in Vogel et al. (1983).

example were taken from the work of Vogel et al. (1983). They considered rich gas condensate containing nitrogen, carbon dioxide, hydrogen sulfide, and the normal and isoparaffins up to C_6 (the identified components) and then a C_7 and heavier fraction of molecular weight 170 and mole fraction of 0.0913. Initially they represented the C_{7+} fraction with 34 empirically chosen cuts, and then, under pressure from their simulation group, with successively 14, then 7, and finally only 2 pseudocomponents chosen by extensive trial and error. They express surprise at finding that the first fraction covered the range from C_7 to C_{16} with an average carbon number of 9.9, and the second C_{17+} fraction had a carbon number of 26.3.

In Figure 4–2 we see how the average molecular weight is used to determine the *m* parameter in the exponential distribution, and then how the two-point Gauss–Laguerre integration scheme is used, with hardly any effort, to identify two pseudocomponents which are almost identical with those that were found by Vogel et al. after a great deal of work. This illustrates the utility of the Gauss–Laguerre method of choosing pseudocomponents.

Another example is given in Figures 4–3, 4–4, and 4–5, all of which deal with retrograde dew point behavior of a reservoir fluid. In Figure 4–3 we show the shift of the predicted dew point curve as the C_{7+} fraction is modeled first by a single component of the measured molecular weight, and then successively by 5, 10, and finally 84 identified components, chosen in such a way that each pseudocomponent is of equal mole fraction. What should be clear from this figure is that the single component representation of the C_{7+} fraction is very poor, and that the convergence to the most accurate 84 component result on addition of more equal mole fraction pseudocomponents is quite slow.

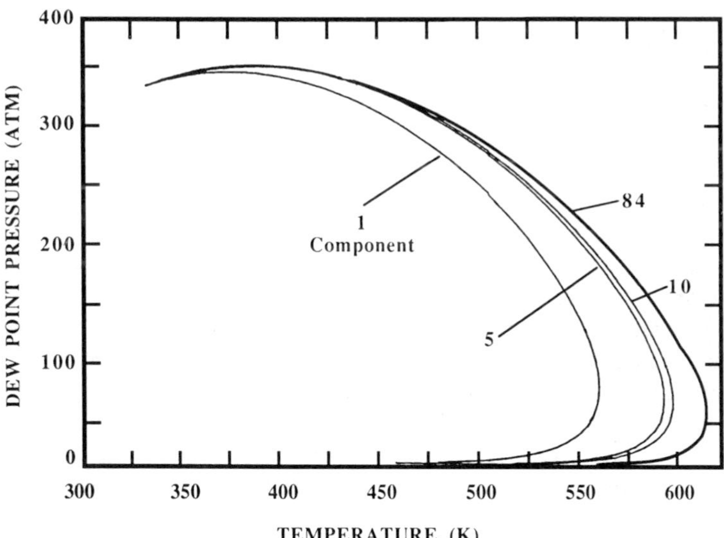

Figure 4–3 Dew point pressure predictions versus temperature 1, 5, 10, and 84 equal mole fraction components.

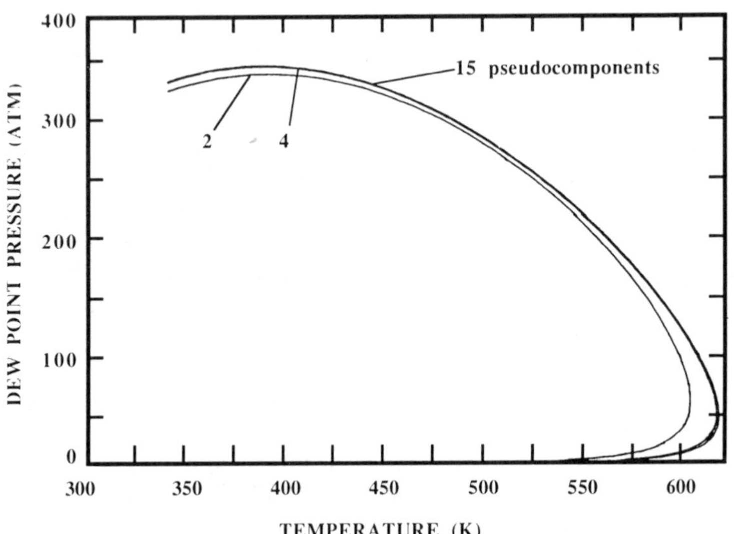

Figure 4–4 Dew point pressure predictions versus temperature with Gauss–Laguerre pseudocomponent lumping.

In Figure 4–4 we consider the use of 2, 4, and 15 pseudocomponents chosen by the Gauss–Laguerre quadrature method discussed here to represent the C_{7+} fraction. We need to point out that the Gauss–Laguerre method leads to a specification of the pseudocomponents, but not their properties. In this work the critical properties of the pseudocomponents were obtained from generalized correlations, though that means using those correlations for fractional carbon numbers. Densities and boiling points were obtained from the correlations of Katz and Firoozabadi (1978); critical pressure and temperature were obtained from the work of Riazi and Daubert (1980), and the acentric factor from the correlation of Kesler and Lee (1976).

We see in Figure 4–4 that the two Gauss–Laguerre pseudocomponent or lumping description chosen in this way leads to more accurate retrograde dew point behavior than 10 components chosen previously, and that four Gauss–Laguerre pseudocomponents leads to predictions that are equivalent to that obtained with 84 components chosen by the equal mole fraction recipe. This is another demonstration of the value of using the Gauss–Laguerre quadrature method as a basis for choosing pseudocomponents or lumps.

Finally, in Figure 4–5, we demonstrate the importance of knowing the average molecular weight of the C_{7+} fraction in pseudocomponent identification. In this figure we see that the 84 component and the Gauss–Laguerre

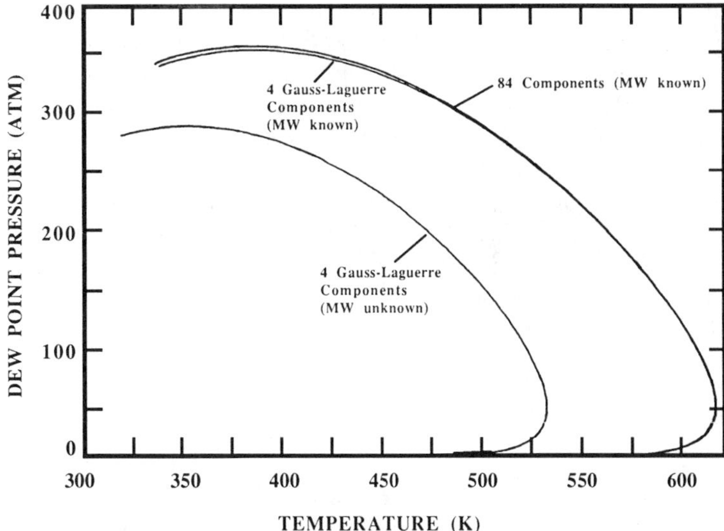

Figure 4–5 Dew point pressure predictions versus temperature.

four-pseudocomponent predictions, both using the known molecular weight, are essentially coincident. The additional line results from a calculation in which we assume the molecular weight of the C_{7+} fraction is unknown and estimate the exponential distribution parameters from the variation of composition with carbon number for the identified components. The large error in these predictions points to the necessity of having molecular-weight information to obtain an accurate tail of the distribution function and an accurate pseudocomponent representation.

A number of other examples of the success of the Gauss–Laguerre method of choosing pseudocomponents appear in Shibata et al. (1987) and Behrens and Sandler (1988). There examples are given of satisfactory phase equilibrium descriptions be obtained for fluids whose C_{7+} fraction is more than 70 mol% with as few as two Gauss–Laguerre components. Also there we consider how the bubble point changes with gas injection as in CO_2-miscible flooding.

GENERALIZATIONS OF THE QUADRATURE-BASED LUMPING PROCEDURE

Here we consider two generalizations to the quadrature-based lumping procedure discussed in the last section. The first generalization is for a finite upper bound to the distribution function. This can occur in two situations. First, for the case of very heavy reservoir fluids and crude oils (flashed reservoir fluid to remove light components), the direct application of the Gauss–Laguerre quadrature method may result in the selection of one or more components with unrealistically high carbon number. In such cases it may be desirable to reduce the large carbon number(s) while still maintaining the mathematical rigor of the quadrature method by assuming an upper bound for the distribution function. A second case occurs when one would like to force the location (or carbon number) of some of the pseudocomponents to lie within specific ranges in order to represent certain refinery streams. In this case one needs to use a collection of distribution functions, each of which covers a nonoverlapping finite range.

In such cases we must evaluate an integral of the form

$$\int_{\eta}^{\varepsilon} e^{-mI}\theta(I)\, dI = \int_{0}^{\Delta} e^{-J}\phi(J)\, dJ, \tag{4–18}$$

where $J = m(I - \eta)$, $\Delta = m(\varepsilon - \eta)$, and $\phi(J) = (e/m^{-m\eta})\theta(J/m + \eta)$. Returning to Figure 4–1, we see that the Gaussian integration scheme is easily generalized to a bounded integral. The main difference which results is that the locations and weights of the pseudocomponents which now must be found numerically, rather than from tables of the Laguerre polynomial.

The analysis here is easily generalized to other distribution functions which occur in other applications. For example, it is common to use the Gaussian distribution

$$F(I) = c\ \exp\left[-\left(\frac{I-\mu}{\sigma}\right)^2\right] \tag{4–19}$$

to represent the molecular-weight distribution in polymer solutions. The mathematically correct method of choosing the optimum location of n pseudocomponents, $z_1, z_2, \ldots, z_n$, to represent this distribution over the unbounded region $0 \leq I \leq \infty$ is the zeroes of the Hermite polynomial $H_n(x)$ with weights equal to Stroud and Secrest (1966)

$$w_k = \frac{n!\sqrt{\pi}(z_k)^{n+1}}{[H_{n+1}(z_k)]^2}. \tag{4–20}$$

Alternatively, the Schultz–Flory (1937) distribution

$$F(I) = CI^a e^{-I} \tag{4–21}$$

is used to represent molecular-weight distribution of polymers and of particulates. In this case the optimum locations of the pseudocomponents are the zeroes of the associated Laguerre polynomial $L_n^{(\alpha)}(x)$ with weights

$$w_k = \frac{n!\Gamma(n + \alpha + 1)z_k}{[L_{n+1}^{(\alpha)}(z_k)]^2}. \tag{4–22}$$

In a similar fashion, the mathematically optimum pseudocomponent locations and weights can be determined for other distribution functions based on the Gaussian quadrature family of integration procedures (Stroud and Secrest 1966).

CONCLUSION

In this chapter we have shown how phase equilibrium calculations involving continuously distributed components reduce to the problem of evaluating certain integrals of the distribution function of concentrations. These integrals can be evaluated analytically or numerically. The numerical evaluation is equivalent to choosing lumped pseudocomponents which can be used in current phase equilibrium computer codes. However, by recognizing that choosing the best pseudocomponents is equivalent to finding the optimum choice of quadrature points, a solved problem, allows us to make the pseudocomponent or lumping choice in a mathematically correct manner. The success of choosing pseudocomponents in this way has been demonstrated by two examples here, and others elsewhere (Behrens and Sandler 1988; Shibata et al. 1987).

We also briefly showed how the procedure for identifying pseudocomponents or lumped components can be generalized to a variety of distribution functions over both semiinfinite and bounded intervals.

ACKNOWLEDGMENTS

This work was supported, in part, by Grant DE-FG-85ER13436 from the United States Department of Energy to the University of Delaware.

REFERENCES

Abramowitz, M. and Stegun, I. A. (1972) *Handbook of Mathematical Functions*, Dover, NY.

Behrens, R. A. and Sandler, S. I. (1988) *SPE Res. Eng.* August, 1041.

Bowman, J. R. (1949) *Ind. Eng. Chem.* **41**:2004.

Bowman, J. R. and Edmister, W. C. (1951) *Ind. Eng. Chem.* **43**:2625.

Cotterman, R. L. and Prausnitz, J. M. (1985) *IEC Proc. Des. Dev.* **24**:434.

Cotterman, R. L., Benter, R., and Prausnitz, J. M. (1985) *IEC Proc. Des. Dev.* **24**:194.

Cotterman, R. L., Chou, G. F., and Prausnitz, J. M. (1986) *IEC Proc. Des. Dev.* **25**:840.

Edmister, W. C. and Buchanan, D. H. (1953) *Chem. Eng. Prog. Symposium Series* No. 6:**67**.

Gualtieri, J. A., Kincaid, J. M. and Morrison, G. (1981) Equilibrium properties of polydisperse systems. In *Proceedings of the Eighth Symposium on Thermophysical Properties*, American Society of Mechanical Engineering.

Gualtieri, J. A., Kincaid, J. M. and Morrison, G. (1982) *J. Chem. Phys.* **77**:521.

Kang, C. H. and Sandler, S. I. (1988) *Macromolecules* **21**:3088.

Katz, D. L. and Firoozabadi, A. (1978) *J. Pet. Tech.* **20**:1649.

Kesler, M. G. and Lee, B. I. (1976) *Hydrocarbon Process.* **55**:153.

Peng, D. Y. and Robinson, D. B. (1976) *Ind. Eng. Chem. Fund.* **15**:59.

Riazi, M. R. and Daubert, T. E. (1980) *Hydrocarbon Process.* **59**:115.

Schultz, G. V. (1937) *Z. Phys. Chem.* **A179**:321.

Shibata, S. K., Sandler, S. I., and Behrens, R. A. (1987) *Chem. Eng. Sci.* **42**:1977.

Stroud, A. J. and Secrest, D. (1966) *Gaussian Quadrature Formulas.* Prentice-Hall, Englewood Cliffs, NJ.

Tables of Functions and Zeros of Function (1964) *Natl. Bur. Standards Appl. Math. Series* **55**, Washington, DC.

Vogel, J. L., Turek, E. A., Metcalfe, R. S., Bergman, D. F., and Morris, R. W. (1983) *Fluid Phase Eq.* **14**:103.

PART 2

STRUCTURE–ACTIVITY RELATIONSHIPS

5

Fundamental Kinetic Modeling of Complex Processes

G. F. FROMENT

The feedstocks processed in petroleum refining and in many petrochemical operations consist of a large number of components, each reacting according to complicated pathways. Selectivity is often an important feature of these commercial processes. In kinetic modeling the actual reaction network is frequently reduced to a single overall reaction or to a limited number of reactions. Excessive lumping may lead to parameters which depend upon the feed composition and even upon the type of reactor in which they were determined. As a consequence extensive experimentation is required for each new feedstock.

This chapter advocates a fundamental approach in which the detailed reaction network of the feed components is retained to a maximum extent and the kinetics are developed in terms of the elementary steps involved in the reactions. True, for a single component the number of rate parameters is significantly higher than with the usual lumping approach, but their fundamental nature gives them a much wider validity: they do not depend upon the chain length and are not affected by a different feed composition. The true benefit of the approach becomes clear when mixtures are dealt with. In that case the lumping approach needs an increasing number of lumps and parameters to retain the significance and accuracy of the modeling. The chapter illustrates to what level of detail the network has to be developed to come to really invariant parameters. Two examples of complex processes will be dealt with. The first is a homogeneous gas phase process: the therma cracking of naphtha for olefins production. The second is the gas phas hydrocracking of paraffins on a zeolite catalyst loaded with a noble metal

KINETICS OF THE THERMAL CRACKING OF NAPHTHA FOR OLEFIN PRODUCTION

Production of olefins and aromatics by thermal cracking of naphtha i carried out in long cracking coils at temperatures of the order of 800°C an residence times of the order of 0.4 s. One element that enters into the desig or simulation of cracking units is the kinetics of the process.

Naphtha is a complex mixture containing 100–200 components, each c which reacts in a complicated way. As always in such a case a choice of th level of complexity of the kinetic equations has to be made. The experimenta data base for the evaluation of various levels of kinetic modeling is provide by the work of Van Damme et al. (1981), who cracked three naphthas, wit

Table 5–1 Main Characteristics of Three Typical Naphthas

	Naphtha		
	I	II	III
Density, 15/4°C	0.6790	0.7056	0.7180
API density	76.89	69.04	65.55
UOP-K factor	12.618	12.322	12.264
ASTM-distillation			
IBP°C	38	46	45
10%	52	61	68
50%	71	88	110
90%	113	138	158
EBP	141	170	195
PONA (wt%)			
Paraffins	84.1	71.2	70.6
Olefins	—	—	—
Naphthenes	12.2	19.7	20.8
Aromatics	3.7	9.1	8.6
av mol wt	85.5	93.0	98.3
H/C molar ratio	2.255	2.177	2.177
Ratio *n*/*i* paraffins	1.177	1.084	1.008

Source: Van Dierendonch et al. (1985).

main characteristics given in Table 5–1, in the pilot plant of the Laboratorium voor Petrochemische Techniek.

In the first level of kinetic modeling all the components of the naphtha were lumped into one single pseudocomponent. In spite of its complexity the pseudocomponent "naphtha" cracks according to a first-order kinetic law, $r_N = kC_N$. Table 5–2 shows the kinetic parameters for the global disappearance of the three naphthas. The table also shows how the rate coefficients at 800°C differ for each naphtha.

Consequently, the kinetic experimentation has to be done over for each new feedstock. This is not an attractive perspective and the question arises if a more refined model would not lead to a set of invariant rate coefficients. An analytical technique that is commonly practiced in industrial laboratories splits the naphtha into normal and iso-paraffins, olefins, naphthenes, and aromatics. Olefins are usually present in negligible quantities only. The so-called PONA analysis (*p*araffins, *o*lefins, *n*aphthenes, *a*romatics) of the three naphthas considered here is also given in Table 5–1. The first-order rate coefficient for the disappearance of these lumps at 800°C, the frequency factors, and activation energies are given in Table 5–3. In this table the aromatics are absent because they not only disappear but are also formed in the cracking process. Again the rate coefficients differ for the three naphthas because of the difference in composition of the lump, which becomes obvious from a gas chromatographic analysis of the three naphthas. The *n*-paraffin lump in naphtha III contains a larger fraction of heavier components than in naphtha I, for example, and these crack faster than the components with low molecular weight.

Clearly, to reach a sufficient level of accuracy in the kinetic modeling and to obtain invariant rate parameters, the breakdown of the naphtha composition has to go as far as the feedstock components themselves. The rate coefficients at 800°C for the thermal cracking of fifteen representative components are given in Table 5–4.

Table 5–2 Kinetic Parameters for the Global Disappearance of Three Typical Naphthas

	A (s^{-1})	E (kJ/kmol)	k (800°C) (s^{-1})
Naphtha I	2.40×10^{11}	212,050	11.4
Naphtha II	1.78×10^{12}	228,500	13.5
Naphtha III	4.71×10^{11}	216,150	14.2

Source: Van Dierendonch et al. (1985).

Table 5–3 Kinetics of Naphtha Cracking Based on Characteristic Lumps

	A (s^{-1})	E (kJ/kmol)	k (800°C) (s^{-1})
		Naphtha I	
n-Paraffins	2.05×10^{11}	210,560	11.6
Isoparaffins	7.20×10^{11}	221,330	12.2
Naphthenes	3.51×10^{10}	196,850	9.2
		Naphtha II	
n-Paraffins	7.22×10^{11}	220,670	13.1
Isoparaffins	4.52×10^{13}	255,700	16.2
Naphthenes	2.02×10^{11}	210,560	11.4
		Naphtha III	
n-Paraffins	3.86×10^{11}	214,090	14.7
Isoparaffins	5.56×10^{12}	237,200	15.85
Naphthenes	9.24×10^{9}	184,010	10.2

Source: Van Damme et al. (1981).

Even with this detailed breakdown some variation in the rate coefficients is still present. It exceeds the inaccuracies associated with this type of experimentation and analysis. The variation can be traced down to interaction between reacting species.

The results obtained thus far in the development outlined above, regardless of their accuracy, are insufficient as a basis for simulation or design of a commercial unit. Indeed, they relate only to the disappearance of the feedstock or its components, but do not provide any means for predicting the effluent composition, which is of capital importance in commercial olefin production. What is required for this is a set of kinetic equations describing the production of hydrogen, methane, ethylene, propylene, butenes, butadiene, and aromatics, which are the main components of the effluent. The way to obtain these kinetic equations is to write detailed reaction schemes for the feed components and the intermediates. In thermal cracking the reactions proceed by radical mechanisms involving initiation, hydrogen abstraction, radical isomerization and decomposition, radical addition on double bonds, and termination.

Yet, up to 10 or 15 years ago the radical based network was approximated

Table 5–4 First-Order Rate Coefficients at 800°C (in s^{-1}) for the Cracking of the Key Components in Three Typical Naphthas

	Naphtha		
	I	II	III
n-Pentane	9.3	—	—
n-Hexane	12.7	11.7	11.85
n-Heptane	14.5	16.7	15.9
n-Octane	16.1	19.0	18.25
n-Nonane	17.9	20.8	24.35
n-Decane	—	23.9	30.6
2-Methylpentane	9.9	12.15	10.25
3-Methylpentane	13.6	15.3	11.7
2-Methylhexane	11.3	18.2	17.85
3-Methylhexane	14.1	14.5	17.35
2-Methylheptane	16.1	17.7	19.5
3-Methylheptane	17.9	17.25	26.6
Methylcyclopentane	10.7	11.95	10.6
Cyclohexane	5.2	8.85	7.0
Methylcyclohexane	11.8	13.1	11.75

Source: Van Damme et al. (1981).

by a molecular "equivalent," partly to limit the size of the set of kinetic equations, partly for circumventing mathematical problems associated with the solution of sets of stiff differential equations. The precision required by present-day olefin producers in the predicted effluent composition of a thermal cracker is extremely high. There are good reasons for this: an increase of the ethylene yield from, for example, 29 to 30 wt% represents a tremendous benefit for the olefin producer. The simulation software should have the capability of predicting the effect of feedstock composition and operating policy within such a narrow limit. Reaching such a degree of accuracy is not a simple matter: the effluent composition is not only determined by averaged values of the operating conditions such as temperature, steam dilution ratio, and total pressure, but also by their evolution along the coil. Molecular approximation schemes do not provide a sufficient level of accuracy. Modern simulation packages for olefins production account in detail for the radical nature of the reaction network. Years ago this looked like an impossible task. By way of example, Figure 5–1 represents the radical reaction scheme for the thermal cracking of 3-methylheptane, one

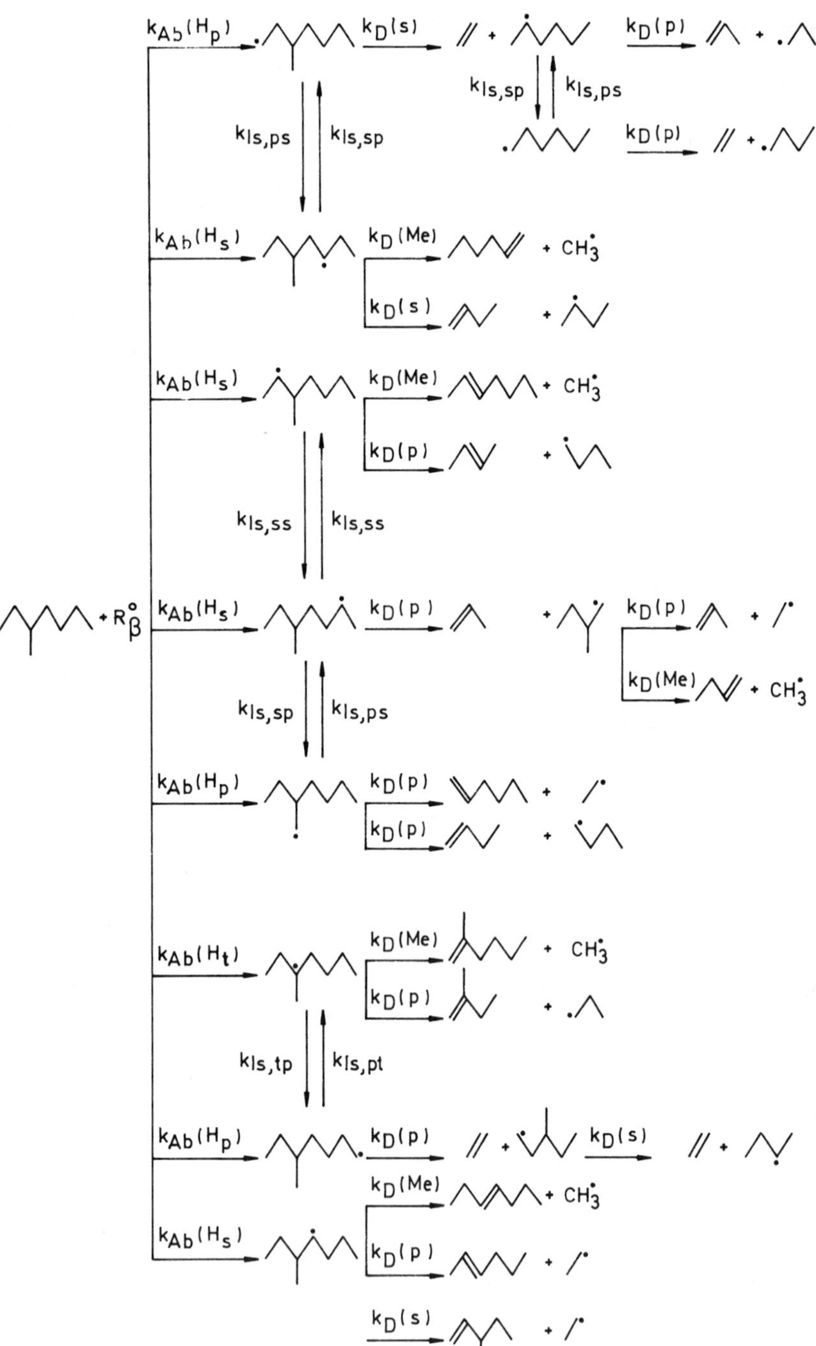

Figure 5–1 Mechanism of thermal cracking of 3-methylheptane (after Clymans and Froment 1984).

f the many components of naphtha (Clymans and Froment 1984). Devel-ping such reaction schemes for all the main components in a naphtha is n overwhelming task in itself. Therefore, Clymans and Froment (1984) and (illewaert et al. (1988) devised software for the computer generation of omplex reaction networks, based on Boolean relation matrices reflecting ıe structure of the hydrocarbons and accounting for the fundamental rules f radical chemistry.

Figure 5–2 represents such a Boolean relation matrix for 3-methylheptane. ı this matrix the existence of a bond between two carbon atoms is indicated y 1 on the intersection of the row and column corresponding to these carbon toms. The number of relations on a row (or column) indicates whether a ırbon atom is primary, secondary, tertiary, or quaternary (Clymans and roment 1984).

The scission of a C–C bond in a radical occurs in β-position with respect the carbon bearing the free electron. The matrix $\mathbf{M}^2 - \mathbf{I}$ (Figure 5–3) ontains all the possible β-positions. The matrix $\mathbf{M}^4 - \mathbf{I}$ yields all the ossible 1 → 5 isomerizations. Relation matrices can also convey information n the products of β-scission in a radical.

The number of reactions and, therefore, kinetic parameters appearing in ıe radical network of Figure 5–1 may seem to be high. What then if a

M =

	1	2	3	4	5	6	7	8
1	0	1	0	0	0	0	0	0
2	1	0	1	0	0	0	0	0
3	0	1	0	1	0	0	0	1
4	0	0	1	0	1	0	0	0
5	0	0	0	1	0	1	0	0
6	0	0	0	0	1	0	1	0
7	0	0	0	0	0	1	0	0
8	0	0	1	0	0	0	0	0

gure 5–2 Boolean relation matrix for 3-methylheptane (after Clymans ıd Froment 1984).

$(\mathbf{M}^2 - \mathbf{I}) =$	1	2	3	4	5	6	7	8
1	0	0	1	0	0	0	0	
2	0	0	0	1	0	0	0	1
3	1	0	0	0	1	0	0	0
4	0	1	0	0	0	1	0	1
5	0	0	1	0	0	0	1	0
6	0	0	0	1	0	0	0	0
7	0	0	0	0	1	0	0	0
8	0	1	0	1	0	0	0	0

Figure 5–3 Boolean relation matrix for β-scission of radicals (after Clyman and Froment 1984).

mixture of components representing a naphtha has to be dealt with? A close look at the network of Figure 5–1 reveals that the reactions—or mor specifically the elementary steps—pertain to a small number of types only initiation; hydrogen abstraction from a primary, secondary, or tertiar carbon; radical decomposition; etc., as mentioned above. The associated rat parameters no longer depend on the chain length, as those shown in Tabl 5–4 do. Also, because of their fundamental nature, the parameters are vali for any naphtha. Evidently, the parameters depend on the nature of th reactants and products. By way of example, Table 5–5 shows the variou rate coefficients for hydrogen abstraction considered by Willems and Fro ment (1988a,b).

The activation energy of a hydrogen abstraction depends on the natur of the abstracting radical and of the produced olefin and radical. The variou activation energies are related to that of a reference elementary step, whic is the abstraction of hydrogen from a primary carbon atom by means of hydrogen radical. This is accomplished by structural contributions. Eac contribution accounts for specific effects of the structure of reactant an product. From a comparison with the group contributions for the calculatio of the standard enthalpies of formation a range can be set for the value c the structural contribution, which helps in the parameter estimation b means of optimization routines. A thermodynamic analysis leads to a numbe of relations between the structural contributions and reduces the number c independent parameters.

The number of molecular components considered in the naphtha crackin network is close to 200, the number of radicals 40. The number of elementar

Table 5.5 Definition and Use of the Structural Contribution for the Calculation of the Activation Energies of Hydrogen Abstractions

1. Initialization	$E = E_{AB}(H_p)(H^\bullet)$
2. The abstracted hydrogen is	
primary	
secondary	$+\Delta E_{Ab}(H_s)$
tertiary	$+\Delta E_{Ab}(H_t)$
3. The abstracted hydrogen is located in the neighborhood of a double bond	
in α-position	$+\Delta E_{Ab}(\alpha)$
in β-position	$+\Delta e_{Ab}{=}(\beta)$
others	
4. The molecule from which hydrogen is abstracted is	
hydrogen	$+\Delta E_{Ab}(H_2)$
methane	$+\Delta E_{Ab}(CH_4)$
others	
5. The abstracting radical is	
hydrogen	
methyl	$+\Delta E_{Ab}(Ch_3^\bullet)$
vinyl	$+\Delta E_{Ab}(C_2H_3^\bullet)$
ethyl	$+\Delta E_{Ab}(C_2H_5^\bullet)$
etc.	etc.

Source: Willems and Froment (1988).

radical steps is of the order of 10,000. The total number of independent rate parameters is 120. This is a number that can be determined by means of modern multiresponse optimization routines from a broad data base derived from carefully selected experiments on the cracking of homologous series of individual hydrocarbons, of various mixtures, and of naphthas with varying compositions. The approach has been extended to the modeling of the thermal cracking of gasoil.

The fundamental radical model briefly outlined here is at present used in the simulation of existing cracking coils and furnaces or in the design of new units (Plehiers et al. 1990).

KINETICS OF HYDROCRACKING, A CATALYTIC PROCESS

Hydrocracking is a catalytic process for converting heavy petroleum fractions into lighter and more valuable fractions. It is often carried out in two stages. In the first, sulfur- and nitrogen-containing compounds are decomposed and aromatics are hydrogenated. The liquid fraction of the first stage is then hydroisomerized and hydrocracked in the second stage on amorphous or crystalline silica–alumina loaded with a noble metal.

The kinetic modeling to be discussed below deals more specifically with the processes taking place in the second stage. It follows very much the lines encountered in the discussion of the kinetic modeling of the thermal cracking of petroleum fractions. The experimental data base required for such a study comprises results on the hydroisomerization and hydrocracking of key components representative for the various types of hydrocarbons, of binary and ternary mixtures of these, and finally of a number of gasoils. Steijns et al. (1981) and Vansina et al. (1983) collected data on the hydroisomerization and hydrocracking of C_8-, C_{10}-, and C_{12}-paraffins on a Pt-US-Y-zeolite in a flow reactor with complete mixing in the temperature range 180–240°C and pressures from 7 up to 100 bar.

Steijns and Froment (1981) and Baltanas et al. (1983) based their kinetic modeling on a lumped reaction scheme. The lumping was based on the observation that the isomers with the same degree of branching reached equilibrium beyond ± 30% feed conversion. At higher conversions di- and tri-branched isomers are also obtained in significant amounts. Normally isomerization involving different degrees of branching does not reach equilibrium. Hydrocracking only starts at conversions of the order of 90%. After discrimination between seven rival networks the following was retained:

$$\begin{array}{ccccc} A & \rightleftharpoons & MB & \rightleftharpoons & MTB \\ & & & & \downarrow \\ & & & & CR \end{array} \qquad (5\text{–}1)$$

where A represents the normal paraffin, MB the monobranched isoparaffins, MTB the multibranched isoparaffins, and CR the lumped cracking products. The lumping in this model is of the type applied by Marin and Froment (1982) to the catalytic reforming of C_6 components on a dual-function Pt-alumina catalyst.

Within network (5–1) a variety of sets of rate equations is possible. These equations were based on the bifunctional character of the catalyst, with (de)hydrogenation taking place on the metal and the rate determining hydroisomerization and hydrocracking on the acid sites. They also accounted for the physisorption inside the zeolite cages. Some 30 sets of rate equations were tested. The parameters were determined by a multiresponse

Marquardt optimization routine. The model and parameters were submitted to a number of statistical and physicochemical tests (Froment and Bischoff 1990). The following rate equations, written in terms of partial pressures, were retained:

Rate of disappearance of *n*-alkane, r_A

$$r_A = \frac{k_{MB}(p_A - p_{MB}/K_1)}{ADSORB}. \tag{5–2}$$

Net rate of formation of the monomethylisomers, r_{MB}:

$$r_{MB} = \frac{k_{MB}(p_A - p_{MB}/K_1) - k_{MTB}(p_{MB} - p_{MTB}/K_2)}{ADSORB}. \tag{5–3}$$

Net rate of formation of multibranched isomers, r_{MTB}

$$r_{MTB} = \frac{k_{MTB}(p_{MB} - p_{MTB}/K_2) - k_{CR}\, p_{MTB}}{ADSORB}. \tag{5–4}$$

Rate of formation of cracked products, r_{CR}

$$r_{CR} = \frac{k_{CR}\, p_{MTB}}{ADSORB} \tag{5–5}$$

with ADSORB $= p_{H_2}[1 + K_L(p_A + p_{MB} + p_{MTB})]$.

K_L in ADSORB is a lump of physi- and chemisorption parameters. The denominator also reflects that the concentration of carbenium ions generated from the olefins by protonation is much smaller than the total concentration of active sites.

Figure 5–4 represents the various rate coefficients and the physisorption coefficient as a function of the number of carbon atoms in the *n*-paraffin feed (Froment 1987b). The rate coefficients increase with chain length, as can be observed also for thermal cracking in Table 5–4.

Although the kinetic modeling outlined above goes quite a way beyond what is usually practiced for complex reactions, its shortcomings are evident. Indeed, for heavier hydrocarbons extrapolation is required and additional experimentation is advisable. Since the model equations (5–2)–(5–5) already contain seven parameters, the modeling of the hydrocracking of complex mixtures would require an overwhelming number of parameters if the effluent composition were to be predicted with some degree of accuracy. This is why

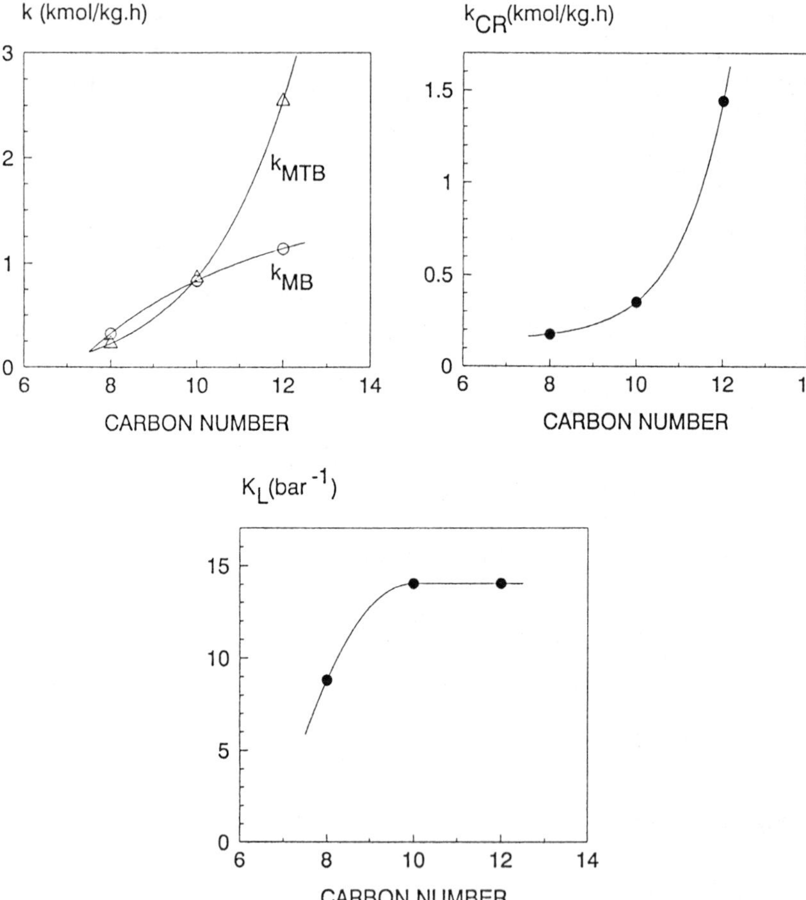

Figure 5–4 Rate and adsorption parameters for hydroisomerization and hydrocracking of *n*-octane, *n*-decane, and *n*-dodecane on Pt/US-Y zeolite at 200°C.

Baltanas et al. (1989) abandoned the lumping of reactions and parameters and switched to a fundamental approach based on the detailed mechanistic description of the reactions and interaction of the chemical species with the active sites of the catalyst.

As mentioned above, hydroisomerization and hydrocracking on metal loaded zeolites is known to proceed on the acid sites of the zeolite, through carbenium ions formed by protonation of olefins produced by dehydrogenation of the paraffins on the metal function. Regardless of the complexity of the feed, the reaction network consists of a limited number of types of elementary steps only. The types of elementary steps to be considered in the hydroisomerization and hydrocracking of paraffins are listed in Table 5–6. These steps are shown in some more detail in Table 5–7. A schematic representation of the resulting reaction network of a paraffin is shown in Figure 5–5.

Evidently, with complex mixtures the reaction network is generated by computer, using Boolean relation matrices introduced in the discussion of the kinetic modeling of thermal cracking (Baltanas and Froment 1985).

A rate coefficient is assigned to each type of elementary step. Its value depends on the nature of the carbenium ion and olefin. It is well known that methyl- and primary carbenium ions are far less stable than secondary and tertiary ions. Therefore, if only the latter are retained in the reaction network, only four rate coefficients have to be considered for each of the isomerization mechanisms listed in Table 5–8, regardless of the chain length of the involved paraffins. These rate coefficients are shown in Table 5–8. For (de)protonation

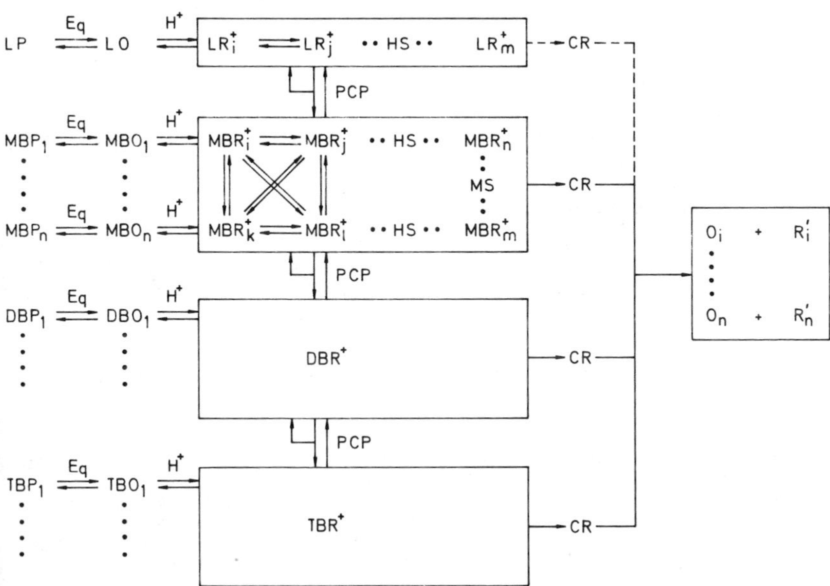

Figure 5–5 Fundamental reaction scheme for hydroisomerization and hydrocracking of alkanes (Baltanas and Froment 1985). LP, linear alkane; MEP_i, DBP_i, TBP_i; mono-, di-, and tribranched isomers; LO, $MBO_i, \ldots,$ corresponding olefins. LR_j^+, MBR_j^+, corresponding carbenium ions; HS, hydride shift; MS, methyl shift; PCP, protonated cyclopropane mechanism; CR, cracking.

Table 5–6 Steps Involved in Bifunctional Hydroisomerization and Hydrocracking of Alkanes

Step	Reaction
Physical adsorption of paraffin in the zeolite pores	$\text{P(g)} + \text{zeolite} \overset{EQ}{\rightleftharpoons} \text{P(ads)}$
Dehydrogenation of paraffin on metal sites	$\text{P(ads)} \overset{EQ}{\rightleftharpoons} \text{O(ads)} + \text{H}_2$
Protonation of olefin on acid sites	$\text{O(ads)} + \text{H}^+ \xrightarrow{k_{\text{PR}}(\text{O}; m)} \text{R}_m^+$
Isomerization of R_m^+ on acid sites	
Hydride shift	$\text{R}_m^+ \underset{k_{\text{HS}}(q; m)}{\overset{k_{\text{HS}}(m; q)}{\rightleftharpoons}} \text{R}_q^+$
Methyl shift	$\text{R}_m^+ \underset{k_{\text{MS}}(t; m)}{\overset{k_{\text{MS}}(m; t)}{\rightleftharpoons}} \text{R}_t^+$
PCP branching	$\text{R}_m^+ \underset{k_{\text{PCP}}(U; m)}{\overset{k_{\text{PCP}}(m; u)}{\rightleftharpoons}} \text{R}_u^+$
Cracking of R_m^+ on acid sites	$\text{R}_m^+ \xrightarrow{k_{\text{Ci}}(m; v, \text{O}')} \text{O}'(\text{ads}) + \text{R}_v^+$
Deprotonation of R_h^+ from sites	$\text{R}_h^+ \xrightarrow{k_{\text{De}}(h; \text{O}'')} \text{O}''(\text{ads}) + \text{H}^+$
Hydrogenation of produced isoolefins on metal sites	$i\text{O(ads)} + \text{H}_2 \overset{EQ}{\rightleftharpoons} i\text{P(ads)}$
Hydrogenation of produced normal olefins	$\text{O(ads)} + \text{H}_2 \overset{EQ}{\rightleftharpoons} \text{P}''(\text{ads})$
Physical desorption of produced isoparaffins	$i\text{P(ads)} \overset{EQ}{\rightleftharpoons} i\text{P(g)} + \text{zeolite}$
Physical desorption of produced normal paraffins	$\text{P(ads)} \overset{EQ}{\rightleftharpoons} \text{P(g)} + \text{zeolite}$

Source: Baltanas et al. (1989).
Note: *EQ* denotes quasi-equilibrium. PCP, protonated cyclopropane.

Table 5–7 Carbenium Ion Reactions in Hydroisomerization and Hydrocracking

Type of Reaction	Example
Type A rearrangement	
Hydride shift	$C-\overset{+}{C}-\overset{H}{\underset{}{C}}-C \rightleftharpoons C-\overset{H}{C}-\overset{+}{C}-C$
Alkyl shift	$C-\overset{+}{C}-\overset{C}{C}-C \rightleftharpoons C-\overset{C}{C}-\overset{+}{C}-C$
Type B rearrangement	
PCP-branching	$C-\overset{+}{C}-C-C-C$ ⇌ [PCP: C—C (a) C (b) C—C, H^{+}] ⇌ (a) $C-\overset{+}{C}-\overset{C}{C}-C$; (b) $C-\overset{C}{C}-\overset{+}{C}-C$
Cracking by β-scission	$C-\overset{+}{C}-C-\underset{C}{C}-C \longrightarrow C-C=C + C-\overset{+}{C}-C$

PCP, protonated cyclopropane.

Table 5–8 Rate Coefficients for Elementary Carbenium Ion Reactions on the Acid Sites of the Zeolite

Elementary Step	Elementary Rate Coefficients	
Olefin protonation	$k_{Pr}(O_j; s)$	$k_{Pr}(O_j; t)$
Carbenium ion deprotonation	$k_{De}(s; O_j)$	$k_{De}(t; O_j)$
Hydride shift	$k_{HS}(s, s)$	$k_{HS}(t; s)$
	$k_{HS}(s; t)$	$k_{HS}(t; t)$
Methyl shift	$k_{MS}(s; s)$	$k_{MS}(t; s)$
	$k_{MS}(s; t)$	$k_{MS}(t; t)$
PCP-branching	$[k_{PCP}(s; s)]$	$k_{PCP}(t; s)$
	$k_{PCP}(s; t)$	$k_{PCP}(t; t)$
Cracking	$k_{Cr}(s; s, O_j)$	$k_{Cr}(t; s, O_j)$
	$k_{Cr}(s; t, O_j)$	$k_{Cr}(t; t, O_j)$

Source: Van Raemdonck and Froment (1989).
PCP, protonated cyclopropane.

and cracking steps that involve olefins, the number of rate coefficients becomes very large when one accounts for the nature of the olefins. Thermodynamic constraints permit the reduction of the number of kinetic parameters to a small number of independent ones. By way of example, constraints will be derived in what follows for the rate coefficients for protonation and deprotonation.

Consider two olefins, O_1 and O_2, which isomerize through one single intermediate secondary carbenium ion

(O_1)

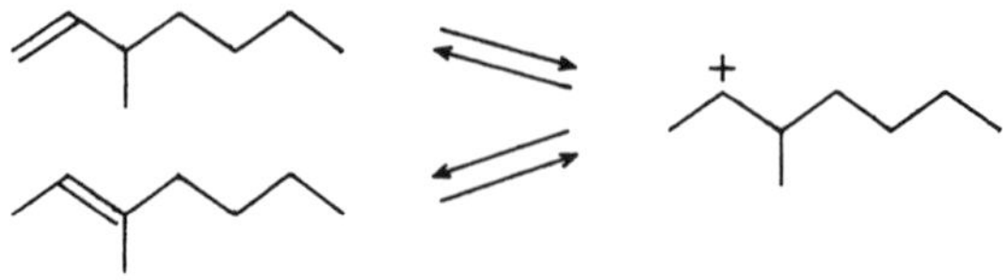

(O_2)

The equilibrium constant for the isomerization can be written

$$K_{is}(O_1 \rightleftarrows O_2) = \frac{k_{Pr}(O_1; s)k_{De}(O_2; s)}{k_{Pr}(O_2; s)k_{De}(O_1; s)}. \tag{5–6}$$

If it is assumed that the activated complex in a protonation reaction has a planar structure resembling the olefin structure, but with the double bond not yet completely broken, it is logical to conclude that the nature of the reacting olefin does not enter into the value of the rate coefficient of protonation, so that Eq. (5–6) becomes

$$K_{is}(O_1 \rightleftarrows O_2) = \frac{k_{De}(s; O_2)}{k_{De}(s; O_1)}. \tag{5–7}$$

This relation shows that there is only one independent rate coefficient for deprotonation of a secondary carbenium ion per carbon number and one for the deprotonation of a tertiary carbenium ion.

It is possible also to derive a relation between the four rate coefficients for respectively hydrogen shifts, methyl shift, and protonated cyclopropane isomerization given in Table 5–8, so that only three of these are independent within each type of isomerization.

If it is assumed that the enthalpy and entropy differences between the activated complexes in two cracking reactions are equal to the summation of the enthalpy and entropy differences between the produced olefins and carbenium ions, there is only one independent cracking coefficient (Van Raemdonck and Froment 1989).

The reaction network representing the hydrocracking of *n*-octane contains 14 octanes, 5 paraffinic cracking products, 49 octenes, 9 olefinic cracking products, 42 octyl-carbenium ions, and 6 carbenium ions with a shorter chain length, when only secondary and tertiary carbenium ions are considered. The number of elementary steps amounts to 383 (75 protonations and 75 deprotonations of C_8 species; 10 deprotonations of species below C_8; 88 hydride shifts of the 1.2 and 1.3 type; 24 methyl shifts; 96 PCP-branching isomerizations, and 15 cracking reactions). The number of isomerization rate coefficients amounts to 12, protonation coefficients to 75, deprotonation coefficients to 85, and cracking coefficients to 14. The total number of rate coefficients amounts to 186; the number of physisorption coefficients amounts to 14.

Introduction of the thermodynamic constraints leads to only 17 independent rate coefficients (2 for protonation; 5 for deprotonation, 3×3 for the isomerizations, and 1 for cracking) (Van Raemdonck and Froment 1989). A further refinement remains to be introduced. An elementary step such as hydrogen or methyl shift can occur through a number of single events. Take, for example, the methyl shift represented by

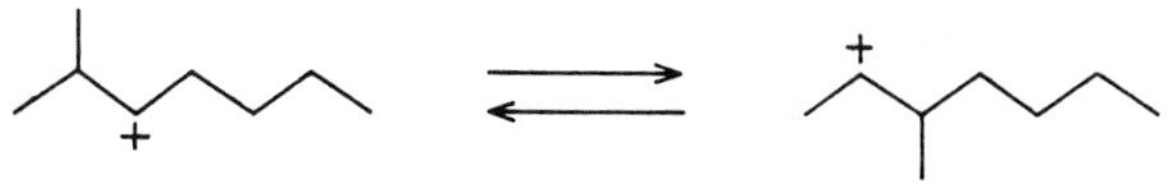

Toward the right there are two methyl groups that can shift; toward the left only one, however. Different values would be deduced for k_{Me} (ss) depending

on the carbenium ion considered. This is why the rate parameters have to be related to single events, a further breakdown of the elementary step. Through the symmetry number concept, an analysis of the transition state leads unambiguously to the number of single events comprised in one elementary step.

Another approach is based on the statistical factors for the forward and reverse step of the equilibrium reaction between reactants and activated complex (Baltanas et al. 1989). Table 5–9 lists the number of single events for various elementary steps.

Finally, the rate equations have to be derived. The first step in the bifunctional hydroisomerization and hydrocracking of a paraffin is the dehydrogenation of the paraffin, P_i, on the metal sites of the catalyst

$$P_i \rightleftarrows O_{ij} + H_2. \tag{5–8}$$

A subscript j is added to the olefin because one paraffin can dehydrogenate to several double-bond isomers. On the acid sites of the catalyst the olefins are protonated on each carbon atom of the double bond, yielding two different carbenium ions of types m_1 and m_2

$$O_{ij} + H^+ \rightleftarrows \begin{cases} R^+_{ij,m_1} \\ R^+_{ij,m_2} \end{cases} \tag{5-9}$$

Olefins can also be formed by carbenium ion cracking on the acid sites

$$R^+_v \rightarrow O_{ij} + R^+_w. \tag{5–10}$$

Table 5–9 Number of Single Events for Different Types of Carbenium Ion Elementary Steps

Type of Reaction	Number of Single Events
Protonation	1
Deprotonation	$n_{\alpha H}$
Hydride shift	$2n_{\alpha H}$
Methyl shift	$2n_{\alpha Me}$
PCP-branching	
β-Cleavage PCP	2
α-Cleavage PCP	$2n_{\beta H}/n_{C^+-H}$

Source: Baltanas et al. (1989).
PCP, protonated cyclopropane.

If the rate-determining step is on the acid sites, the global reaction rate of a paraffin is equal to the summation of the net rates of formation on the acid sites, of the olefins that are dehydrogenation products of the given paraffin

$$R_{P_i} = \sum_j R_{O_{ij}}. \tag{5–11}$$

Since, on the acid sites, the olefins are involved in the protonation, deprotonation, and cracking reactions given by Eqs. (5–9) and (5–10), the expression for the net rate of olefin formation should contain contributions from each of these reactions. Equation (5–11) then becomes

$$\begin{aligned} R_{P_i} = \sum_j \{&(k_{De}(m_1; O_{ij})C_{R^+_{ij,m_1}} + k_{De}(m_2; O_{ij})C_{R^+_{ij,m_2}} \\ &- [k_{Pr}(m_1) + k_{Pr}(m_2)]C_{O_{ij}}C_{H^+} \\ &+ k_{Cr}(v; w, O_{ij})C_{R_v^+})\}. \end{aligned} \tag{5–12}$$

Equation (5–12) expresses the global reaction rate of a paraffin as a function of elementary rate coefficients, each of which should be interpreted as the product of the single-event rate coefficient and the number of single events, of concentrations of carbenium ions, free acid sites, and physisorbed olefins. Since quasi-equilibrium is assumed for the dehydrogenation step, the concentrations of the physisorbed olefins can be easily calculated in terms of the physisorbed paraffin concentrations and the partial pressure of hydrogen, via

$$C_{O_{ij}} = \frac{C_{P_i} K_{DH,ij}}{p_{H_2}}. \tag{5–13}$$

The concentrations of the carbenium ions and the free acid sites are eliminated through the use of the pseudo-steady-state approximation for the carbenium ions

$$R_{R_m^+} = 0, \tag{5–14}$$

and the total active sites balance

$$C_t = C_{H^+} + \sum_m C_{R_m^+}. \tag{5–15}$$

Since the carbenium ions are involved in isomerization and cracking reactions, as well as in protonation and deprotonation reactions, the general form of the rate expression to be substituted in Eq. (5–14) is

$$R_{\mathrm{R}_m^+} = \left\{ \sum_{\mathrm{O}} k_{\mathrm{Pr}}(m) C_{\mathrm{O}} C_{\mathrm{H}^+} + \sum_{q} k_{\mathrm{HS}}(q; m) C_{\mathrm{R}_q^+} + \sum_{r} k_{\mathrm{MS}}(r; m) C_{\mathrm{R}_r^+} + \sum_{u} k_{\mathrm{PCP}}(u; m) C_{\mathrm{R}_u^+} + \sum_{v} k_{\mathrm{Cr}}(v; m, \mathrm{O}') C_{\mathrm{R}_v^+} \right\}$$
$$- \left\{ \sum_{\mathrm{O}} k_{\mathrm{De}}(m, \mathrm{O}) + \sum_{q} k_{\mathrm{HS}}(m; q) + \sum_{r} k_{\mathrm{MS}}(m; r) + \sum_{u} k_{\mathrm{PCP}}(m; u) + \sum_{z} k_{\mathrm{Cr}}(m; z, \mathrm{O}'') \right\} \cdot C_{\mathrm{R}_m^+}. \qquad (5\text{–}16)$$

The equations for the global reaction rates of the paraffins that result from the above eliminations still contain the concentrations of the physisorbed paraffins. Therefore, quasi-equilibrium is assumed for the physical adsorption, allowing the concentrations of the physisorbed paraffins to be eliminated in favor of the observable partial pressures. In previous lumped kinetic studies (Baltanas et al. 1983; Steijns and Froment 1981) a Langmuir-type isotherm was found most successful in describing the physisorption of the paraffins inside the zeolite cages

$$c_{\mathrm{P}_i} = \frac{c_{s,\mathrm{sat}} K_{L,i} \mathrm{p}_i}{1 + \sum_m K_{L,m} \mathrm{p}_m}. \qquad (5\text{–}17)$$

Equations (5–12) and (5–17) express the global reaction rates of the paraffins in terms of the observable partial pressure of the paraffins and hydrogen and contain a number of single-event rate and physisorption coefficients. The elementary rate coefficients are the products of the single-event rate coefficient and the number of single events. The kinetic parameters are independent of the chain length of the paraffins and are preferably determined from experiments in which all the isomers can still be identified. This is why Van Raemdonck and Froment (1989) determined the rate parameters from the hydrocracking of n-octane. The estimation problem was partitioned into two parts: hydroisomerization only and hydroisomerization plus hydrocracking. Only low conversion data were used in the first stage. The rate parameters for isomerization of a n-paraffin into a monomethyliso-mer were then used as first estimates for the parameter estimation of the second stage, comprising all the data, both at low and at high conversion.

For the rate coefficients the following trends, generally accepted in carbenium ion chemistry, are observed: hydride shifts is the fastest type of isomerization; branching isomerization between carbenium ions of the same type is much slower than isomerization without change in degree of branching; isomerization between secondary carbenium ions is much faster than isomerization between tertiary carbenium ions; olefin protonation is a fast reaction with low activation energy; and deprotonation of tertiary carbenium ions is much slower than deprotonation of the less stable secondary ion.

At present, the approach is extended to naphthenic components and to mixtures of key components.

CONCLUSION

A fundamental kinetic modeling of processes involving complex feedstocks has been outlined and illustrated. By decomposing the reactions of the network into elementary steps the approach keeps the number of parameters within tractable limits. Also, by preserving the full reaction network it enables the prediction of a detailed effluent composition, which is normally not possible with lumped models.

When a complete component analysis of feedstock and effluent is not possible, a certain degree of lumping is inevitable. The kinetic behavior of the lumps can be correctly predicted when correlations are available for the distribution of the components of the lump. Such a correlation might be based on extrapolation from fractions that are easier to analyze or constructed from information obtained by different analytical techniques.

It should be emphasized that the experimental database from which the kinetic and physisorption parameters of the fundamental model were calculated was exactly the same as that used in the lumped approach. In other words, if the experiments are carefully designed, the fundamental approach does not require a larger experimental effort.

Finally, the kinetics of other complex processes such as catalytic reforming and catalytic cracking can be modeled along the lines illustrated here for hydrocracking and thermal cracking.

NOMENCLATURE

A	preexponential factor (s^{-1}) (in thermal cracking)
A	normal paraffin (in hydrocracking)
ADSORB	denominator of catalytic kinetic expression (bar)
C_{H^+}	surface concentration of vacant acid sites ($kmol/kg_{cat}$)
C_N	concentration of "naphtha" pseudocomponent ($kmol/m_r^3$)
$C_O, C_{O_{ij}}$	olefin surface concentration ($kmol/kg_{cat}$)
C_{P_i}	paraffin surface concentration ($kmol/kg_{cat}$)
CR	lumped cracking products

$C_{R_m^+}, C_{R_{i,m}^+}$	Carbenium ion surface concentration ($kmol/kg_{cat}$)
$C_{s,sat}$	saturation surface concentration of physisorbed hydrocarbons ($kmol/kg_{cat}$)
C_t	total surface concentration of acid sites ($kmol/kg_{cat}$)
E	activation energy (kJ/kmol or kcal/kmol)
$E_{AB}(H_p)(H^\bullet)$	initialization of the activation energy for hydrogen abstraction from a primary carbon atom by a hydrogen radical (kJ/kmol or kcal/kmol)
ΔE_{AB}	structural contribution to the activation energy of hydrogen abstraction (kJ/kmol or kcal/kmol)
H^+	vacant acid site
k	rate coefficient for naphtha cracking (s^{-1})
$\tilde{k}$	single-event rate coefficient
k_{Cr}	rate coefficient for cracking of MTB ($kmol/kg_{cat} \cdot s$)
$k_{Cr}(m; n; O)$	rate coefficient for cracking of a carbenium ion of type m into a carbenium ion of type n and olefin O (s^{-1})
$k_{De}(m; O_{ij})$	rate coefficient for deprotonation of a carbenium ion of type m, with formation of olefin O_{ij}(h^{-1})
$K_{DH_{ij}}$	equilibrium constant for dehydrogenation of paraffin P_i to olefin O_{ij} (bar)
$K_{HS}(m; n)$	rate coefficient for hydride shift between carbenium ions of type m (h^{-1})
K_L	adsorption parameter (bar^{-1})
$K_{L,i}$	Langmuir physisorption equilibrium constant for hydrocarbon i (bar^{-1})
k_{MB}	rate coefficient of formation of MB from A ($kmol/(kg_{cat}$-hr))
$k_{MS}(m; n)$	rate coefficient for methyl shift between carbenium ions of type m and n (h^{-1})
k_{MTB}	rate coefficient of formation of MTB from MB ($kmol/(kg_{cat}$-hr))
$k_{PCP}(m; n)$	rate coefficient for PCP branching between carbenium ions of type m and n (h^{-1})
$k_{Pr}(m)$	rate coefficient for protonation of an olefin to a carbenium ion of type m (kg_{cat}/(kmol-h))
K_1	lumped thermodynamic equilibrium constant of $A \rightleftarrows MB$ (bar/bar)
K_2	lumped thermodynamic equilibrium constant of $MB \rightleftarrows MTB$ (bar/bar)
M	Boolean relation matrix
MB	monobranched paraffin
Me	methyl
MTB	multibranched paraffin
n_{C^+-H}	number of hydrogen atoms on the charge-bearing carbon atom

n_e	number of (single) events involved in a reaction
$n_{\alpha H}$, $n_{\beta H}$	number of hydrogen atoms on the carbon atom in the positions respectively α and β with respect to the charge-bearing carbon atom
$n_{\alpha Me}$	number of methyl groups on the carbon atom in α-position with respect to the charge-bearing carbon atom
O	olefin, type not specified
O_i	ith olefin of reaction network
O_{ij}	olefin with hydrocarbon structure i and double-bond subscript j
p_{H_2}	partial pressure of hydrogen (bar)
P_i	paraffin with hydrocarbon structure i
R	ideal gas constant (kJ/(kmol-K))
r_A	rate of disappearance of A (kmol/(kg$_{cat}$-hr))
r_{Cr}	rate of cracking of MTB (kmol/kg($_{cat}$-hr))
$R^+_{ij,m}$	carbenium ion of type m, formed by protonation of olefin O_{ij}
R^+_m	carbenium ion of type m, without reference to related olefin
r_{MB}	rate of formation of MB (kmol/kg$_{cat}$-hr))
r_{MTB}	rate of formation of MTB (kmol/(kg$_{cat}$-hr))
r_N	rate of naphtha cracking (kmol/(m$_r^3$-s))
$R_{O_{ij}}$	net rate of formation of olefin O_{ij} (kmol/(kg$_{cat}$-s))
R_{P_i}	net rate of formation of paraffin P_i (kmol/(kg$_{cat}$-s))
$R_{R^+_m}$	net rate of formation of carbenium ion R^+_m (kmol/(kg$_{cat}$-s))
	secondary
	tertiary
T	temperature (K)

Subscripts

Cr	cracking
De	deprotonation
DH	dehydrogenation
HS	hydride shift
MS	methyl shift
PCP	PCP-branching
Pr	protonation

REFERENCES

Baltanas, M. A. and G. F. Froment (1985) Computer generation of reaction networks and calculation of product distributions in the hydroisomerization and hydrocracking of paraffins on Pt-containing bifunctional catalysts. *Comput. Chem. Engin.* **9**:71.

Baltanas, M. A., H. Vansina, and G. F. Froment (1983) Hydroisomerization and hydrocracking. 5. Kinetic analysis of rate data for *n*-octane. *Ind. Engin. Chem. Prod. Res. Dev.* **22**:531.

Baltanas, M. A., K. K. Van Raemdonck, G. F. Froment, and S. R. Mohedas (1989) Fundamental kinetic modeling of hydroisomerization and hydrocracking on noble-metal-loaded faujasites. 1. Rate parameters for hydroisomerization. *Ind. Engin. Chem. Res.* **28**:899.

Clymans, P. J. and G. F. Froment (1984) Computer-generation of reaction paths and rate equations in the thermal cracking of normal and branched paraffins. *Comput. Chem. Engin.* **8**:137.

Froment, G. F. (1987a) The kinetics of complex catalytic reactions. *Chem. Engin. Sci.* **42**:1073.

Froment, G. F. (1987b) Kinetics of the hydroisomerization and hydrocracking of paraffins on a platinum containing bifunctional Y-zeolite. *Catalysis Today* **1**:455.

Froment, G. F. and K. B. Bischoff (1990) *Chemical Reactor Analysis and Design*, 2nd Ed. John Wiley, New York.

Hillewaert, L. P., J. L. Dierickx, and G. F. Froment (1988) Computer generation of reaction schemes and rate equations for thermal cracking. *A.I.Ch.E.J.* **334**:17.

Marin, G. B. and G. F. Froment (1982) Reforming of C_6 hydrocarbons on a Pt-Al_2O_3 catalyst. *Chem. Engin. Sci.* **37**:759.

Plehiers, P. M., G. C. Reyniers, and G. F. Froment (1990) Simulation of run length of ethane cracking furnaces. *Ind. Engin. Chem. Res.* **29**:636.

Steijns, M. and G. F. Froment (1981) Hydroisomerization and hydrocracking. 3 Kinetic analysis of rate data for *n*-decane and *n*-dodecane. *Ind. Engin. Chem. Prod. Res. Dev.* **20**:660.

Steijns, M., G. F. Froment, P. Jacobs, J. Uytterhoeven, and J. Weitkamp (1981) Hydroisomerization and Hydrocracking. 2. Product distributions from *n*-decane and *n*-dodecane. *Ind. Engin. Chem. Prod. Res. Dev.* **20**:654.

Van Damme, P. S., G. F. Froment, and W. B. Balthasar (1981) Scaling up of naphtha cracking coils. *I.E.C. Proc. Des. Dev.* **20**:366.

Van Dierendonck, L. L., H. G. M. Egberink, and G. F. Froment (1985) *Proceedings of the 6th Symposium Large Chemical Plants*, K.V.I.V., Antwerpen, October 1985.

Van Raemdonck, K. and G. F. Froment (1989) Fundamental kinetic modeling of hydroisomerization and hydrocracking of paraffins on noble metal loaded zeolites. II. The elementary cracking steps. Presented at *A.I.Ch.E. National Meeting*, San Francisco, November 5–10.

Vansina, H., M. A. Baltanas, and G. F. Froment (1983) Hydroisomerization and hydrocracking. 5. Kinetic analysis of rate data for *n*-octane. *Ind. Engin. Chem. Prod. Res. Dev.* **22**:531.

Willems, P. A. and G. F. Froment (1988a) Kinetic modeling of the thermal cracking of hydrocarbons. 1. Calculation of frequency factors. *Ind. Engin. Chem. Res.* **27**:1959.

Willems, P. A. and G. F. Froment (1988b) Kinetic modeling of the thermal cracking of hydrocarbons. 2. Calculation of activation energies. *Ind. Engin. Chem. Res.* **27**:1966.

6

Structural Models of Catalytic Cracking Chemistry: A Case Study of a Group Contribution Approach to Lumped Kinetic Modeling

DAVID T. ALLEN
DIMITRIS LIGURAS

The kinetic modeling of mixtures containing many components is a complex problem that has received considerable attention. One of the traditional approaches to modeling the reactions of such mixtures is to group molecules that react at similar rates together into compound classes. The compound classes, or kinetic lumps, are then treated as pseudocomponents in the modeling. Such lumped kinetic models have been very successful in a variety of applications (Weekman 1979); however, new reaction engineering challenges are stretching the limits of lumped models. A notable example is the use of size and selective catalysts in cracking complex petroleum feedstocks. The size and shape selective features of zeolite catalysts can give molecules of similar structure very different reactivities. The consequence of these differences in reactivity is that lumped kinetic models must contain scores to hundreds of lumps. Determining the rates of reaction of this many kinetic lumps presents significant difficulties. Direct experimental evaluation of the rate parameters would be exceedingly time consuming and difficult. Further, if the rate parameters for the lumps are considered adjustable parameters, it is unlikely that a unique, invariant set of rate parameters could be identified. Clearly, we must adopt a predictive approach to estimating rate parameters for kinetic models containing large numbers of lumps.

Our approach to solving this problem utilizes group contribution concepts. To illustrate the approach we will use the case study of cracking of petroleum feedstocks over an amorphous catalyst. Based on analytical data available on the feedstock, several hundred kinetic lumps are chosen. Each lump is represented by a chemical structure. These hundreds of pseudocomponents vary widely in structure, but they can all be constructed from a small number of functional groups. Using model compound studies and group contribution concepts we have developed algorithms that use functional group distributions to estimate the reaction rates and pathways for a broad spectrum of pseudocomponent structures. We thus *predict* the pathways and rates of reaction of our hundreds of kinetic lumps. Using this modeling approach, we significantly reduce the number of parameters in a model containing many lumps and the parameters (group contributions) can be determined using relatively simple model compound experiments.

There are a number of approximations inherent in this approach. Chief among the approximations is the selection of the pseudocomponents (perhaps several hundred) used to represent the thousands of components in the actual feedstock. If the approximations are successful, however, this type of model can be very powerful. One of the most significant features of this type of model is its utility in predicting the properties of the reaction products. Because the model predicts the concentrations of product molecules, it is possible to use group contribution methods to evaluate and optimize a wide variety of physical, chemical, and thermodynamic property values.

This chapter will demonstrate the development and use of this new generation of lumped kinetic models. Our case study will be the catalytic cracking reactions of petroleum feedstocks over amorphous silica alumina catalysts. We present sections on pseudocomponent (lump) selection, rate parameter estimation, interfacing the model with property estimation methods, and finally, the information required to extend the model to shape and size selective catalysts.

PSEUDOCOMPONENT SELECTION

The overall goal of this work is to examine the feasibility of a new type of kinetic model for the catalytic cracking reactions of petroleum. The model uses a large number of pseudocomponents to characterize a petroleum feedstock. This section will describe methods for the selection of the pseudocomponents. The pseudocomponent selection is based on analytical data commonly available on petroleum feedstocks.

A vast array of analytical tools are available for petroleum characterization. In this work, however, we will focus on using the results of a few primary analysis tools: compound class separations, gas chromatography, mass spectrometry, and nuclear magnetic resonance (NMR) spectroscopy. Detailed descriptions of each of these techniques is beyond the scope of this chapter. The focus of this discussion will be the information provided by these techniques and the use of the data in selecting pseudocomponents. Summarized below are brief descriptions of the data obtained from each of the analytical tools.

Compound Class Analysis by Chromatographic Separation/Mass Spectrometry

In this type of analysis, a petroleum feedstock is run through a series of resin columns. These columns sequentially separate neutrals, acids, and bases (e.g., see Allen et al. 1985; Jewell et al. 1974). Alternatively one can use liquid and gas chromatography to separate and determine the weight fractions of paraffins, isoparaffins, naphthenes of varying ring size, and aromatics of varying ring size (Later et al. 1981). Separations can be done in tandem with mass spectrometry, yielding a carbon number distribution for each of the compound classes.

NMR Spectroscopy

While compound class separations coupled with mass spectrometry can yield information about carbon number distributions, NMR spectroscopy yields information on the carbon type distribution in the feedstock. A typical proton NMR spectrum, shown in Figure 6–1, can be divided into the eight bands listed in Table 6–1 (Clutter et al. 1972). The relative intensities of these resonances provide the distribution of hydrogen types in the samples. A typical ^{13}C NMR spectrum, shown in Figure 6–2, is much more detailed than the ^{1}H spectrum of Figure 6–1. The carbon spectrum can be separated into more than 30 bands (Thiel and Gray 1988). Typical band assignments are listed in Table 6–2.

Taken together, the ^{1}H and ^{13}C NMR spectra yield approximately 40 independent pieces of structural data. Spectra can be taken on the whole oil or on each of the compound classes. In most of our work, we have collected NMR spectra only on the whole oil.

Synthesis of Characterization Data

If we assume that a mass spectrum is available on each of five compound classes (*n*-paraffins, isoparaffins, olefins, naphthenes, aromatics) and that ^{1}H and ^{13}C NMR spectra are available on the whole oil, then there are roughly 190 constraints on our pseudocomponent selection (five compound classes with 30 carbon numbers per class plus 40 NMR-based constraints). All of these constraints are linear in the pseudocomponent concentrations. As an example of the form of these constraints, consider the C_{12} isoparaffins. The concentration of these pseudocomponents must be related to the constraints

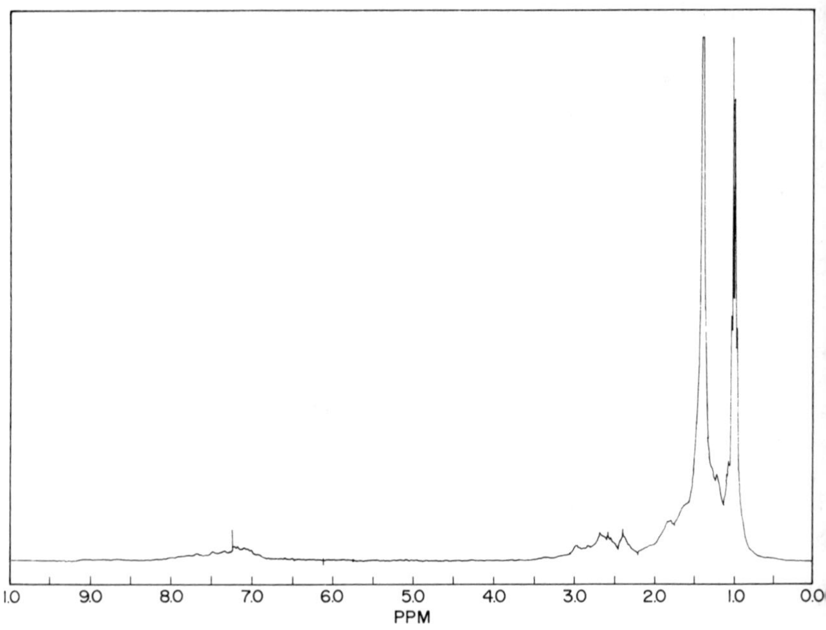

Figure 6–1 ^{1}H NMR spectrum of a gas oil.

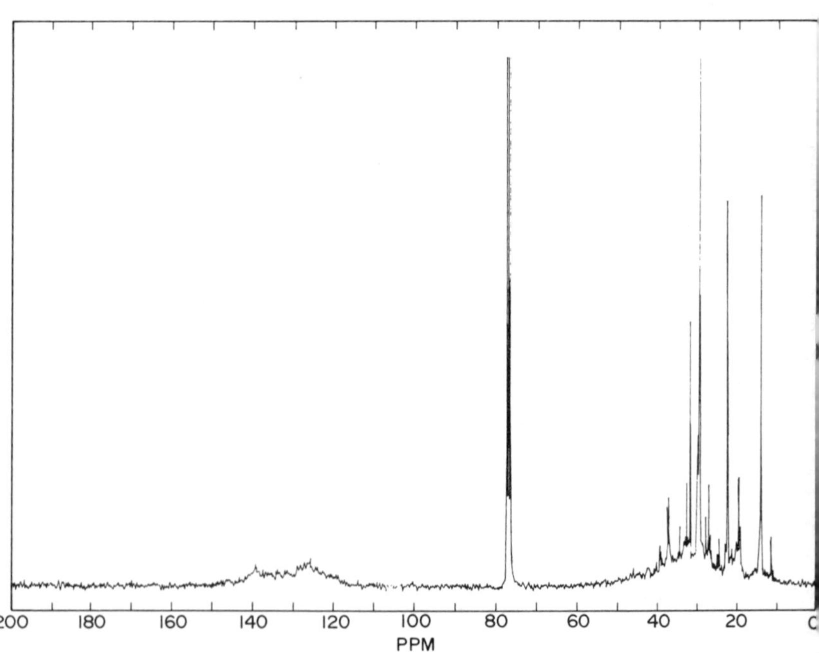

Figure 6–2 ^{13}C NMR spectrum of a gas oil.

Table 6–1 Band Assignments for ^{1}H NMR Spectra

	Chemical Shift Range (ppm)
H in tri+ aromatic rings	10–8.15
H in diaromatic rings	8.15–6
H in monoaromatic rings	7.9–6
H in CH_2 groups alpha to an aromatic ring	4.0–2.6
H in methyl groups alpha to an aromatic ring	2.6–1.9
H in naphthenic rings beta for farther from an aromatic ring (including saturated molecules)	1.9–1.65
H beta or farther from an aromatic ring (including saturated molecules)	1.9–1.0
H in methyl groups not alpha to an aromatic ring (including saturated molecules)	1.0–0

Source: Clutter et al. (1972).

imposed by the analytical data. Some possible C_{12} isoparaffin structures are shown in Figure 6–3. The constraint based on the carbon number distribution is

$$x_1 + x_2 + x_3 + x_4 + x_5 + x_6 + x_7 = C_{12}^{IP}, \tag{6–1}$$

where the x_i are the concentrations of the seven pseudocomponents used to model the C_{12} isoparaffins and C_{12}^{IP} is the total concentration of C_{12} isoparaffins. This equation reflects the fact that the sum of the concentrations of the C_{12} pseudocomponents must equal the total concentration of C_{12} isoparaffins. An additional series of constraints can be written based on the carbon type distribution. For example, from the ^{1}H NMR spectrum and the total elemental composition of the oil, we can determine the concentration of methyl hydrogens. This constraint can be written as

$$H_{\text{methyl}} = 9*(x_1 + x_2 + x_3 + x_4 + x_5) + 12(x_6 + x_7) + \sum_{\substack{\text{all other} \\ \text{pseudocomponents}}} S_{ij}x_i, \tag{6–2}$$

Table 6–2 ^{13}C NMR Peak Assignments

	Chemical Shift (ppm)	Assignment
Aromatics	115–135	Aromatic carbon
Aliphatics		
Peak 1	11.4	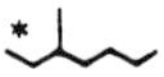 * R (R_{ar})
Peak 2	12.85	* R (ethyl branch)
Peak 3a	14.1	 * R (R_{ar})
Peak 3b	14.5	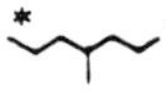 * R (R_{ar})
Peak 4	15.7	* AR
Peak 5a	19.3	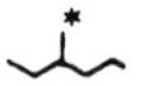 * R (R_{ar})
Peak 5b	19.8	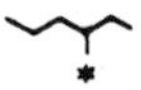 * R (R_{ar})
Peak 5c	20.4	 * R (R_{ar})
Peak 6	21.5	$*CH_3–R_{ar}$
Peak 7	23	 * * $(CH_2)_n$ R (R_{ar})
Peak 8a	24.6	C_8 of C_5–C_9 farnesane subunit
Peak 8b	24.8	C_1 of C_1–C_4 farnesane subunit
Peak 9a	26.7	
Peak 9b	27.1	
Peak 9c	27.4	 * R (R_{ar})
Peak 11	28.1	 -CH* $(CH_2)_n$ R (R_{ar})
Peak 12a	29.4	 * R (R_{ar})
Peak 12b	29.9	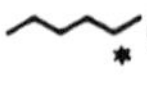 * $(CH_2)_n$ R (R_{ar})

Table 6–2 *(continued)*

	Chemical Shift (ppm)	Assignment
Aliphatics		
Peak 12c	30.1	$\sim\sim\overset{*}{(CH_2)}_n\frown R\ (R_{ar})$
Peak 13	32.1	$\underset{*}{\sim\sim}(CH_2)_n\frown R\ (R_{ar})$
Peak 14a	32.5	
Peak 14b	32.8	-(-$\underset{*}{\overset{\vert}{C}}H$⌣)- of C_5–C_8 farnesane subunit
Peak 14c	33.2	
Peak 15a	34.5	-(-$\underset{*}{\overset{\vert}{C}}H$⌣) of C_9–C_{12} farnesane subunit
Peak 15b	34.7	⌣$\underset{*}{\overset{\vert}{C}}H$⌢R
Peak 16a	37.2	R⌣⌢$\underset{*}{\overset{\vert}{C}}H$⌢⌣R (isolated methyl branch)
Peak 16b	37.5	–(* ⅄ *)– of C_5–C_8 farnesane subunit
Peak 17a	39.1	⅄ * ∽∽R (R_{ar})
Peak 17b	39.4	–(⅄ *)– of C_1–C_4 farnesane subunit
Peak 18	33–38	AR–$\overset{*}{C}H_2$–; AR–CH <
Peak 19	42.0–42.4	

Source: Thiel and Gray (1988).

Farnesane: $[C_1-\underset{\underset{C_1}{\vert}}{C_2}-C_3-C_4-][-C_5-\underset{\underset{C_{13}}{\vert}}{C_6}-C_7-C_8-][-C_9-\underset{\underset{C_{14}}{\vert}}{C_{10}}-C_{11}-C_{12}]$

C_1—C_4 subunit C_5—C_8 subunit C_9—C_{12} subunit

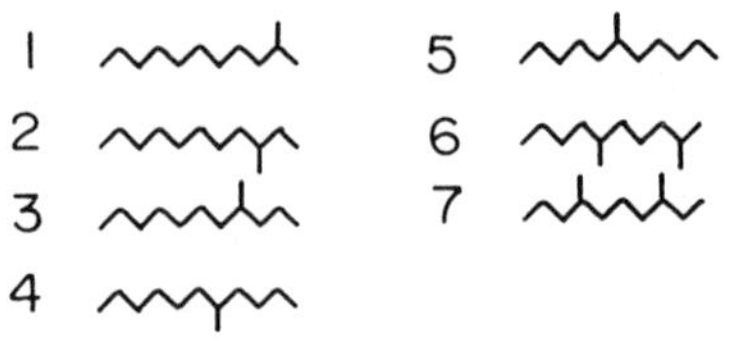

1 2 3 4 5 6 7

Figure 6–3 C_{12} isoparaffin structures used as potential pseudocomponents.

where H_{methyl} is the total concentration of methyl hydrogen in the feedstock and S_{ij} is a stoichiometric coefficient accounting for the number of methyl hydrogens in pseudocomponent i.

The two simple examples given in Eqs. (6–1) and (6–2) are typical of the linear constraints that bound the pseudocomponent selection. The constraints can be written in the matrix form

$$\mathbf{Sx} = \mathbf{b}, \tag{6–3}$$

where **S** is the matrix of stoichiometric coefficients, **x** is the vector of pseudocomponent concentrations, and **b** is the vector of concentrations derived from the analytical data.

These equations define a feasible space of allowed pseudocomponent concentrations. Typically, there are several hundred possible pseudocomponent structures and their concentrations are bounded by roughly 200 constraints. At this point, with a defined space of allowed pseudocomponent concentrations, a variety of strategies for selecting pseudocomponents can be employed. One strategy is to select many possible sets of pseudocomponents. Variations in the predictions of the model over a number of pseudocomponent sets can be averaged. If done over enough pseudocomponent sets, this is essentially a Monte Carlo approach. A second strategy, which is similar to the first, is to use a very large ((0)10^4) number of pseudocomponents. Both of these strategies are computationally intensive. A somewhat less computationally demanding route is to choose a single set of pseudocomponents that minimizes the heat of formation or Gibbs free energy of the set of pseudocomponents. The relative merits of these various strategies for selecting pseudocomponents have not been tested. In this work, we apply the first strategy. Several sets of pseudocomponents are selected. All the sets of pseudocomponents represent the analytical data equally well. Each is used to model the mixture, and variations in the predictions of the model are monitored.

We now examine the problem of identifying valid sets of pseudocomponents. The problem is simply to find a set of concentrations, **x**, that are non-negative

$$x_i \geq 0, \quad i = 1, n \tag{6–4}$$

and that satisfy the analytical constraints

$$\sum_{i=1}^{n} S_{ij}x_i = b_j, \quad j = 1, \ldots, m, \tag{6–5}$$

where n is the number of pseudocomponents and m is the number of constraints. Several well-known algorithms exist for finding the maxima or minima of equations that are linear or quadratic in **x**, subject to constraints of the type given in Eqs. (6–4) and (6–5). The simplest of these algorithms are the SIMPLEX methods for finding the extrema of a linear function subject to linear equality and inequality constraints and subject to the non-negativity of **x** (see, e.g., Strang 1980). We use SIMPLEX algorithms to select our pseudocomponents, maximizing the objective P, where

$$P = \sum_{i=1}^{n} w_i x_i. \tag{6–6}$$

The w_i are weighting factors. In our current work, all weighting factors are assumed to be unity. An alternative approach would be to use a heat of formation as the weighting factor, minimizing P.

In summary, analytical data are used to generate roughly 200 constraints on the pseudocomponent selection. These constraints, along with the physical constraints of non-negative concentrations, idenitify a feasible space of pseudocomponent concentrations. A number of alternative pseudocomponent sets are chosen from the feasible space using the SIMPLEX linear programming algorithm. As an example of pseudocomponent selection, Figure 6–4 shows two different sets of C_{12} structures that equally well represent a petroleum feedstock described by Liguras and Allen (1989b).

OVERVIEW OF THE STRUCTURAL KINETIC MODEL

There are two basic elements in the structural model of catalytic cracking kinetics. The first element is the characterization of the petroleum feedstock in terms of a reasonable number of pseudocomponents. The issue of pseudocomponent selection was addressed in the previous section. The second element of the model is the definition of reaction pathways and the estimation of reaction rate parameters. This is accomplished using group contribution principles. The details of the approach are described elsewhere (Liguras and Allen 1989a,b). In this chapter, we will review the basic principles, focusing on the number and type of adjustable parameters in the model.

Group contributions are determined for each of the major reaction pathways (cracking, isomerization, cyclization, dehydrogenation, ring open-a)

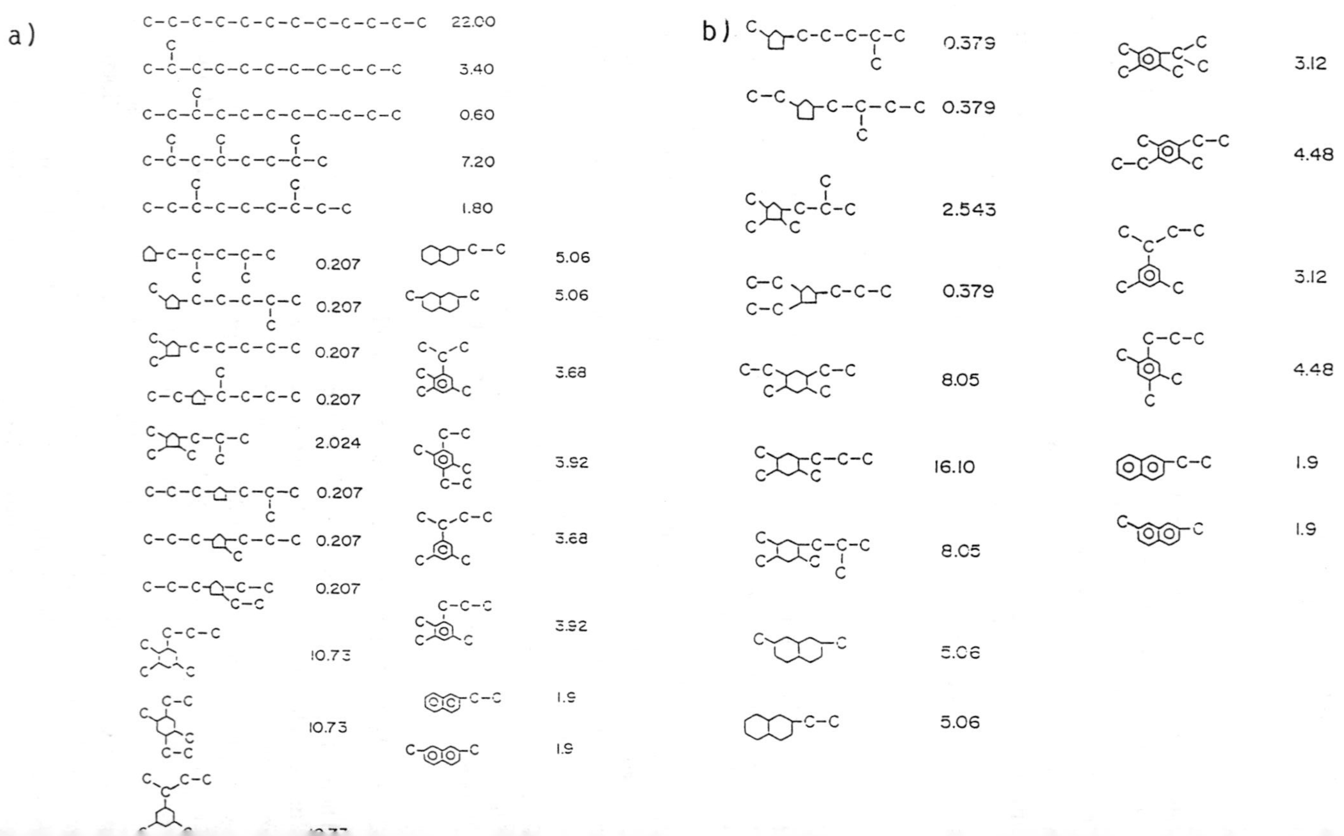

C–C–C–C–C–C–C–C–C–C–C–C 22.00

C–C–C–C–C–C–C–C–C–C–C 3.40

C–C–C–C–C–C–C–C–C–C–C 0.60

C–C–C–C–C–C–C–C 7.20

C–C–C–C–C–C–C–C–C 1.80

0.207

0.207

0.207

0.207

2.024

0.207

0.207

0.207

10.73

10.73

5.06

5.06

3.68

3.92

3.68

3.92

1.9

1.9

b)

0.379

0.379

2.543

0.379

8.05

16.10

8.05

5.06

5.06

3.12

4.48

3.12

4.48

1.9

1.9

ing, dealkylation, and coke formation) for each of the major compound classes (n-paraffins, isoparaffins, olefins, naphthenes, aromatics). The group contribution principles are best demonstrated through an example, so we will describe the methods for estimating the overall rates and product distributions for cracking reactions.

We begin by separating the problems of estimating overall cracking rates and predicting the product distributions. To accomplish the separation, we express the rate of change of a given product concentration C_p and the rate of change of a given reactant concentration C_r as

$$\frac{dC_p(t)}{dt} = KC_r(t)\Phi(t)\mathrm{SP}_i\mathrm{SPN}_k, \qquad (6\text{–}7)$$

$$\frac{dC_r(t)}{dt} = -\sum_i \sum_k KC_r(t)\Phi(t)\mathrm{SP}_i\mathrm{SPN}_k, \qquad (6\text{–}8)$$

where K is the overall rate of cracking of the reactant, $\Phi(t)$ is the catalyst activity decay function, SP_i is the probability of formation of a reactive carbenium ion on carbon number i of the reactant, and SPN_k is the probability that bond k in the reactant cracks, given that a carbenium ion exists on carbon i. Consider the application of these equations to the cracking of n-paraffins and isoparaffins. The overall cracking rate constant K, at 500°C is, a function of carbon number and the types of carbon centers in the paraffin. Based on model compound data for cracking over amorphous silica alumina catalysts, we conclude that the overall cracking rate for n-paraffins is given by a quadratic function of carbon number

$$a(\mathrm{CN})^2 + b(\mathrm{CN}) + c = K, \qquad (6\text{–}9)$$

where CN is the carbon number of the n-paraffin, and K is the pseudo-first-order rate constant at 500°C in units of s^{-1}. The overall cracking rate for isoparaffins is estimated by multiplying the rate of cracking for the

---

Figure 6–4 Two alternative sets of C_{12} pseudocomponents; both sets are consistent with the gas oil structural data given by Liguras and Allen (1989b) (n-paraffin and isoparaffin structures in set b are the same as in set a). Reprinted with permission from Liguras, D. K. and D. T. Allen, Structural models for catalytic cracking. 2. Reactions of simulated oil mixtures. *Ind. Engin. Chem. Res.* **28**:674 (1989). Copyright 1989 American Chemical Society.

n-paraffin of the same carbon number by the following factor:

$$F = (0.62P + 1.29S + 12.3T)/(0.62P' + 1.29S'), \qquad (6\text{–}10)$$

where P is the number of primary hydrogens, S is the number of secondary hydrogens, and T is the number of tertiary hydrogens in the isoparaffin; P' and S' are the number of primary and secondary hydrogens in the n-paraffin of the same carbon number. This correlation is based on model compound data and on the principle that carbenium ions (the reactive intermediate in cracking reactions) are formed at different rates on primary, secondary, and tertiary carbons. The performance of the correlation, for a variety of model compounds, is shown in Table 6–3.

It is instructive at this point to consider how many adjustable parameters have been introduced to model the overall rate of cracking for paraffins and isoparaffins. The quadratic dependence on carbon number introduces three parameters and the isoparaffin correction factor F introduces an additional three parameters to bring the count to six.

We now examine the calculation of product distributions, using the case of 2,7-methyloctane to illustrate our procedures. From Eqs. (6–9) and (6–10) we know the overall rate of cracking for 2,7-dimethyloctane. The molecule has 10 carbon atoms, 12 primary hydrogens, 8 secondary hydrogens, and 2 tertiary hydrogens. Its overall cracking rate K, at 500°C, is thus

$$0.556\ \mathrm{s}^{-1} = (0.032\ \mathrm{s}^{-1})[0.62(12) + 1.29(8) + 12.3(2)]/[0.62(6) + 1.29(16)].$$

To predict the product distributions we rely on our understanding of the cracking mechanism. Cracking proceeds through a reactive intermediate known as a carbenium ion. This ion can form on any of the carbons; however the relative stability of the ion depends on whether the ion is formed on a

Table 6–3 Comparison of Experimental and Estimated Rate Constants for Overall Cracking Rate (in Inverse Seconds)

Compound	Experimental[a]	Estimated
3-Methylpentane	0.0095	0.0090
2,3-Dimethylbutane	0.0115	0.0126
2,2,4-Trimethylpentane	0.0130	0.0139
2,7-Dimethyloctane	0.0550	0.0556
2,2,4,6,6-Pentamethylheptane	0.0580	0.0640

[a] Data of Greensfelder et al. (1949).

primary, secondary, or tertiary carbon. The relative stabilities of tertiary, secondary, and primary carbons are 0:14:21 kcal/mol (Franklin 1949). At 500°C these energy differences translate to relative probabilities of 1:1.1×10^{-4}:1.2×10^{-6}. For the case of 2,7-dimethyloctane the probability that the ion forms on carbon number 7 is thus

$$SP_7 = [2(1) + 8(1.1 \times 10^{-4}) + 12(1.2 \times 10^{-6})]^{-1}. \qquad (6\text{–}11)$$

Note that the terms in the denominator of Eq. (6–11) reflect that there are three potential ion formation sites per primary carbon, two potential ion formation sites per secondary carbon, and one potential ion formation site per tertiary carbon. Similar probabilities for ion formation (SP_i) can be calculated for each carbon position. These calculations assume an equilibrium distribution of carbenium ions. This is a major assumption in the model. Given that an ion forms on carbon number 7, cracking can occur only at the 5–6 bond. In general, however, more than one cracking site will be possible and the relative probabilities for each pathway must be estimated. Our model assumes that the relative probability of the pathways depends on the relative stability of the carbenium ion formed by the cracking reaction. The relative probabilities of bond cleavage are 1:2:20 depending on whether a primary, secondary, or tertiary carbenium ion is formed. Predictions of this model are compared to experimental data in Table 6–4 and Figure 6–5.

In this formulation the model for cracking has eight adjustable parameters. Six of the parameters are used in modeling the overall cracking rate. The remaining two parameters, the relative probabilities of pathways involving primary, secondary, and tertiary carbenium ion products, are used in modeling product distributions. Note that in our count of adjustable param-

Table 6–4 Compound Class Analysis of 2,7-Dimethyloctane Cracking Products

	Experimental[a]		Predicted	
Carbon No.	Olefins	Paraffins + Naphthenes	Olefins	Paraffins + Naphthenes
C_2	83	17	2	98
C_3	81	19	69	31
C_4	40	60	38	62
C_5	43	57	39	61

[a] Data of Greensfelder et al. (1949).

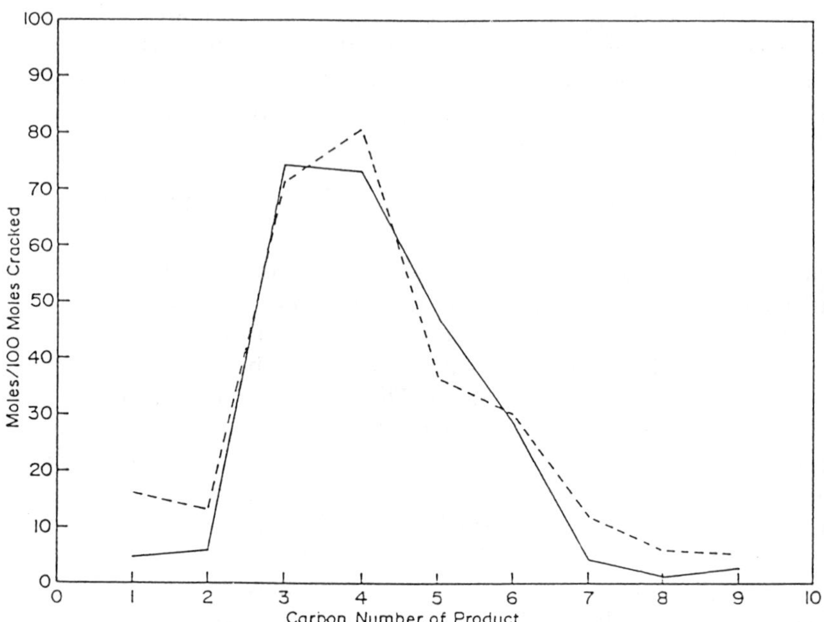

Figure 6–5 Carbon number distribution of the cracking products of 2,7-dimethyloctane. (------) Experimental data of Greensfelder and Voge (1945) and Greensfelder et al. (1949). (——) This work.

eters we have not included the relative stabilities of primary, secondary, and tertiary carbenium ions. These thermodynamic quantities are independent of the kinetic model.

Thus, a molecular level model has been built for the cracking reactions of a large number of pseudocomponents, based on eight adjustable parameters. Isomerization, cyclization, dehydrogenation, ring opening, dealkylation, and coke formation reactions can be modeled in a similar manner (Liguras and Allen 1989a). The comprehensive model contains approximately 25 adjustable parameters. These parameters were determined using a model compound data base of roughly 20 components. Since the 20 model compounds generated hundreds of reaction products, the data were more than sufficient to estimate the 25 adjustable parameters. Using this group contribution approach, each reaction pathway can be followed for each of the hundreds of pseudocomponents that represent a typical oil. The aggregated behavior of the pseudocomponents represents the behavior of the oil. Results obtained when this approach is applied to a typical oil have been described by Liguras and Allen (1989b). The data of Figure 6–6 and Table

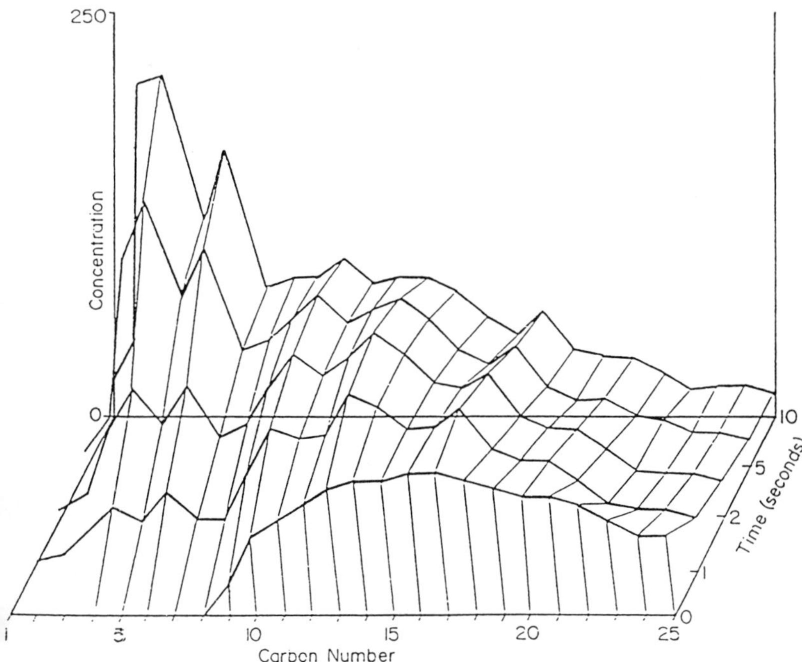

Figure 6–6 Carbon number distributions of the cracking products of the simulated oil described by Liguras and Allen (1989b) (concentrations in mol/100 mol cracked). Reprinted with permission from Liguras, D. K. and D. T. Allen, Structural models for catalytic cracking. 2. Reactions of simulated oil mixtures. *Ind. Engin. Chem. Res.* **28**:674 (1989). Copyright 1989 American Chemical Society.

–5 are representative of the information provided by the model. It is important to note, however, that Figure 6–6 and Table 6–5 are *summaries* of the predictions of the model. The model predicts the concentrations of the initial pseudocomponents and all product structures. Such information can be extremely valuable in interfacing the kinetic model with property estimation methods. This interface is the topic of the next section.

Before moving on to the interface between kinetic and thermodynamic modeling, we will use the structural kinetic model to address the following questions:

- How sensitive is the model to the uncertainties in the pseudocomponent selection?
- How many pseudocomponents are necessary?

Table 6–5 Carbon Center Distributions in Catalytic Cracking Products: Base Cases

Time (sec)	Carbon Number	Mole Fraction of Cut	Carbon Center Distribution (mole fraction in mixture)					
			Aliphatic			Olefinic		
			CH_3	CH_2	CH	Terminal	Nonterminal	Aromatic
1	C_1–C_3	0.0246						
	C_4–C_9	0.1718	0.0146	0.0225	0.0136	0.0023	0.0023	0.0037
	C_{10}–C_{13}	0.2735	0.0474	0.0942	0.0603			0.0190
	C_{14}–C_{17}	0.2563	0.0584	0.1239	0.0661			0.0329
	C_{18}–C_{23}	0.2305	0.0585	0.1361	0.0650			0.0696
	C_{23+}	0.0431	0.0115	0.0255	0.0107			0.0198
2	C_1–C_3	0.0494						
	C_4–C_9	0.2690	0.0295	0.0433	0.0254	0.0052	0.0052	0.0068
	C_{10}–C_{13}	0.2651	0.0576	0.1104	0.0704			0.0274
	C_{14}–C_{17}	0.2129	0.0590	0.1195	0.0594			0.0403
	C_{18}–C_{23}	0.1722	0.0522	0.1052	0.0430			0.0821
	C_{23+}	0.0307	0.0106	0.0181	0.0069			0.0226
5	C_1–C_3	0.0991						
	C_4–C_9	0.3803	0.0568	0.0776	0.0411	0.0108	0.0108	0.0132
	C_{10}–C_{13}	0.2245	0.0633	0.1135	0.0698			0.0343
	C_{14}–C_{17}	0.1523	0.0520	0.0905	0.0404			0.0474
	C_{18}–C_{23}	0.1230	0.0444	0.0624	0.0224			0.0940
	C_{23+}	0.0208	0.0099	0.0108	0.0043			0.0251
10	C_1–C_3	0.1450						
	$C_4{=}C_9$	0.4167	0.0710	0.0951	0.0468	0.0123	0.0122	0.0198
	C_{10}–C_{13}	0.1885	0.0585	0.1000	0.0595			0.0321
	C_{14}–C_{17}	0.1371	0.0500	0.0803	0.0367			0.0496

To address the first question, we selected two sets of pseudocomponents. The two sets represent the analytical data on a typical oil equally well. The differences in the model predictions using the two sets of pseudocomponents provide a measure of the inherent uncertainty involved in using a pseudocomponent approach. We can make this comparison quantitative using a deviation measure

$$D = \sum_{i=1}^{5} \sum_{j=1}^{6} (C_{ij} - C_{ij}^{1})^2,$$

where the first summation is performed over the various carbon types and the second summation is taken over the cuts reported in Tables 6–5 and 6–6. Note that the relative concentrations are normalized over the entire mixture

$$\sum_{i=1}^{5} \sum_{j=1}^{6} C_{ij} = 1.$$

At reaction times of 1, 2, 5, and 10 s, this measure of deviation had values of 0.000370, 0.000189, 0.000229, and 0.000029 for the comparison of the base case (Table 6–5) and the equivalent set of starting pseudocomponents (Table 6–6). This helps establish the accuracy of the pseudocomponent approach. Equally valid sets of pseudocomponents result in product distributions that deviate by these amounts. These deviations are orders of magnitude less than deviations resulting from differences in the initial feedstock and deviations due to perturbations of the rate parameters.

A second issue that can be examined using the structural kinetic model is the effect of the number of pseudocomponents. The base case of Table 6–5 uses 325 pseudocomponents. The effect of successively reducing number to 250, 200, 125, 95, and 75 is examined in Table 6–7. Table 6–7 shows the overall deviation parameter, at various reaction times, as the number of pseudocomponents is reduced. The deviation measure goes through a dramatic change when the number of pseudocomponents is reduced below about 150. This number corresponds to one pseudocomponent per carbon number per compound class. Reducing the number of pseudocomponents below about 150 requires that several carbon numbers be represented by a single pseudocomponent structure. As a result of this lumping, the product distributions are spiked at certain carbon numbers. Based on these results, we come to the general conclusion that the model requires at least one pseudocomponent structure per carbon number for each compound class. This conclusion is based on results for an amorphous catalyst. It is conceivable that the minimum number of pseudocomponents could increase for modeling cracking catalyzed by zeolites.

Table 6–6 Carbon Center Distributions in Catalytic Cracking Products: Effect of Pseudocomponent Selection

Time (sec)	Carbon Number	Mole Fraction of Cut	Carbon Center Distribution (mole fraction in mixture)					
			Aliphatic			Olefinic		
			CH_3	CH_2	CH	Terminal	Nonterminal	Aromatic
1	C_1–C_3	0.0231						
	C_4–C_9	0.1662	0.0147	0.0226	0.0149	0.0021	0.0021	0.0030
	C_{10}–C_{13}	0.2570	0.0458	0.0906	0.0583			0.0183
	C_{14}–C_{17}	0.2664	0.0630	0.1302	0.0723			0.0314
	C_{18}–C_{23}	0.2397	0.0639	0.1446	0.0721			0.0682
	C_{23+}	0.0415	0.0137	0.0325	0.0168			0.0187
2	C_1–C_3	0.0470						
	C_4–C_9	0.2591	0.0285	0.0417	0.0265	0.0048	0.0048	0.0054
	C_{10}–C_{13}	0.2593	0.0543	0.1034	0.0667			0.0255
	C_{14}–C_{17}	0.2279	0.0629	0.1233	0.0643			0.0383
	C_{18}–C_{23}	0.1748	0.0550	0.1083	0.0461			0.0793
	C_{23+}	0.0314	0.0110	0.0201	0.0091			0.0208
5	C_1–C_3	0.0964						
	C_4–C_9	0.3666	0.0533	0.0742	0.0414	0.0101	0.0101	0.0109
	C_{10}–C_{13}	0.2260	0.0613	0.1088	0.0683			0.0321
	C_{14}–C_{17}	0.1700	0.0575	0.0980	0.0466			0.0458
	C_{18}–C_{23}	0.1203	0.0478	0.0669	0.0245			0.0944
	C_{23+}	0.0207	0.0095	0.0102	0.0041			0.0239
10	C_1–C_3	0.1456						
	C_4–C_9	0.4081	0.0696	0.0947	0.0463	0.0124	0.0124	0.0162
	C_{10}–C_{13}	0.1914	0.0588	0.1010	0.0602			0.0310
	C_{14}–C_{17}	0.1378	0.0512	0.0791	0.0365			0.0500
	[illegible]	[illegible]	[illegible]	[illegible]	[illegible]			[illegible]

Table 6–7 Effect of Number of Pseudocomponents on Predictions of Fraction Yields

Number of Pseudocomponents	Deviation from Base Case (Reaction Time = 2 s)	Deviation from Base Case (Reaction Time = 10 s)
325	1.9×10^{-4}	2.9×10^{-5}
230	2.2×10^{-4}	4.1×10^{-5}
200	3.1×10^{-4}	3.5×10^{-5}
125	117.0×10^{-4}	ND
95	125.0×10^{-4}	ND
75	124.0×10^{-4}	60.0×10^{-5}

ND, not determined.

Interfacing the Structural Kinetic Model with Thermodynamic and Property Estimation Models

The previous sections have described the formulation of a catalytic cracking model based on molecular structure. This section will demonstrate one of the primary applications of such models: investigating the links between reaction pathways and the properties of product oil fractions. This link between kinetic model and property estimation is possible because the kinetic model predicts the concentrations of molecular structures. With molecular level product characterization available, it becomes completely straightforward to estimate the chemical, physical, and thermodynamic properties of the products using group contribution methods (see, e.g., Benson 1976; Fredenslund et al 1975; Vajdi and Allen 1989). As a case study, the output of the structural kinetic model was interfaced with a method for estimating gasoline range research octane number (RON) and gasoline range motor octane number (MON) (Cotterman and Plumlee 1989). The sensitivity of RON, MON, and gasoline yield to changes in the values of model parameters were examined quantitatively. The results are reported in Table 6–8. Two types of model parameters were varied in the calculations. In the first type of calculation, the rate constants for major reaction pathways were perturbed. In the second type of calculation, the structure of the initial feedstock was perturbed.

In evaluating the results of Table 6–8, consider first the effect of varying the rates of major reaction pathways. In these sensitivity studies the rate for a particular reaction pathway (e.g., isomerization) was changed for all compounds by the same percentage. Examination of Table 6–8 reveals that

Table 6–8 Sensitivity of Gasoline Yield and Octane Number to Kinetic Model Parameters

Case Study	Gasoline Yield (wt%)	Percent Change from Base Case	RON	Percent Change from Base Case	MON	Percent Change from Base Case	Gasoline Yield × MON	Percent Change from Base Case
Base case								
Reaction time, 1 s	24.04	—	66.60	—	65.62	—	1,578	—
Reaction time, 10 s	39.75	—	75.02	—	72.43	—	2,879	—
Primary paraffin cracking rate increased by 50%								
Reaction time, 1 s	26.65	+10.9	65.76	−1.3	64.70	−1.4	1,724	+9.3
Reaction time, 10 s	42.07	+5.8	74.98	−0.1	72.24	−0.3	3,039	+5.5
Primary paraffin cracking rate decreased by 50%								
Reaction time, 1 s	19.35	−19.5	66.90	+0.5	66.48	+1.3	1,286	−18.5
Reaction time, 10 s	33.56	−15.6	76.34	+1.7	73.51	+1.5	2,467	−14.3
Secondary paraffin cracking rate increased by 50%								
Reaction time, 1 s	24.08	+0.2	67.04	+0.7	65.86	+0.4	1,586	+0.5
Reaction time, 10 s	39.78	+0.1	75.04	0.0	72.18	−0.4	2,871	−0.3
Secondary paraffin cracking rate decreased by 50%								
Reaction time, 1 s	23.87	−0.7	66.05	−0.8	65.38	−0.4	1,561	−1.1
Reaction time, 10 s	40.37	+1.5	74.86	−0.2	72.29	−0.2	2,918	+1.3

increased by 50%								
Reaction time, 1 s	23.99	−0.2	66.27	−0.5	65.43	−0.3	1,569	−0.6
Reaction time, 10 s	39.56	−0.5	74.54	−0.6	72.23	−0.3	2,857	−0.8
Olefin cracking rate decreased by 50%								
Reaction time, 1 s	24.11	+0.3	67.09	+0.7	65.92	+0.5	1,589	+0.7
Reaction time, 10 s	39.85	+0.3	76.18	+1.5	72.91	+0.7	2,905	+0.9
Aromatization rate increased by 50%								
Reaction time, 1 s	23.14	−3.7	67.12	+0.8	66.05	+0.7	1,528	−3.2
Reaction time, 10 s	37.63	−5.3	76.62	+2.1	73.51	+1.5	2,766	−3.9
Aromatization rate decreased by 50%								
Reaction time, 1 s	24.69	+2.7	65.97	−0.9	65.12	−0.8	1,608	+1.9
Reaction time, 10 s	41.29	+3.9	73.74	−1.7	71.49	−1.3	2,952	+2.5
Isomerization rate increased by 50%								
Reaction time, 1 s	24.07	+0.1	67.68	+1.0	66.78	+1.8	1,607	+1.8
Reaction time, 10 s	39.77	+0.1	75.74	+1.0	73.09	+0.9	2,906	+0.9
Feedstock made more paraffinic								
Reaction time, 1 s	22.16	−7.8	66.31	−0.4	65.13	−0.7	1,443	−8.6
Reaction time, 10 s	38.22	−3.8	76.67	+1.9	72.94	+0.7	2,788	−3.2
Feedstock made more aromatic								
Reaction time, 1 s	23.11	−3.9	68.24	+2.5	66.83	+1.8	1,544	−2.2
Reaction time, 10 s	38.78	−2.4	76.42	+1.9	73.38	+1.3	2,845	−1.2

the effect of the reaction rates on the absolute magnitude of gasoline yield exhibits the following order

$$\begin{array}{ccccccccc} \text{primary} & & & & \text{secondary} & & & & \\ \text{paraffin} & \gg & \text{aromatization} & > & \text{paraffin} & \cong & \text{olefin} & > & \text{olefin} \\ \text{cracking} & & & & \text{cracking} & & \text{cracking} & & \text{isomerization.} \end{array}$$

That is, the effect of a change in primary paraffin cracking rate far exceeds the effect of a change in aromatization rate, which in turn exceeds the effects of secondary paraffin cracking, olefin cracking, and isomerization. For octane number, all pathways are of comparable importance, but the following minor differences were observed:

$$\begin{array}{ccccccccc} & & & & \text{primary} & & \text{secondary} & & \\ \text{aromatization} & > & \text{olefin} & \cong & \text{paraffin} & > & \text{paraffin} & \cong & \text{olefin} \\ & & \text{isomerization} & & \text{cracking} & & \text{cracking} & & \text{cracking} \end{array}$$

Finally, for the octane-barrel yield (gasoline yield multiplied by MON) we observe the following ordering

$$\begin{array}{ccccccccc} \text{primary} & & & & \text{secondary} & & & & \\ \text{paraffin} & \gg & \text{aromatization} & > & \text{paraffin} & \cong & \text{olefin} & > & \text{olefin} \\ \text{cracking} & & & & \text{cracking} & & \text{cracking} & & \text{isomerization.} \end{array}$$

Although this sensitivity analysis is somewhat simplistic in that it assumes that each compound's reaction rate can be altered by the same factor, it does reveal that the two most important reaction pathways in catalytic cracking are formation of the initial carbenium ion on paraffins and aromatization. The validity of this conclusion will of course depend on the nature of the feedstock, because primary paraffin cracking clearly cannot be of great importance in a feedstock with no paraffins. To examine the importance of feedstock variations relative to rate constant variations, several sensitivity calculations were performed. In one study, 10% of the cyclic fraction (roughly 1% of the oil) was converted to aromatics of the same carbon number. This had a minor overall effect when compared to changes in reaction rate constant. In a second set of calculations, 30% of the cyclic fraction (roughly 3% of the oil) was converted to *n*-paraffins and isoparaffins of the same carbon number (2 mol of *n*-paraffin/mol of isoparaffin). This change in feedstock structure had a major impact on gasoline yield and a moderate impact on octane number.

The sensitivity results reported in Table 6–8 assess the impact of incremental changes in rate constants on model predictions. Additional sensitivity

Table 6–9 Sensitivity of Gasoline Yield and Octane Number to Elimination of Isomerization Pathways

Case Study	Gasoline Yield (wt%)	Percent Change from Base Case	RON	Percent Change from Base Case	MON	Percent Change from Base Case	Gasoline Yield × MON	Percent Change from Base Case
Isomerization rate decreased by a factor of 10^6								
Reaction time, 1 s	24.07	+0.1	50.18	−24.7	47.53	−27.6	1,144	−27.5
Reaction time, 10 s	39.77	+0.1	61.25	−18.4	58.12	−14.3	2,311	−19.7
Isomerization rate decreased by a factor of 5×10^{-3}								
Reaction time, 1s	24.07	+0.1	50.62	−24.0	48.00	−26.9	1,155	−26.8
Reaction time, 10 s	39.77	+0.1	65.29	−13.0	62.52	−13.7	2,486	−13.7

calculations were performed to address the issue of eliminating reaction pathways from the model. Reaction pathways could be eliminated for the sake of model simplification or because of the shape selectivity of non-amorphous catalysts (e.g., dehydrocyclization reactions would be eliminated for paraffins cracking in a ZSM-5 zeolite). Table 6–9 shows the effect of eliminating isomerization reactions. Gasoline yield is not impacted; however, octane number yields are dramatically reduced. In general, it was found that none of the reaction pathways reported in Table 6–8 could be eliminated from the model without sacrificing the quality of the model's predictions.

Structural Kinetic Models for Zeolites

The structural model of catalytic cracking chemistry described to this point is based on model compound reactions over an amorphous catalyst and assumes no mass transfer limitations. Because of these restrictions, the model is not directly applicable to the current generation of catalytic cracking processes which employ zeolite catalysts. The goal of this section is to describe briefly some problems that will need to be addressed in extending the cracking model for amorphous catalysts to cracking on and within zeolites.

Zeolites affect overall reaction rates and product selectivities in at least two ways. The small channel sizes of the zeolites can result in large differences in diffusivities of structurally similar molecules, creating a size selectivity. This size selectivity is a fairly conventional mass transfer limitation that can be modeled using an effectiveness factor approach. In addition to this size selectivity, zeolites can also have a shape selectivity. The size and shape of the channels in the zeolite may tend to favor the transition states required by some reaction pathways over those required by other pathways. This transition state selectivity affects the intrinsic rate of reaction and cannot be incorporated into a standard effectiveness factor.

Extending the model described in this chapter to zeolites will require, as a minimum, a structurally based model for diffusivity in zeolites. In addition, the group contributions used to predict intrinsic reaction rates will be highly dependent on catalyst properties due to transition state selectivities. Developing structural models for size and transition state selectivities will be a challenging problem.

ACKNOWLEDGMENT

This work was supported by W. R. Grace & Co. and the National Science Foundation through Presidential Young Investigator Award CBT 86 57289.

REFERENCES

Allen, D. T., D. W. Grandy, K. M. Jeong, and L. Petrakis (1985) Heavier fractions of shale oils, heavy crudes, tar sands and coal liquids: comparison of structural profiles. *Ind. Engin. Chem. Process Des. Dev.* **24**:737–742.

Benson, S. (1976) *Thermochemical Kinetics*, 2nd ed. John Wiley, New York.

Clutter, D. R., L. Petrakis, R. L. Stenger, Jr., and R. K. Jensen (1972) Nuclear magnetic resonance spectrometry of petroleum fractions: carbon-13 and proton nuclear magnetic resonance characterizations in terms of average molecule parameters. *Anal. Chem.* **44**:1395.

Cotterman, R. and K. Plumlee (1989) Effects of gasoline composition on octane number. *Prepr. Am. Chem. Soc. Div. Petr. Chem.* **34:**756.

Franklin, J. L. (1949) Prediction of heat and free energies of organic compounds. *Ind. Engin. Chem.* **41**:1070.

Fredenslund, A., R. L. Jones and J. M. Prausnitz (1975) Group contribution estimation of activity coefficients in non-ideal liquid mixtures. *A.I.Ch.E.J.* **21**:1086.

Greensfelder, B. S. and H. H. Voge (1945) Catalytic cracking of pure hydrocarbons. Cracking of paraffins. *Ind. Engin. Chem.* **37**:514.

Greensfelder, B. S., H. H. Voge, and G. M. Good (1949) Catalytic and thermal cracking of pure hydrocarbons. Mechanisms of reaction. *Ind. Engin. Chem.* **41**:2573.

Jewell, D. M., E. W. Albaugh, B. E. David, and R. G. Ruberto (1974) Integration of chromatographic and spectroscopic techniques for the characterization of residual oils. *Ing. Engin. Chem. Fundam.* **13**:178.

Later, D. W., M. L. Lee, K. W. Bartle, R. C. Kong, and D. L. Vassilaros (1981) Chemical class separation and characterization of organic compounds in synthetic fuels. *Anal. Chem.* **53**:1612.

Liguras, D. K. and D. T. Allen (1989a) Structural models for catalytic cracking: 1. Model compound reactions. *Ind. Engin. Chem. Res.* **28**:665.

Liguras, D. K. and D. T. Allen (1989b) Structural models for catalytic cracking. 2. Reactions of simulated oil mixtures. *Ind. Engin. Chem. Res.* **28**:674.

Strang, G. (1980) *Linear Algebra and Its Applications*, 2nd ed. Academic Press, New York.

Thiel, J. and M. R. Gray (1988) NMR spectroscopic characteristics of Alberta bitumens. *AOSTRA J. Res.* **4**:63.

Vajdi, L. and D. T. Allen (1989) Vapor pressures of coal liquids estimated using a group contribution equation of state. *Fuel* **68**:1388.

Weekman, V. W., Jr. (1979) Lumps, models and kinetics in practice. *A.I.Ch.E. Monogr. Ser.* **75**:3–24.

7

Monte Carlo Modeling of Complex Reaction Systems: An Asphaltene Example

MICHAEL T. KLEIN • MATTHEW NEUROCK
ABHASH NIGAM • CRISTIAN LIBANATI

The practice of "lumping" a larger number N of descriptor species into a smaller number N/p of species has been of great utility in process modeling. In reaction modeling, for example, this can reduce the number of controlling differential equations to a numerically or even analytically tractable level. Moreover, lumped models are relevant. Historical limitations in the analytical chemistry of reactants (e.g., petroleum) forced modeling at a global lumped level, and many commercially relevant products (e.g., gasoline) are defined as boiling point or solubility-based product fractions. However, relatively recent advances in analytical chemistry and computational capability, and a growing interest in product quality information suggest discovery and retention of molecular detail will often be useful.

The problem of model development for heavy oil upgrading processes allows illustration of the relationship between molecular detail and lumping. Modeling of heavy oil reactions requires three conceptual components: the reactant, the reaction, and the description of the products.

Figure 7–1 summarizes limiting-case and hybrid approaches. At the global level described in the upper part of Figure 7–1, petroleum resids are defined as distillation cuts. Classic lumped models provide the kinetics of the interconversion of boiling point lumps to a global product state. Models of this kind are highly relevant and computationally straightforward, but also reflect little of the controlling chemical fundamentals.

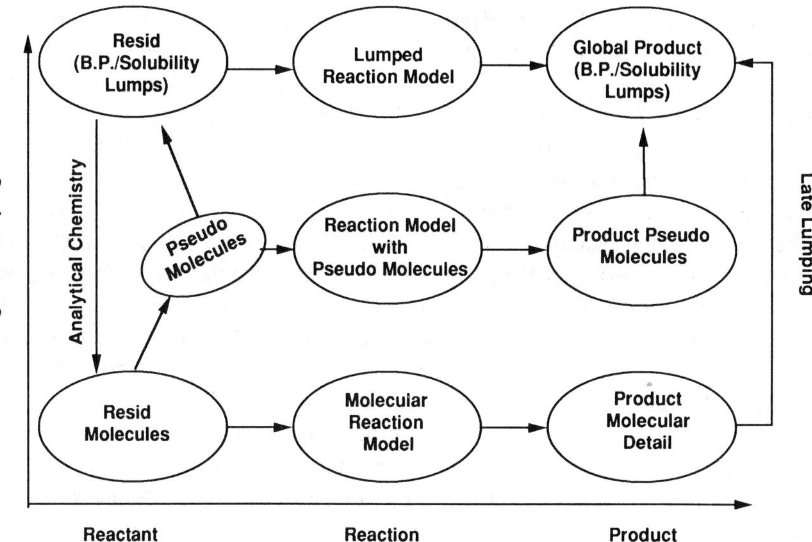

Figure 7–1 Early, late, and hybrid lumping paths in reaction modeling.

Modern analytical chemistry provides the ability to model at a molecular level. Techniques such as nuclear magnetic resonance spectroscopy (NMR), gas chromatography/mass spectrometry (GC/MS), and FTIR spectroscopy provide an indirect glimpse of the molecular components of a resid. The reaction of the resid can be phrased in terms of the reactions of the molecules it comprises. This provides an indication of product quality since full molecular detail is retained. Lumping by thermodynamic methods can transform product molecules into global lumps.

The hybrid approaches in Figure 7–1 confirm that lumping (averaging) entails a loss of information. For example, lumping reactant molecules into pseudospecies followed by reaction modeling provides a product spectrum containing less information than a reaction model of the reactant molecules. The appropriate path of Figure 7–1 to follow, therefore, depends upon the information sought from a model. Global lumped models are appropriate for linear chemistry where only product yields are sought; these models need not be directly connected to the controlling chemistry. On the other hand, models with molecular detail may be appropriate when the chemistry is highly nonlinear and product structure and quality information assumes significance.

Chemical engineers are generally interested in commercial processes. Thus, irrespective of the path taken in Figure 7–1, reaction models should ultimately predict the kinetics of the relevant global lumps. Therefore, it is worth noting that all of the paths of Figure 7–1 will ultimately involve some lumping. The only difference in each path is the point or time of lumping. This motivates the concept of "early" and "late" lumping. The top path in Figure 7–1 is an early lumping example: the reactants, reaction, and product models are all of global lumps. The bottom path of Figure 7–1 is the complementary molecular approach. Full molecular detail is retained in the reactant, reaction, and product models. Global product information is obtained after a molecularly explicit reaction model. The vertical transformations in Figure 7–1 depicting departure from molecular detail also depict loss of information. Thus lumped models represent loss of information only when analytical chemistry provides molecular detail. Lumping is a loss of information. Lumped models starting with lumped reactants do not represent information loss.

The early and late lumping dichotomy is reminiscent of the issues of maximum mixedness and maximum segregation in reactor theory. As long as the nonlinearities of the kinetics are significant, the two extremes will yield different results.

CHEMICAL MODELING

The foregoing issues motivated the development of the chemical modeling approach to the analysis of the reactions of complex feedstocks. This approach involves the three steps outlined in Figure 7–2 and involves stochastic computational methods in support of analytical, reaction, and computational chemistry.

The uppermost horizontal transformation in Figure 7–2 shows the early lumping extreme. The reactants, reaction, and products are all described globally. Chemical modeling is a complementary route to the same and additional information involving delumping, molecular reaction, and thermodynamic lumping steps.

The delumping step in Figure 7–2 is aimed at construction of a molecular representation of the real feed. Analytical characterization techniques (e.g., NMR, GC/MS, elemental analysis, etc.) allow summary of the structural attributes of the molecules of the complex feedstock in terms of cumulative probability distribution functions (pdf) for each attribute. A stochastic construction technique is used to assemble an ensemble of different molecules from these pdf's. This renders the complex feedstock "structure explicit."

The reactions of molecules in the complex feed representation describe the actual feed reaction. A Monte Carlo simulation approach (McDermott et al. 1990; Neurock et al. 1989) is a convenient framework for accounting for heterogeneities (e.g., molecular-weight distribution, phase behavior, kinetic coupling) of real feedstocks. The intrinsic chemistry that forms the basis

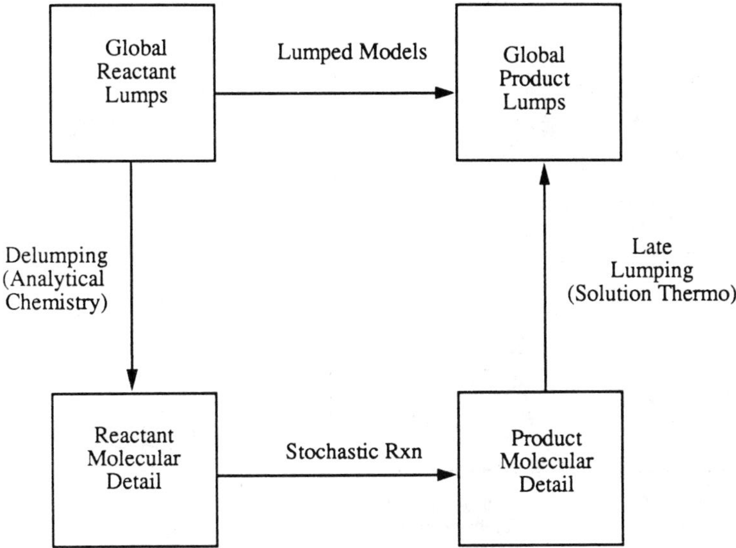

Figure 7–2 Lumping and delumping in reaction models.

for the probability of reaction of the reactive moieties in the representation is obtained via model compound experiments and computational chemistry.

The late lumping step of the chemical modeling approach turns the product molecular detail obtained from the stochastic reaction into commercially relevant pseudospecies. This involves primarily thermodynamic modeling for the assignment of each molecule into a boiling point or solubility-defined product fractions. For reaction models, then, early lumping is primarily a kinetics issue, whereas late lumping is a thermodynamics issue.

The object of the present chapter is to illustrate this approach through an asphaltene pyrolysis example. This alkane-insoluble, aromatic-soluble portion of heavy oils has a major effect on the processibility of the oil. In brief, structural probability distribution functions, assembled from analytical chemistry in the open literature (Speight 1980), were used to construct an ensemble of 10,000 molecules. These were, in turn, stochastically reacted to obtain a spectrum of product molecules. The late lumping technique, based on solution thermodynamics, places the product molecules into different solubility classes.

STOCHASTIC CONSTRUCTION

Asphaltenes, strictly a solubility class, are composed of an ensemble of molecules with the following structural attributes: each asphaltene molecule

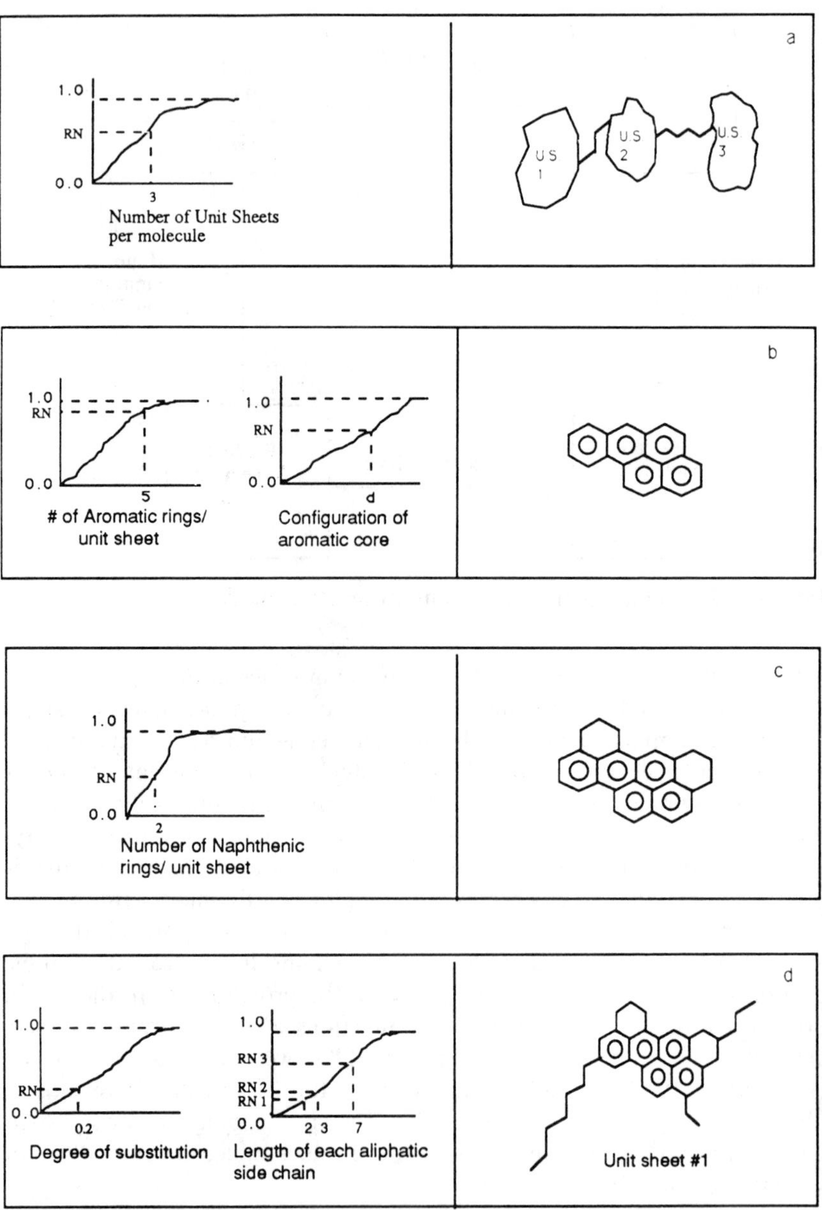

Figure 7–3 Initial structural distributions for a generic asphaltene feed used in the simulation. (*a*) Number of unit sheets per molecule. (*b*) Number and arrangement of the aromatic core. (*c*) Number of naphthenic rings. (*d*) Number and length of aliphatic substituents.

s characterized by a degree of polymerization, a number of aromatic rings, number of naphthenic rings, a degree of substitution, and a chain length or each aliphatic substituent. This information about the asphaltene mole-ule is expressed in terms of pdf's that give the probability that the attribute hould take on the value of the abscissa or less. The distributions used in he present simulation are shown in Figure 7–3 in the boxes on the left-hand ide. These are composite pdf's derived from many investigations of asphal-ene structure reported in the literature, i.e., they are not specific to a articular asphaltene.

Each of these distributions were sampled 10,000+ times to construct 0,000 different reactant molecules that represent the generic asphaltene eedstock. The procedure, depicted in Figure 7–3, involves the sequential etermination of the random variates which correspond to each probability istribution (Neurock et al. 1989).

TOCHASTIC REACTION

Iaving the structure of each molecule explicit allows for the assignment of reaction probability to each potential reaction site. The present simulation of asphaltene reactions in resid included both thermal and catalytic reactions.

The thermal reactions resulted in the cleavage of carbon–carbon bonds f alkylaromatic and alkylnaphthenic moieties. The literature provides ample nodel compound results for the pathways and kinetics of these thermolysis eactions (Savage and Klein 1988a,b).

The catalytic reactions included were the hydrogenation of aromatic noieties and the dehydrogenation of naphthenic moieties. These model ompound pathways and kinetics are described elsewhere (Bhinde 1978; Girgis 1988; Lapinas et al. 1987; Qadar 1973).

This intrinsic information for the catalytic reactions was summarized in erms of a linear free-energy relationship. Figure 7–4 shows a good correla-ion of the experimentally measured kinetics for the hydrogenation of romatic compounds (Bhinde 1978; Girgis 1988; Lapinas et al. 1987; Qadar 973) with the π-electron density of the site on each ring with the highest alue of π. The usefulness of reducing model compound information in this nanner is in the implied predictive ability. For example, the kinetics for sphaltene or asphaltene product molecules not studied experimentally can e determined by performing a simple molecular orbital calculation to etermine the π-electron density at each site and utilizing the linear free-nergy relationship of Figure 7–4.

The reactions of the asphaltene were simulated by tracking the reaction f each reactive moiety using a variable time step Monte Carlo simulation pproach. The transitions (reaction paths) and the transition probabilities

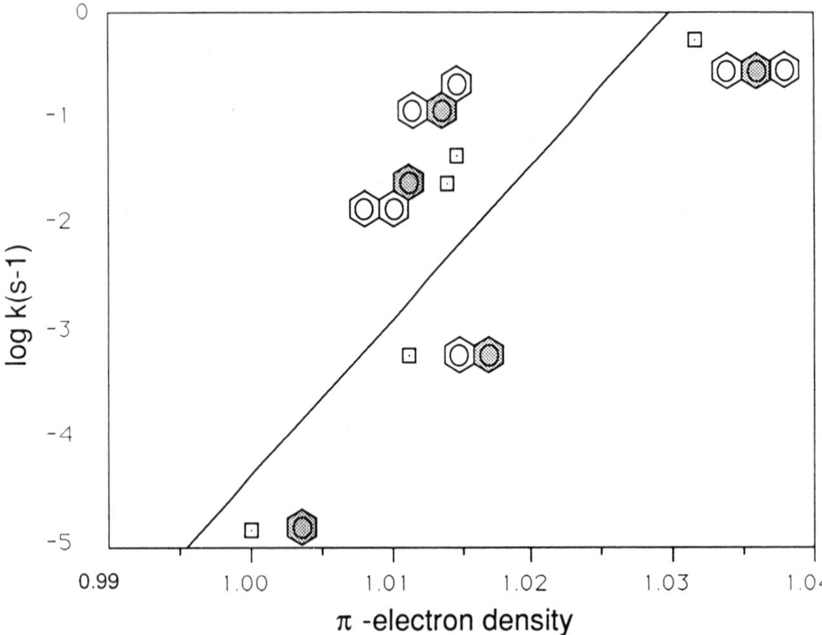

Figure 7–4 Linear-free-energy relationship for the hydrogenation of aromatic compounds.

(kinetics) for each reactive asphaltene moiety were accumulated on a cumulative probability distribution function for any possible reaction time and event. A first-drawn random number was placed on the pdf for reaction time shown in Eq. (7–1), to determine the time increment (Δt) between the previous and the next as yet-unidentified reaction:

$$\Delta t = -\ln(1 - \mathrm{RP})/\textstyle\sum k_i. \tag{7–1}$$

This time is a function of a random probability (RP) between 0 and 1 and the summation of all potential reactive probabilities ($\sum k_i$). A second-drawn random number was placed on the cumulative probability distribution for the likelihood of individual events. Since each reaction changed the molecular structure, the process was repeated by constructing new pdf's for reaction time and events and drawing additional random numbers. The reaction of each asphaltene molecule proceeded event by event until the final simulation time was reached. A new molecule was then constructed and reacted stochastically. This process was repeated until each of the 10,000 molecules were reacted.

The Monte Carlo simulation therefore provides the identity of the 10,000+ molecules derived from the starting 10,000 as a function of reaction time. Analytical results, such as H/C, MWD, M_n, and M_w are explicit and are determined through a simple accounting scheme. More interesting or relevant product information, such as boiling point- or solubility-based product lumps, requires further thermodynamic analysis of each of the 10,000+ molecules.

LATE LUMPING: THE THERMODYNAMIC GUIDELINES

It is traditional to represent the asphaltene reaction products in terms of solubility classes. The following solubility classes are defined based on the addition of, by convention, 40 volumes solvent per volume of reaction products:

Coke: insoluble in aromatic (e.g., benzene) and aliphatic solvents (e.g., hexane),
Asphaltene: soluble in aromatic but insoluble in aliphatic solvents,
Maltene: soluble both in aromatic and aliphatic solvents,
Gases.

Thus, the object of the thermodynamic or late lumping model is to assign 10,000+ product molecules, as generated in the stochastic reaction model, to one of the four coke, asphaltene, maltene, or gas product fractions. The thermodynamic model developed herein views oil as a mixture of molecules that precipitate depending upon the thermodynamic state of the mixture. The model, therefore, performs solid–liquid–vapor equilibrium calculations to predict the quantity and structural attributes of each of the solubility classes.

The three approaches under study are outlined in Figure 7–5. The first approach is to use the 10,000+ molecules to obtain product pdf that can in turn provide optimal quadrature points (Cotterman and Prausnitz 1985; Shibata et al. 1988). The second approach is to subject the 10,000+ molecules to the multivariate statistical technique of cluster analysis to obtain pseudospecies. The final approach is to use Monte Carlo methods to assign each of the 10,000+ molecules to a product fraction. The first two approaches involve lumping *prior* to the thermodynamic analysis and therefore involve a loss of information. The Monte Carlo approach retains molecular detail throughout. The modeling described herein is common to the first two approaches. We describe the Monte Carlo model in a forthcoming communication.

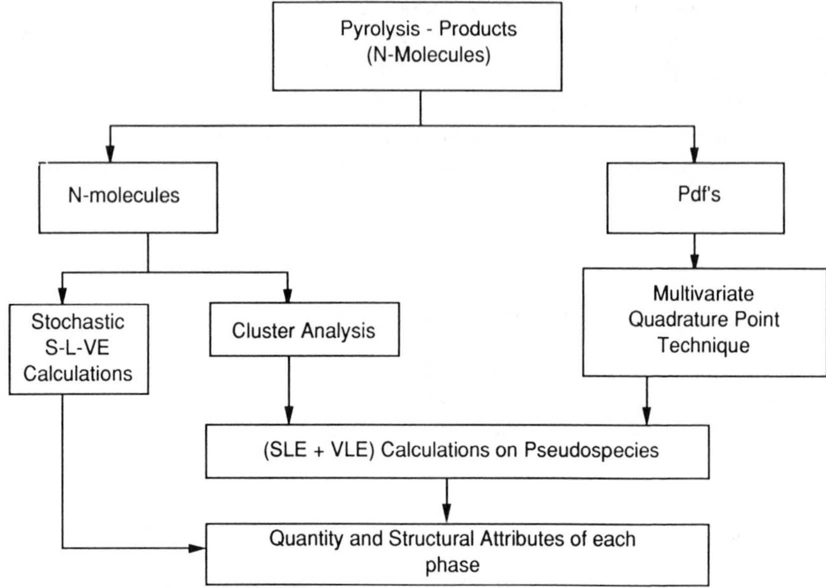

Figure 7–5 Solid–liquid–vapor equilibrium calculations.

The thermodynamic lumping described here involves three subtasks:

1. formulation of pseudospecies,
2. solid–liquid–vapor equilibrium calculations,
3. estimation of structural attributes.

We describe each subtask in turn.

Formulation of Pseudospecies

The input to the thermodynamic model may be represented in terms of either structural probability density functions or a collection of structure explicit molecules. The formulation of pseudospecies renders this input more accessible to the phase equilibrium calculations. In the case of structural probability density functions, optimal quadrature points were chosen following methods described elsewhere (Cotterman and Prausnitz 1985; Shibata et al. 1988). When the model input is in terms of a large collection of product molecules, the multivariate statistical technique of cluster analysis is used on the structural attributes to determine optimal pseudospecies (Anderberg 1973). Cluster analysis identifies similarity classes based on the minimization of intragroup Euclidean error.

Thermodynamic Calculations

Solid–liquid–vapor equilibrium calculations were performed on each of the pseudocomponents to determine the phase split. This work assumed the presence of a single homogeneous solid solution in equilibrium with a liquid solution and a gas phase.

The equilibrium among solid, liquid, and vapor phases is dictated by the isofugacity criterion for each pseudocomponent in each phase

$$\bar{f}_i^{\text{Liquid}} = \bar{f}_i^{\text{Solid}} = \bar{f}_i^{\text{Vapor}}. \tag{7–2}$$

For solid–liquid equilibrium , the isofugacity criterion may be rewritten as

$$K_i^{\text{SL}} = \frac{x_i^{\text{L}}}{x_i^{\text{S}}} = \frac{\gamma_i^{\text{S}}}{\gamma_I^{\text{L}}} \exp\left[-\frac{\Delta H_{\text{im}}}{\text{RT}}\left(1 - \frac{T}{T_{\text{im}}}\right)\right]. \tag{7–3}$$

This neglects the difference between molar heat capacities of components in the subcooled liquid and solid state.

In the case of vapor–liquid equilibrium, the isofugacity criterion may be written as

$$K_i^{\text{VL}} = \frac{x_i^{\text{V}}}{x_i^{\text{L}}} = \gamma_i^{\text{L}} \frac{P_i^{\text{sat}}}{P}. \tag{7–4}$$

The two equilibrium relations were solved in conjugation with the mass balance for each pseudocomponent in the presence of SOL moles of solvent per mole of the reaction products

$$(1 + \text{SOL}) = S + L + V, \tag{7–5}$$

$$(1 + \text{SOL})x_i^{\text{F}} = \text{S}x_i^{\text{S}} + \text{L}x_i^{\text{L}} + \text{V}x_i^{\text{V}} \tag{7–6}$$

to provide the amount and composition of each phase.

This three-phase flash calculation problem was solved using Rachford–Rice method (Cotterman and Prausnitz 1985) as modified by Won (1986) to yield S, V, and L. A detailed block diagram for thermodynamic calculations is given in Figure 7–4.

This required calculation of the activity coefficient in each phase, which in turn required a suitable thermodynamic model for each phase.

Liquid Phase Modeling

The regular solution formalism was employed to model the liquid phase activity coefficients. The multicomponent version of the regular solution

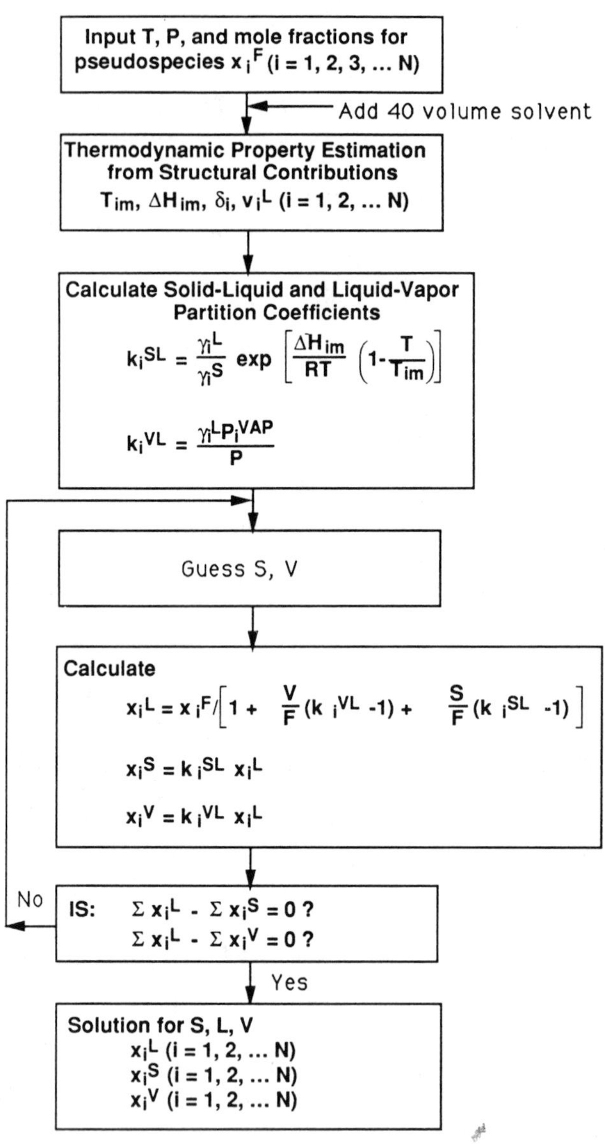

Figure 7–6 Algorithm for solid–liquid–vapor equilibrium calculations.

activity coefficient may be written as

$$\ln \gamma_i^{\mathrm{L}} = \frac{V_i}{RT}(\delta_i - \bar{\delta})^2. \qquad (7\text{–}7)$$

Since the calculations herein followed the conventional addition of 40 volumes of solvent, the infinite dilution assumption of

$$\bar{\delta} \approx \delta_{\mathrm{solvent}} \qquad (7\text{–}8)$$

was invoked.

The molar volumes for the macromolecular asphaltene molecules were determined from structural parameters (Hirsch 1970). Small's (1953) molar attraction constants allowed calculation of solubility parameters solely from the structural parameters.

The regular solution theory is usually valid only for molecules of similar sizes and interaction energies. However, the generalized Flory–Huggins theory due to Lichtenthaler (Hirschberg et al. 1983) provides the rationalization for the use of the regular solution assumption for asphaltene molecules. These molecules tend to be bulky, and in nonpolar solvents the entropy of mixing may be approximated by the ideal entropy of mixing despite a large size difference in solvent and asphaltene molecules.

Solid Phase Modeling

Without experimental data to the contrary, we modeled the solid phase as an ideal mixture of constituent components. Thus $\gamma_i^{\mathrm{S}} = 1$ for every pseudocomponent in the solid mixture. Higher-order phase transitions and contribution from subcooled state were ignored.

The solid–liquid equilibrium calculation required estimates of the melting point and enthalpy of fusion for each pseudocomponent. The melting point was modeled as a function of the molecular weight, structural features, and symmetry of the molecule (Bondi 1968). Linear regression was applied to a set of 70 aromatic components with side chains to develop the following correlation for the melting point:

$$\begin{aligned} T_{\mathrm{M}} = {} & 284.9 + 7.66 \times 10^{-2}\mathrm{MW} - 15{,}684/\mathrm{MW} \\ & + 40.83\ (\text{number of aromatic rings}) \\ & + 11.73\ (\text{number of naphthenic rings}) \\ & + 76.64^{*} \ln\ (\text{symmetry factor}). \end{aligned} \qquad (7\text{–}9)$$

This correlation has an average error of 22.1°C for alkyl substituted aromatics. Trouton's rule, as given by Walden (Bondi 1968) was used to estimate the enthalpy of fusion:

$$\Delta S_{\text{fus}} = \frac{\Delta H_{\text{fus}}}{T_{\text{M}}} = 56.5 \text{ J/mol K}.$$

Vapor Phase Modeling

The isofugacity criterion reduces to

$$\frac{x_i^{\text{V}}}{x_i^{\text{L}}} = \frac{x_i^{\text{L}} P_i^{\text{VAP}}}{P}. \qquad (7\text{–}10)$$

The vapor phase was assumed to be ideal. The Cotterman–Prausnitz (Cotterman and Prausnitz 1985) correlation was used to estimate the vapor pressures.

Estimation of Structural Attributes

The phase split provides the quantity of each of the solubility class. The late lumping approach facilitates the estimation of the structural attributes in each of the solubility class. These structural attributes are indicative of the product quality. Simple accounting of the pseudospecies provided the mean and higher moments of each structural attribute (i.e., degree of polymerization, number of aromatic rings, number of naphthenic rings, degree of substitution, average alkyl chain length, average carbon number of paraffins) for all the solubility classes, namely coke, asphaltene, maltene, and gas.

RESULTS AND DISCUSSION

In illustration of the approach, the catalytic and thermal reactions of the asphaltenes, in a hypothetical inert maltene oil, were simulated by the Monte Carlo methods. Simulation of the combined residual oil is in progress.

The simulation provides analytical and global results. Figure 7–7 is a comparison of the initial and product molecular-weight distribution after 10,000 s (2.78 hours) at 450°C reaction temperature. There is a significant shift in the number of molecules with molecular weights of the order of 2,000 to molecules with molecular weights less than 1,000. The number average molecular weight (M_n) of 1,700 drops to an asymptote of 500 after about 2 hours. These data combine to suggest that the asphaltene molecules undergo complete depolymerization to form individual unit sheets.

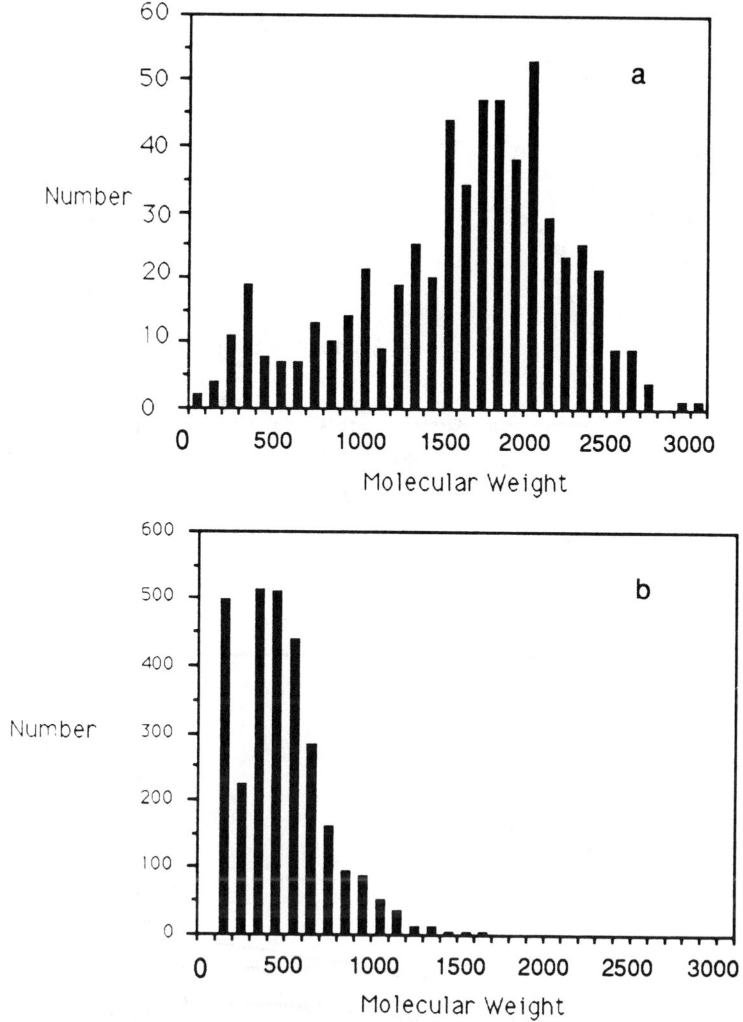

Figure 7–7 Molecular-weight distributions for the simulation of asphaltene pyrolysis at 450°C. (*a*) Initial distribution ($t = 0$ s). (*b*) Final distribution ($t = 10{,}000$ s).

The predicted global product yields (for a literature-averaged, generic asphaltene) from reaction at 450°C compare quite well with the laboratory experiments of Savage et al. (1985) on a California-derived asphaltene. This is shown in Figure 7–8. Quantitative differences might be accounted for by using structural distribution functions derived from the actual California-derived asphaltene instead of generic distributions.

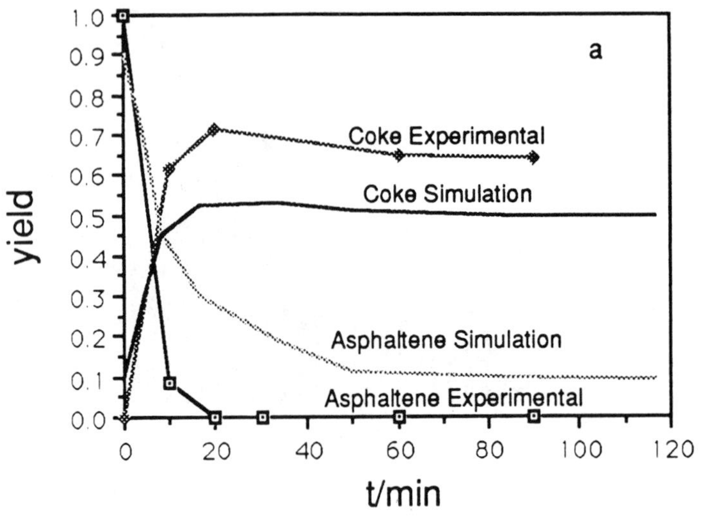

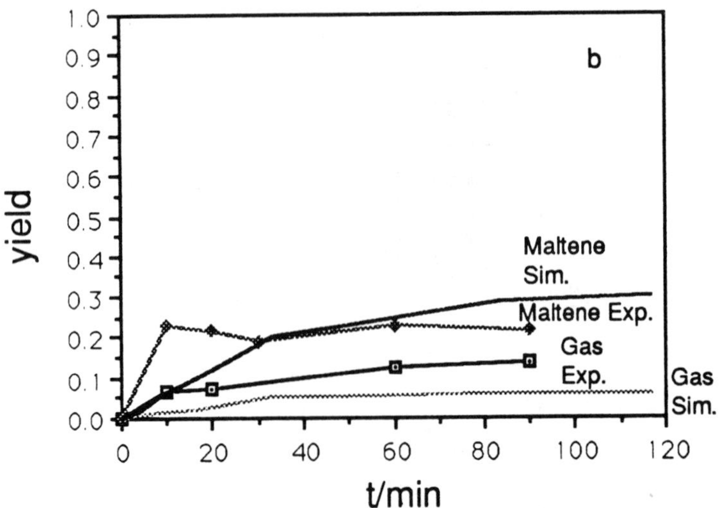

Figure 7–8 Simulation and experimental results for asphaltene pyrolysis a 450°C. (*a*) Asphaltene and coke yields. (*b*) Maltene and gas yields.

CONCLUSIONS

1. A molecularly explicit Monte Carlo simulation allowed modeling of asphaltene hydroprocessing pathways in terms of model compound reaction pathways and kinetics.
2. A solution thermodynamics model was developed for lumping product molecular detail into practical identifiable solubility classes, e.g., asphaltene, maltene, gas, and coke.
3. Temporal yields of global product fractions, obtained from the coupling of reaction and thermodynamic models, compared favorably with experimental data.

NOMENCLATURE

K_i^{SL}	solid–liquid partition coefficient for pseudocomponent i
K_i^{VL}	vapor–liquid partition coefficient for pseudocomponent i
x_I^L	mole fraction of pseudocomponent i in liquid phase
x_i^S	mole fraction of pseudocomponent i in solid phase
y_i^V	mole fraction of pseudocomponent i in vapor phase
γ_i^L	activity coefficient of pseudocomponent i in liquid phase
γ_i^S	activity coefficient of pseudocomponent i in solid phase
ΔH_{im}	enthalpy of fusion for pseudocomponent i
T_{im}	melting point temperature for pseudocomponent i
P_i^{VAP}	vapor pressure for pseudocomponent i
P	total pressure of system
RP	random number corresponding to the cumulative probability
SOL	moles of solvent/moles of solute
z_i^F	mole fraction of component i in feed
S	fraction of solid
L	fraction of liquid
V	fraction of vapor
δ_i	solubility parameter for pseudocomponent i
$\bar{\delta}$	volume fraction averaged solubility parameter of mixture $= \sum \phi_j \delta_j$
ϕ_j	volume fraction of pseudocomponent $i = x_j \phi_j / \sum x_i \phi_j$
Δt	time between event i and event $i + 1$
$\sum k_i$	cumulative reaction probability distribution

REFERENCES

Anderberg, M. R. (1973) *Cluster Analysis for Applications.* Academic Press, New York.

Bhinde, M. V. (1978) Quinoline hydrodenitrogenation kinetics and reaction inhibition. Ph.D. thesis, University of Delaware.

Bondi, A. (1968) *Physical Properties of Molecular Crystals, Liquids, and Glasses.* Jo Wiley, New York.

Cotterman, R. L. and J. M. Prausnitz (1985) Flash calculations for continuous semicontinuous mixtures using an equation of state. *Ind. Engin. Chem. Proce Des. Dev.* **24**:434.

Girgis, M. (1988) Reaction networks and kinetics and inhibition in the hydroproces ing of simulated heavy coal liquids. Ph.D. thesis, University of Delaware.

Hirschberg, A., L. N. J. De Jong, B. A. Schipper, and J. G. Meijers (1983) *Influen of Temperature and Pressure on Asphaltene Flocculation.* Publication 6 Koninklije Shell Exploratie en Produktie Laboratorium Rijswijk, The Nethe lands.

Hirsch, E. (1970) Relation between molecular volume and structure of hydrocarbo at 20 C. *Anal. Chem.* **42**:1326.

Lapinas, A. T., M. T. Klein, B. C. Gates, A. Macris, and J. E. Lyons (1987) Cataly hydrogenation and hydrocracking of fluoranthene: reaction pathways and kine ics. *Ind. Engin. Chem. Res.* **26**:1026–1033.

Lichtenthaler, R. N., D. S. Abrams, and J. M. Prausnitz (1973) Combinatorial entro of mixing for molecules differing in size and shape. *Can. J. Chem.* **51**:3071.

McDermott, J. B., C. Libanati, C. La Marca, and M. T. Klein (1990) Monte Car simulation of the reactions of complex macromolecules: quantitative use of mod compound information. *Ind. Engin. Chem. Res.*

Mansoori, G. A., T. S. Jiang, and S. Kawanka (1988) Asphaltene deposition and role in petroleum production and processing. *Arab. J. Sci. Engin.* **13**:17–34.

Neurock, M., C. Libanati, and M. T. Klein (1989) Modelling asphaltene reacti pathways: intrinsic chemistry, *A.I.Ch.E. Symp. Ser. Fundam. Resid Upgrad.* 27 **85**.

Qadar, S. A. (1973) Hydrocracking of polynuclear aromatic hydrocarbons ov silica-alumina based dual functional catalysts. *J. Int. Pet.* **59**:568.

Rachford, H. H. and J. D. Rice (1952) Procedure for use of electronic digital compute in calculating flash vaporization hydrocarbon equilibrium. *Petrol. Technol.* se 1, p. 19, sect. 2, p. 3 (Oct).

Savage, P. E. and M. T. Klein (1988a) Asphaltene reaction pathways 2. Pyrolysis *n*-pentadecylbenzene. *Ind. Engin. Chem. Res.* **26**:488–496.

Savage, P. E. and M. T. Klein (1988b) Asphaltene reaction pathways 4. Pyrolysis tridecylcyclohexane and 2-ethyltetralin. *Ind. Engin. Chem. Res.* **27**:1348–1356.

Savage, P. E., M. T. Klein, and S. G. Kukes (1985) Asphaltene reaction pathways Thermolysis. *Ind. Engin. Chem. Process Des. Dev.* **24**:1169–1174.

Shibata, S., S. I. Sandler, and R. L. Behrens (1988) Phase equilibrium calculatio for continuous and semicontinuous mixtures. *Chem. Engin. Sci.* **42**:8, 1977–1988.

Small, P. A. (1953) Some factors affecting the solubility of polymers. *J. Appl. Che* **3**:71–80.

Speight, J. G. (1980) *The Chemistry and Technology of Petroleum.* Marcel Dekke New York.

Won, K. W. (1986) Thermodynamics for solid solution-liquid-vapor equilibria: w phase formation from heavy hydrocarbons. *Fluid Phase Equilibria* **30**:265–279.

8

Olefin Oligomerization Kinetics over ZSM-5

R. J. QUANN
F. J. KRAMBECK

Oligomerization of light olefins to higher molecular-weight olefins in the presence of acid catalysts via carbenium ion mechanisms is an area of chemistry that has been extensively studied for many years (Oblad et al. 1958; Pines 1981). Recently, Mobil Research designed and developed commercial process technology for the conversion of light (C_3–C_6) olefins to high quality, higher boiling, and higher valued gasoline and distillate fuels (Garwood 1983; Tabak et al. 1986; Yurchak et al. 1990). This process technology, called Mobil Olefins to Gasoline and Distillate (MOGD), is based on shape-selective catalysis using the Mobil synthetic zeolite ZSM-5 to effect acid-catalyzed olefin oligomerization reactions.

A simplified schematic of the process flow diagram is shown in Figure 8–1. The product distribution with respect to carbon number, or relative yield of gasoline, kerosene, and heavier distillate, is affected by many process variables including operating conditions, feedstock purity (i.e., paraffin content), recycle cut point, and recycle ratio. The gasoline recycle stream also serves as interreactor quench for a multireactor unit to control the exotherm of oligomerization reactions.

During the development phase of the process, it became apparent that a kinetic model of the process would help isolate the impact of a number of process variables. However, the kinetic model had to be of sufficient complexity to represent the effects of a number of process variables or operating

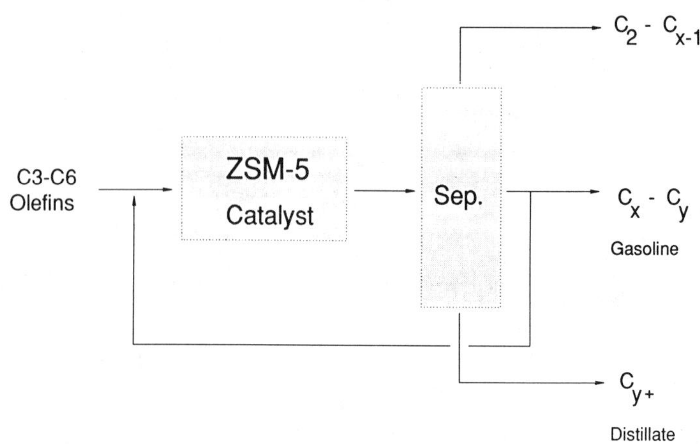

Figure 8–1 Schematic overview of the MOGD process.

strategies on the detailed carbon number distribution of the product streams. This chapter describes the development of the MOGD kinetic process model which utilizes a detailed carbon number distribution approach, incorporates complex oligomerization and cracking reaction networks, and is supported by thermodynamic studies, model compound studies of basic catalytic chemistry, molecular structure investigations, and real (mixed olefin and paraffin) feed pilot plant work.

OLIGOMERIZATION CHEMISTRY WITH ZSM-5

The complex reaction sequence for olefin oligomerization and cracking is shown in Figure 8–2. Any two olefins of carbon number x and y may condense to form a single larger olefin of carbon number $x + y$. Oligomers are the dimers and incremental additions of feed olefins to the dimers and represent the initial products of the reaction sequence, i.e., C_6, C_9, C_{12}, etc., for a propylene (C_3) feed. In addition, any olefin may crack to two olefins of lower carbon number. Olefins may also react to disproportionate, rather than condense, to two olefins of different carbon number via a carbenium ion intermediate which can isomerize and crack prior to desorption from an acid site. These latter two reactions result in a randomization of the carbon number distribution. When these reactions are catalyzed in the narrow pore channels of the ZSM-5 catalyst (Figure 8–3), shape selectivity imposes constraints on the structure of the molecules. More specifically, the degree of isomerization and branching of product molecules is limited by pore diffusion considerations. In this section the phenomenological aspects of

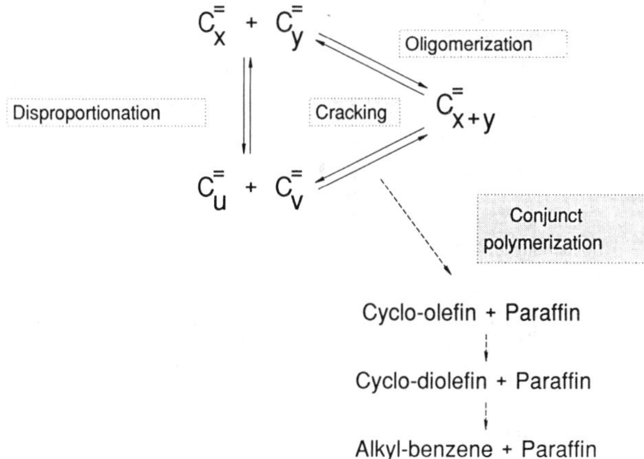

Figure 8–2 Acid-catalyzed olefin reaction pathways (Quann et al. 1988).

olefin reactions in ZSM-5 are briefly reviewed. A more detailed discussion along with experimental methods has been reported elsewhere (Quann et al. 1988).

Olefin carbon number distributions from field ionization mass spectroscopy (FIMS) analysis of products from propene and hexene feeds are shown in Figure 8–4 and reveal the predominance of oligomers but also the definite presence of "randomized" product from cracking and disproportionation reactions. A small fraction of the olefin pool is also continuously and irreversibly lost to cyclization and hydrogen transfer reactions that yield cycloolefins, alkylbenzenes, and paraffins in what has been called "conjunct

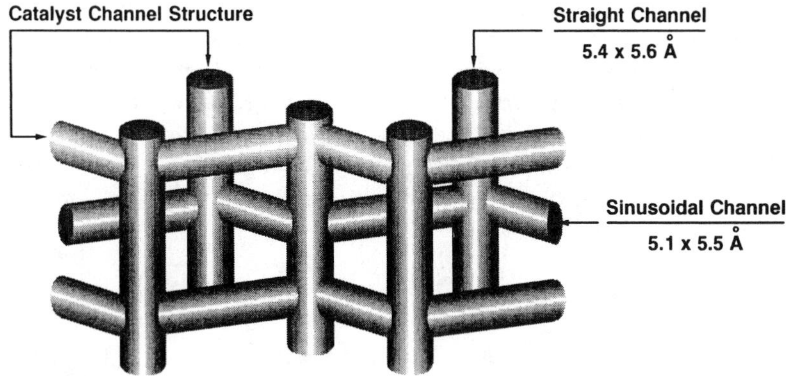

Figure 8–3 Schematic of ZSM-5 pore structure.

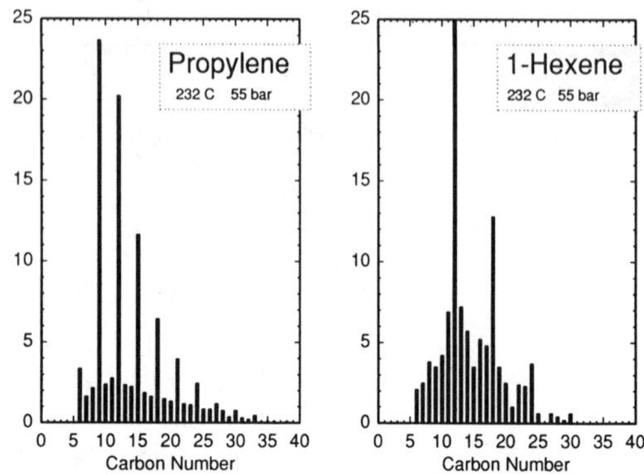

Figure 8–4 Comparison of FIMS analysis of propylene and 1-hexene oligomerization products revealing the oligomers and intermediate olefins.

polymerization" (Pines 1981; Quann et al. 1988). However, these reactions are minimized by the shape-selective constraints imposed by the ZSM-5 catalyst.

Olefin isomerization reactions, both double bond and skeletal, occur rapidly and simultaneously with oligomerization and cracking reactions. The FIMS and gas chromatography (GC) analyses shown in Figure 8–5, and additional H-nuclear magnetic resonance (H-NMR) analysis of the products from 1-hexadecene oligomerization experiments reveal that double bond isomerization is complete prior to any measurable degree of oligomerization or cracking, and that skeletal isomerization of normal hexadecenes to branched hexadecenes has been completed prior to 30% conversion of hexadecenes. Structural analysis of propylene oligomerization products by H-NMR reveals that although isomerization is rapid, it is also limited due to the shape-selective constraints imposed by the ZSM-5 catalyst. As shown in Figure 8–6, the degree of branching of larger olefins is substantially less than that expected for carbenium ion driven oligomerization of propylene without isomerization (polypropylene). Evidently, ZSM-5 will tend to nearly linearize or crack highly branched structures formed from oligomerization in order to minimize diffusion resistance. Larger, highly branched molecules will simply not escape the pore system without isomerizing to less branched isomeric forms.

KINETIC MODEL FORMULATION

In formulating the kinetic model, it was anticipated that a detailed approach, one that would represent the complexity of the reaction sequence in Figure 8–2, would be required to represent properly the effects of wide ranging

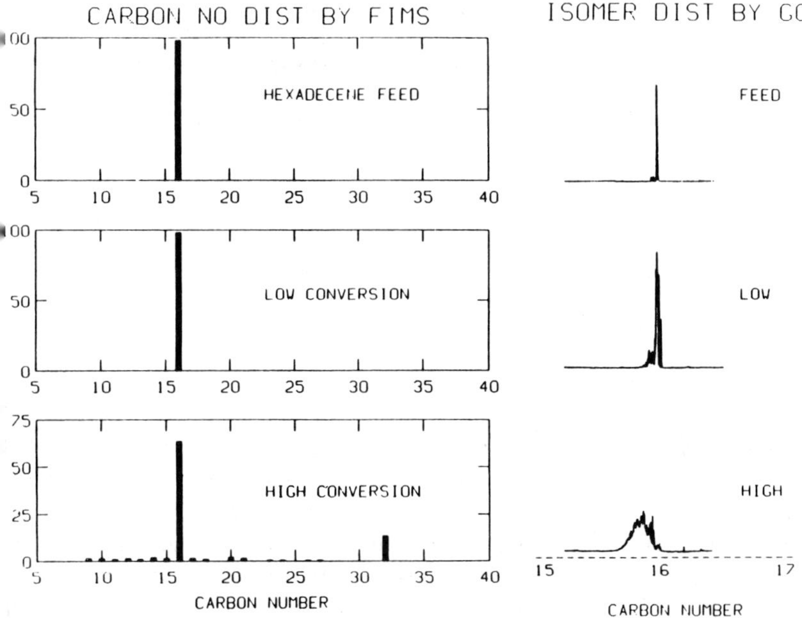

gure 8–5 Hexadecene isomerization and cracking at 232°C and 55 bar uann et al. 1988).

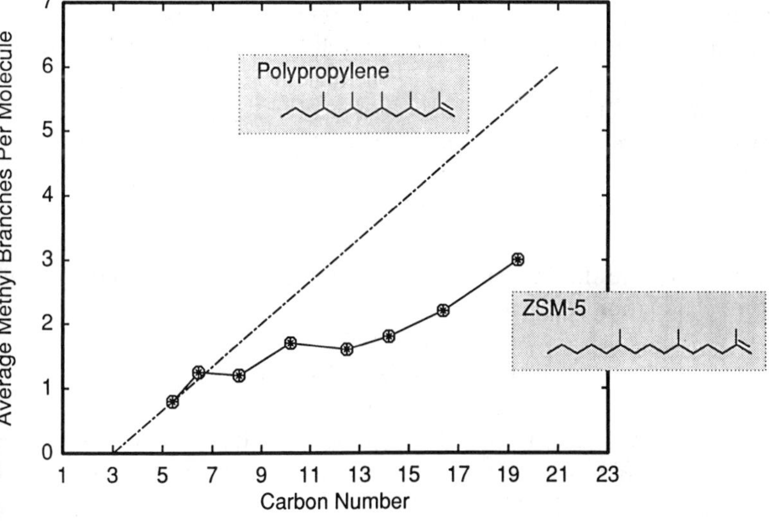

gure 8–6 Branching versus olefin carbon number for MOGD products.

Table 8–1 Olefin Mixture Complexity

Carbon Number	Number of Olefin Isomers
3	1
6	18
10	895
15	185,310
20	46,244,031
25	12,704,949,506

process variables, thermodynamic constraints, and recycle design on tl product distribution. It is quite obvious from Table 8–1, however, that tl system is far too complex to consider every molecular species and the oligomerization, cracking, and isomerization reactions. To reduce comple ity, all isomers of a given carbon number were lumped into a sing component. This approach is consistent with the observation that doub bond and skeletal isomerization reactions were significantly faster tha oligomerization or cracking reactions, and retains the necessary detail of tl carbon number distribution needed to model recycle strategy. A detaile carbon number distribution approach was also necessary for prediction product properties. The viscosity of the distillate product, which is depende on the carbon number distribution of the distillate, can be an importa product property specification in a commercial application. The conjun polymerization mechanism leading to cyclic compounds and saturates neglected.

The kinetics of the MOGD reaction sequence can be described by a simp reversible oligomerization reaction

$$C_i + C_j \leftrightarrows C_{i+j}, \qquad (8–$$

where i and j are the carbon numbers of the two reactants, and a parall disproportionation reaction

$$C_i + C_j \leftrightarrows C_{i+n} + C_{j-n}, \qquad (8–$$

where two olefins react to produce two olefins of different carbon numbe The rate expression for the reversible oligomerization reaction is

$$r_{ij} = Pk_{ij}^{f} x_i x_j - k_{ij}^{r} x_{i+j}, \qquad (8–$$

where k_{ij}^f and k_{ij}^r are the oligomerization and cracking rate constants of olefins i and j, and $i + j$, respectively, and P is total pressure.

The disproportionation reaction can be represented as a two-step reaction involving the formation and splitting of a carbenium ion complex $[C_z]$ having a carbon number z:

$$C_i + C_j \rightarrow [C_z] \rightarrow C_u + C_v \tag{8–4}$$

with

$$i + j = u + v = z. \tag{8–5}$$

The rate of complex formation r'_{ij} is assumed to be proportional to the oligomerization rate by a relative rate parameter α

$$r'_{ij} = \alpha P k_{ij}^f x_i x_j. \tag{8–6}$$

Assuming a pseudo steady state in the concentration of the complex $[C_z]$, the total rate of splitting of the complex $[C_z]$ is equal to the rate of formation of the complex. The complex splitting rate however must be divided among all possible combinations of u and v whose sum is equal to z. This is accomplished by using a factor q_{uv} to scale the complex splitting to the cracking rate of an olefin of carbon number z to two olefins of carbon numbers u and v

$$q_{uv} = \frac{k_{uv}^r}{\sum\limits_{l+m=z} k_{lm}^r} \tag{8–7}$$

with

$$\sum_{l+m=z} q_{lm} = 1. \tag{8–8}$$

The splitting rate of complex $[C_z]$ to olefins u and v is proportional to q_{uv} and the sum of all complex formation rates for $[C_z]$. The net rate of formation of olefins u and v due to disproportionation, r''_{uv}, is then

$$r''_{uv} = q_{uv} \sum r'_{lm} - r'_{uv}. \tag{8–9}$$

It can readily be shown that this net rate is zero at the same equilibrium composition that brings all the r_{ij} to zero in Eq. (8–3).

At equilibrium the net rate of oligomerization for reaction (8–1) is zero and Eq. (8–3) reduces to

$$\frac{k_{ij}^{f}}{k_{ij}^{r}} = \frac{x_{i+j}}{P x_i x j} = K_{ij}^{P}, \tag{8–10}$$

where K_{ij}^{P} is the equilibrium constant for the reaction. At high pressure a more appropriate representation of equilibrium would employ fugacities rather than fugacity partial pressures. However, in the absence of reliable fugacity coefficient information, the ideal gas case is assumed. Equation (8–10) is used in the model to determine the cracking rate constants relative to the forward rate constants. Determination of the equilibrium constant K_{ij}^{P} for the lumped isomer approach is discussed in the next section. Assignment of forward rate constants for oligomerization is discussed below.

Based on isomerization, adsorption, and diffusion considerations, as well as observations from model compound studies, the rate for oligomerization is expected to decrease with increasing carbon number. As the number of possible olefin isomers increases exponentially with carbon number, the steric factors that inhibit diffusion within the zeolite, exclude reactants from pores, and separate reaction centers for the double bond–carbenium ion condensation process of adsorbed species, increase accordingly. Furthermore the capacity for adsorption into the zeolite in terms of molecules per unit volume also decreases with increasing carbon number. Thus a strong variation of the rate constant for oligomerization with carbon number is expected.

Since the forward oligomerization reaction is a bimolecular process, it is logical to assume that the forward rate constant must be dependent on the properties of both reactants i and j. In the absence of extensive model compound studies, a simple functional form of the rate constant was chosen

$$k_{ij}^{f} = k_i k_j k_0 e^{-E/RT} \tag{8–11}$$

with k being a preexponential factor and with the constants k_i and k_j being probability factors that are functionally dependent on the carbon number of the olefin

$$k_i = i^{-\omega}. \tag{8–12}$$

The factor k_i (and k_j) will decrease with increasing carbon number for any positive value of ω. The forward rate constant then has the form

$$k_{ij} = k_0 (ij)^{-\omega} e^{-E/RT}. \tag{8–13}$$

The exception to this functional form is ethylene, which is very unreactive compared to other olefins.

The reactor model is assumed to be an idealized plug flow system. The molar equation of change for an olefin of carbon number n can be expressed as

$$\frac{dy_n}{dM} = \sum_{i+j=n} r_{ij} - \sum_{\substack{i=n \\ \text{or} \\ j=n}} r_{ij} + \sum_{\substack{u=n \\ \text{or} \\ v=n}} \left(q_{uv} \sum_{l+m=z} r'_{lm} - r'_{uv} \right) \tag{8–14}$$

where dy_n is the change in the molar flow y_n through an incremental mass dM of catalyst. The first term of Eq. (8–14) accounts for the (reversible) formations of n from the condensation of all olefins i and j whose carbon number sum is n. The second term accounts for the generation of n from the (reversible) cracking of all olefins larger than $n+1$ *to* n. The last term in this equation is the gain and loss of olefin n from disproportionation. The total mass flow F through the reactor is constant

$$F = \sum y_n \text{MW}_n = \text{constant}, \tag{8–15}$$

where MW_n is the molecular weight of olefin n. The weight hourly space velocity (WHSV) follows as

$$\text{WHSV} = F/M. \tag{8–16}$$

The parameters for the model are the oligomerization rate constant k, the carbon number reactivity parameter ω, and the relative rate parameter α for disproportionation. Reverse rate constants k^r_{ij} are obtained from the forward rate constants using the thermodynamic relationship of Eq. (8–10). The next section will discuss how the equilibrium constant is obtained.

OLIGOMERIZATION THERMODYNAMICS

Thermodynamics provides the driving force for light olefin condensation to heavier olefins and dictates the conditions under which the MOGD process should be operated. Thermochemical quantities of olefins are also required for the estimation of rate parameters for reversible reactions, as noted above. Incorporating thermodynamics into the kinetic model provides the necessary constraint on the system as chemical equilibrium is approached. The two main difficulties in equilibrium computations for reactive olefin systems, however, are the astronomical number of isomers involved and the lack of thermodynamic data on olefins. As shown in Table 8–1, equilibrium computations become impractical with the inclusion of all isomers above C_8 or C_9. In any case, the thermodynamic properties for all olefin isomers are known only through hexene with additional data on linear 1-olefins available through C_{20} (Stull et al. 1969). Approximation techniques are therefore required.

An assumption that reduces the complexity of the problem, and one that is consistent with the kinetic model formulation, is that the isomer distributions of a given carbon number are at equilibrium and that the isomer group can be lumped and treated as a single component. The basis behind this assumption is that the relative distribution of isomers in a carbon number group depends only on temperature and not on the overall composition of the system. The standard Gibbs free energy of the isomer group is then given by the summation over all individual isomers N and a free energy of mixing term

$$\Delta G^0_{f\mathrm{I}} = \sum_i r_i \Delta G_{fi} + RT \sum_i r_i \ln r_i = -RT \ln \sum_i e^{-\Delta G_{fi}/RT} \qquad (8\text{–}17)$$

where r is the fraction of isomer I in the group and $\Delta G^0_{f\mathrm{I}}$ is the standard Gibbs free energy of formation of isomer I. Again the difficulty is that this approach requires thermodynamic data on all individual isomers. Alberty (1986) developed an approximation method to estimate the group thermodynamic properties by assuming a linear extrapolation of the form

$$\Delta G^0_{fi} = A + B\,n, \qquad (8\text{–}18)$$

where A and B are constants at a given temperature and n is the carbon number of the isomer group. The constants A and B are determined from the known properties of olefin isomer groups through hexene. The enthalpy of formation of an isomer group is obtained by an analogous expression

$$\Delta H^0_{f\mathrm{I}} = A_h + B_h\,n, \qquad (8\text{–}19)$$

The standard Gibbs free energy of formation for an isomer group at any temperature T other than T_0 can now be approximated by the relation

$$\Delta G^0_{f\mathrm{I}}(T) = \Delta G^0_{f\mathrm{I}}(T_0) + \int_{T_0}^{T} \frac{\Delta H^0_{f\mathrm{I}}}{RT^2}\, dT. \qquad (8\text{–}20)$$

Using a free energy minimization program developed at Mobil, olefin equilibrium distributions were computed using the above extrapolation method of Alberty and compared to previously published experimental data (Garwood 1983) on olefin equilibration over ZSM-5 using five different starting olefins. This comparison is shown in Figure 8–7 along with an

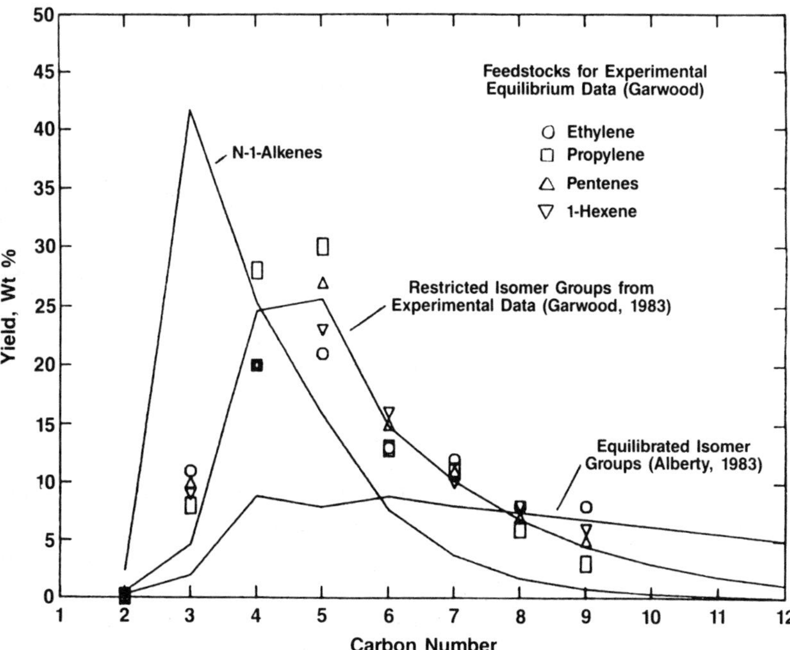

Figure 8–7 Olefin equilibration for different feedstocks (Quann et al. 1988).

additional computation that assumes only 1-alkenes in the system. It seems unquestionable that equilibrium has been obtained given that the net direction of the reaction for ethylene, propylene, and butene is molecular weight increase, and that of hexene and decene is molecular weight decrease. The conditions chosen for the experimentation were low pressure and high temperature to ensure equilibration. At MOGD high-pressure conditions equilibrium, with an average carbon number in the thousands, is never reached because of kinetic limitations.

A possible explanation for the lack of agreement between predicted and measured equilibrium distributions is that not all isomers are formed because of the shape selectivity restrictions imposed by ZSM-5, as discussed earlier and shown in Figure 8–6. Highly branched isomers generally have a lower Gibbs free energy of formation and their exclusion would increase the isomer group free energy and shift the apparent equilibrium distribution toward lower carbon numbers. The extrapolation constants A and B for free energy were adjusted to fit the experimental data for the case of the restricted isomer equilibration set for ZSM-5. The new isomer group parameters were then employed for the estimation of reversible reaction rate parameters. The equilibrium constant for the reaction in Eq. (8–1) is given by

$$K_{ij}^{P} = \exp[-(\Delta G_{f1}^{i+j} - \Delta G_{f1}^{i} - \Delta G_{f1}^{j})/RT]. \quad (8\text{–}21)$$

COMPARISON OF MODEL PREDICTIONS WITH PILOT PLANT RESULTS

The initial application of the model was on single-pass pilot plant experiments using single-component, model compound feedstocks. Comparisons of these model results with detailed carbon number distributions of products obtained from FIMS analysis of pilot plant products for 1-hexene and 1-hexadecene feeds are shown in Figures 8–8 and 8–9. The fits are obtained by adjustment of the three parameters k, ω, and α. Also shown in Figure 8–8 is a comparison of the model prediction without incorporating the disproportionation reaction (i.e., $\alpha = 0$). It is clear from this latter case that the reversible oligomerization/cracking reaction is insufficient with respect to randomizing the carbon number distribution even though the ratio of cracking to oligomerization rates is enhanced by use of restricted thermodynamics. If isomerization reactions are much faster than oligomerization or cracking, then, in principle, disproportionation cannot be a significant reaction pathway because the chemisorption, and desorption, of olefins to

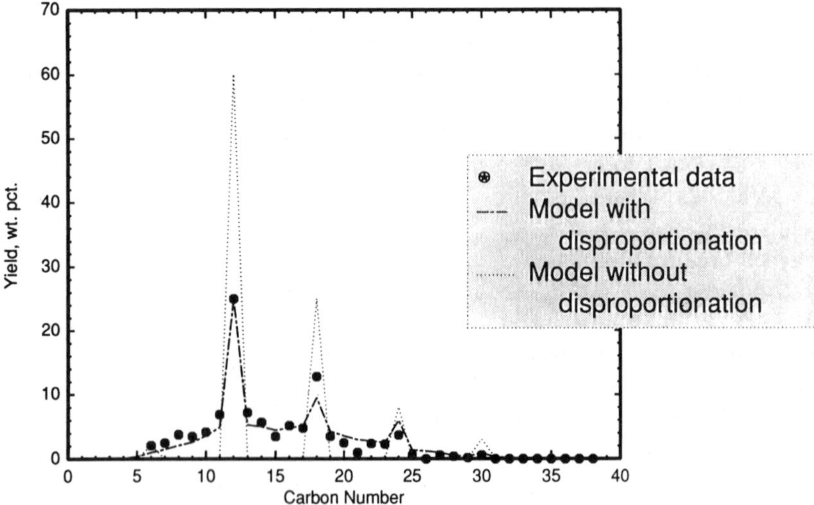

Figure 8–8 Comparison of model predictions of the olefin carbon number distribution with experimental data for a 1-hexene feed.

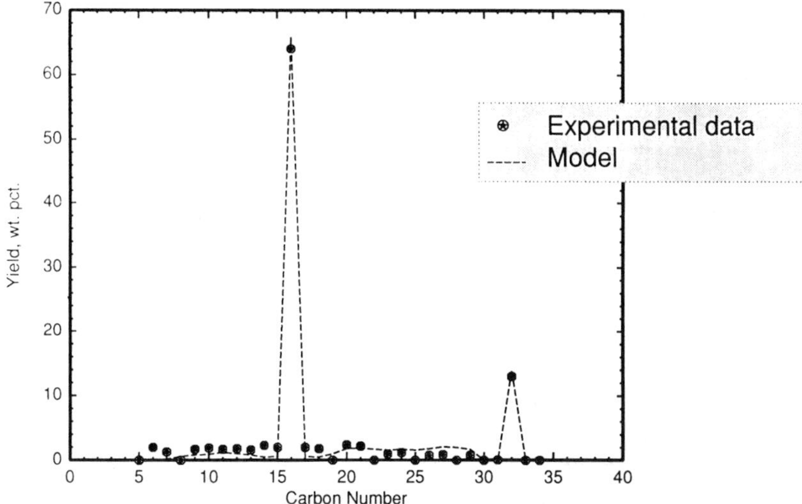

Figure 8–9 Comparison of model prediction with experimental feed for 1-hexadecene feed.

carbenium ions is fast and at equilibrium. In fact, isomerization is faster than cracking, but not much faster. Furthermore using the thermodynamic lump of all isomers of a given carbon number is not necessarily consistent as a kinetic lump. The cracking rate of carbenium ions may be strongly dependent on the isomer. Hence, the use of lumped thermodynamics to predict the reverse reaction rate fails to give satisfactory agreement and a disproportionation reaction must be invoked.

Additional single-pass experiments published previously (Tabak et al. 1986) at different pressures using propylene as a feedstock indicate that Eq. (8–3) has the correct pressure dependence, see Figure 8–10.

A summary of the effects of process variables on olefin reactions in ZSM-5 for single-pass conditions is shown in the predicted surface plot of Figure 8–11. The product distribution, as characterized by average carbon number, reflects the relative rate of oligomerization, cracking, and disproportionation at each temperature and pressure, as well as thermodynamic constraints. High pressure is the driving force for condensation to higher molecular-weight olefins. At low temperatures, oligomerization is favored thermodynamically but molecular-weight growth is kinetically limited. As temperature is increased, all reaction rates increase with cracking eventually dominating kinetically, resulting in thermodynamic limitation to molecular-weight growth.

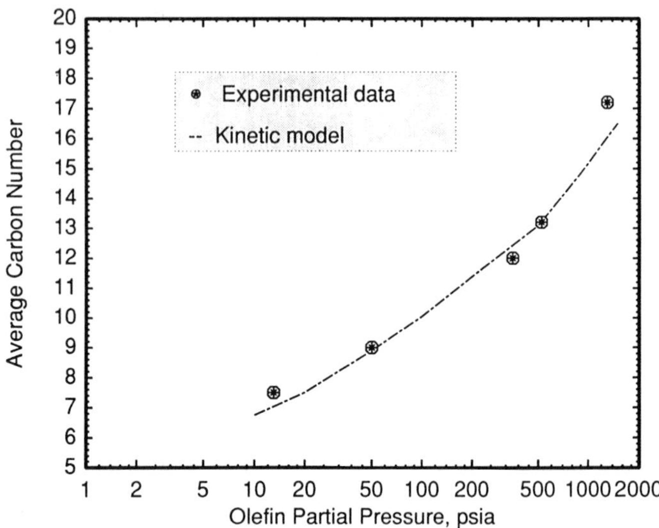

Figure 8–10 Effect of pressure on average carbon number of olefin product.

RECYCLE EFFECTS

The MOGD process employs product recycle to control reactor exotherms and the carbon number distribution of the product. The commercial design incorporates multiple flash separators to isolate product and recycle cuts and multiple reactors with recycle as cofeed to each reactor (Yurchak et al. 1990). Pilot plant development work was performed on single reactor units using a continuous still (as in Figure 8–1) to isolate a recycle cut for the purpose of studying the effects of recycle strategy on product distribution. A comparison of the carbon number distribution from FIMS analysis of the total product from a recycle pilot plant experiment with the model in the similar recycle configuration is shown in Figure 8–12. The feedstock was a pentene/hexene blend that contained unreactive C_5/C_6 paraffins. The recycle range for both pilot plant and model was C_5–C_{13} with a recycle/fresh feed ratio of 2:1. The model clearly predicts the carbon number profile of the entire product.

The recycle model was used to examine the combined effects of recycle process variables, including recycle cutpoints and recycle ratio, and feedstock purity on the product distribution. Since the recycle model in the configuration of Figure 8–1 (one reactor and one recycle stream cofeed with fresh feed) only approximates a commercial design, the observations presented here are meant only to illustrate trends that can be predicted by a complex

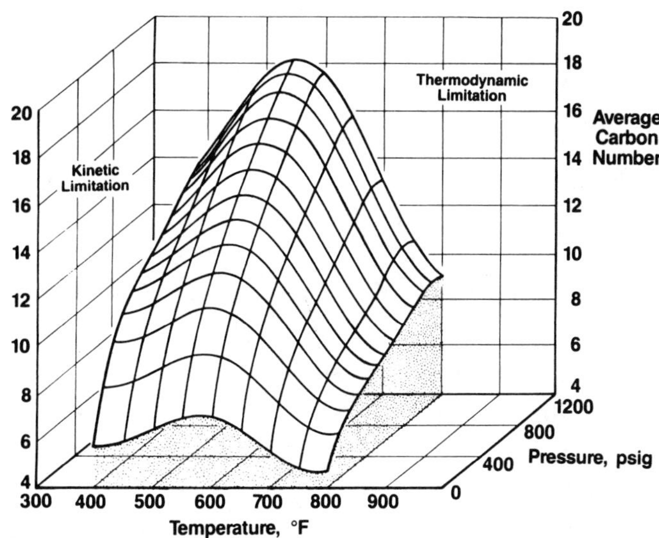

Figure 8–11 Model prediction of temperature and pressure effects on the average carbon number distribution of the product.

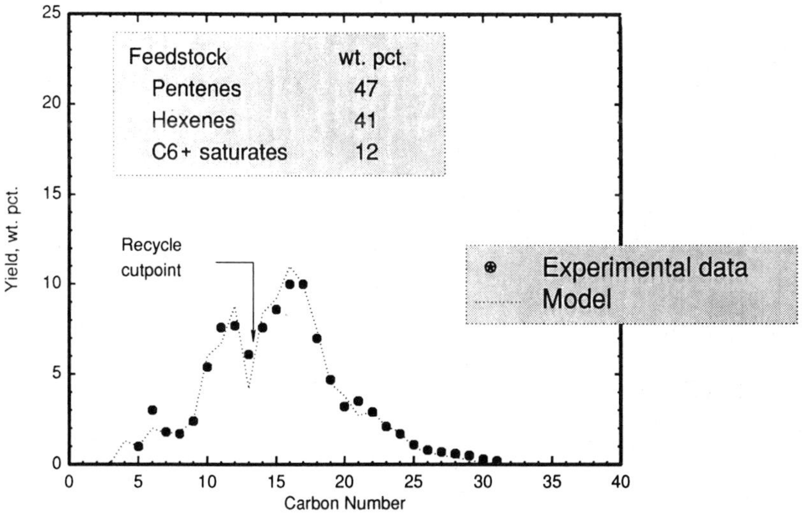

Figure 8–12 Comparison of model and pilot plant data for recycle operation.

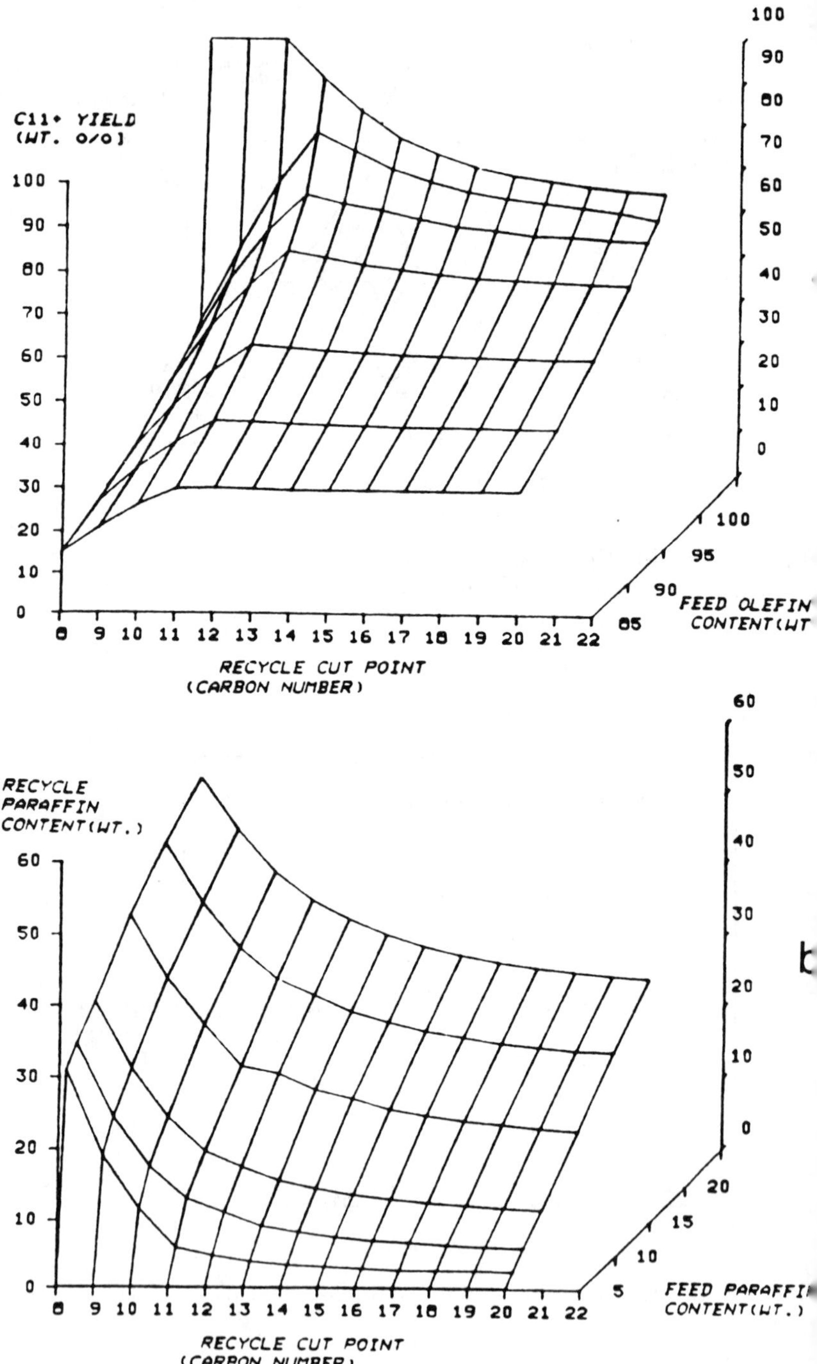
C11+ YIELD
(WT. O/O)
100
90
80
70
60
50
40
30
20
10
0
8 9 10 11 12 13 14 15 16 17 18 19 20 21 22
RECYCLE CUT POINT
(CARBON NUMBER)
85
90
95
100
FEED OLEFIN
CONTENT(WT
RECYCLE
PARAFFIN
CONTENT(WT.)
60
50
40
30
20
10
0
8 9 10 11 12 13 14 15 16 17 18 19 20 21 22
RECYCLE CUT POINT
(CARBON NUMBER)
5
10
15
20
FEED PARAFFI
CONTENT(WT.)

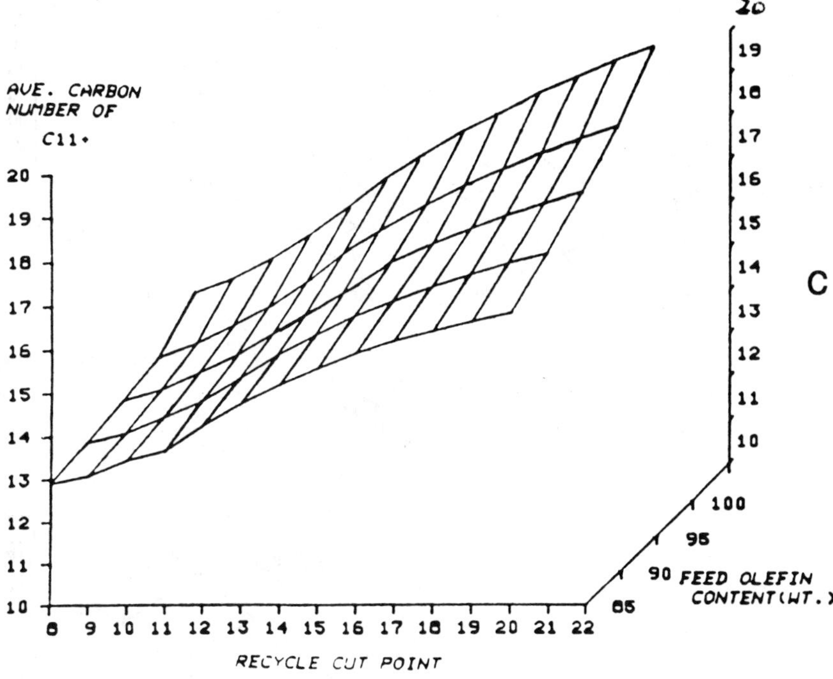

Figure 8–13 Model predictions for the effects of feed olefin purity and recycle cutpoint for recycle system.

kinetic model. The predicted effects of recycle cutpoint and olefin feed purity on C_{11+} distillate yield are shown in Figure 8–13a for a C_3–C_6 olefin feed blend and a recycle ratio of 2:1. For a pure olefin feed, increasing the recycle cut point decreases C_{11+} as a consequence of the reversibility of the reaction. When unreactive feed components such as paraffins are present, increasing the recycle cutpoint can increase C_{11+} yield. Feed paraffins can dramatically decrease C_{11+} yield by lowering olefin partial pressure. As shown in Figure 8–13b, paraffin concentration builds in the recycle stream, particularly at low recycle cutpoints. Increasing recycle cutpoint dilutes the paraffin content of the recycle stream, raising overall olefin pressure in the reactor. The average carbon number of the distillate also increases substantially with recycle cutpoint as seen in Figure 8–13c. Recycle ratio has very little impact on distillate yield or average carbon number as shown in Figures 8–14a and 8–14b. This is due to a space velocity effect. Increasing recycle ratio increases overall space velocity in the reactor at constant fresh feed space velocity with no net gain in distillate.

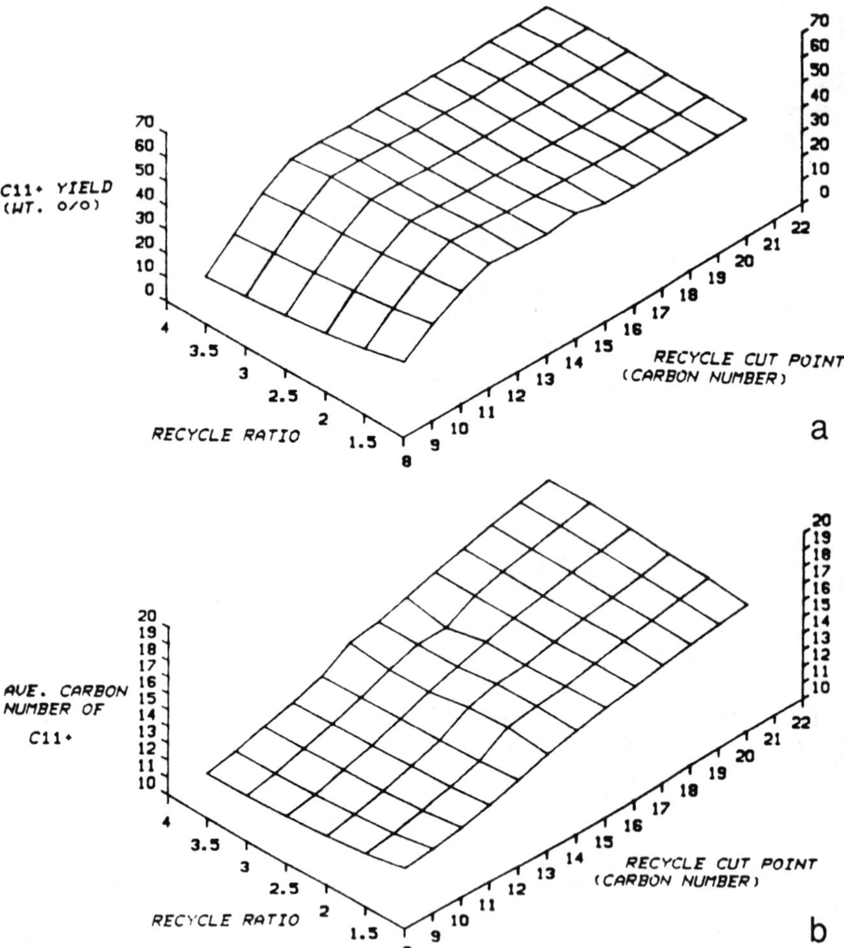

Figure 8–14 Model predictions for the effect of recycle ratio on distillate yield and molecular weight for the recycle system.

CONCLUSION

A kinetic model of Mobil's MOGD process has been developed. The model is based on a complex sequence of olefin oligomerization, cracking, and disproportionation reactions and predicts the detailed carbon number distribution of the product. The model has provided valuable insight into the dynamics of the process and has been used extensively in the design and development of the process.

ACKNOWLEDGMENTS

The authors are grateful to the many individuals who were involved in the development of the MOGD process, including S. A. Tabak, J. H. Beech, J. D. Kushnerick, W. E. Garwood, and L. A. Green.

REFERENCES

Alberty, R. A. (1986) J. Phys. Chem. **87**:4999.

Garwood, W. E. (1983) In *Intrazeolite Chemistry* (Stucky, G. D. and F. G. Dwyer, Eds.), *ACS Symposium Series 218.*

Oblad, A. G., G. A. Mills, and H. Heinemann (1958) *Catalysis.* Van Nostrand Reinhold, New York.

Pines, H. (1981) *Chemistry of Catalytic Hydrocarbon Conversions.* Academic Press, New York.

Quann, R. J., L. A. Green, S. A. Tabak, and F. J. Krambeck (1988) *IEC Res.* **27**:.

Read, R. C. (1976) In *Chemical Applications of Group Theory* (Balaban, A. T., Ed.), Chapter 4. Academic Press, New York.

Stull, D. R., E. F. Westrum and G. C. Sinke (1969) *The Chemical Thermodynamics of Organic Compounds.* John Wiley, New York.

Tabak, S. A., F. J. Krambeck and W. E. Garwood (1986) *A.I.Ch.E.J.* **32**:1526.

Yurchak, S., J. E. Child, and J. E. Beech (1990) Paper presented at the 1990 Annual NORA Meeting.

PART 3

KINETIC ANALYSIS

9

Some Current Issues in Kinetic Lumping of Discrete Mixtures

KENNETH B. BISCHOFF

Lumping analysis of very complex reaction networks, such as in petroleum processing, has been extensively developed over the years, primarily by heuristic means based on a large amount of data. A comprehensive survey was provided by Weekman (1979), focusing on one of the most complicated such networks—catalytic cracking of gasoil or heavier feedstocks. Weekman also considered other processes such as catalytic reforming [a very thorough review of this process was recently given by Ramage et al. (1987)], steam cracking, and even a brief description of biochemical pathways. Many practical engineering aspects were also discussed.

The modern rigorous mathematical framework for lumping analysis was provided by Wei and Kuo (1969) for monomolecular first-order kinetics. It is worthwhile briefly recapitulating their results, since much that will follow is based on their analysis. For an n-vector of species compositions (say mole fractions), x, the general kinetic scheme can be represented as in Wei and Prater (1962)

$$\frac{dx(t)}{dt} = K\ x(t), \quad x(0) = x_0. \tag{9–1}$$

where K is the $n \times n$ rate constant matrix. Now we wish to describe the system with a far smaller lumped system of l-lumped pseudospecies, $\hat{x}$, related to the original species by the $\hat{n} \times n$ lumping matrix, L:

$$\hat{x}(t) = L\, x(t). \tag{9–2}$$

They defined exact lumping as one that satisfies invariant response, which means that precisely the same results will be obtained (a) by also solving the monomolecular lumped kinetics,

$$\frac{d\hat{x}}{dt} = \hat{K}\hat{x}, \qquad \hat{x}(0) = Lx_0, \tag{9–3}$$

where $\hat{K}$ is the $\hat{n} \times \hat{n}$ lumped rate constant matrix, (b) as would be found by solving the original full system, and then lumping the appropriate species by Eq. (9–2). They derived the necessary and sufficient conditions for this by noting that the two above approaches to obtaining the lumped solution can be written

$$\frac{d\hat{x}}{dt} = \hat{K}\hat{x} = \hat{K}Lx$$

$$= L\frac{dx}{dt} = LKx.$$

For these to be identical for the nonzero vector x

$$LK = \hat{K}L. \tag{9–4}$$

This, then, is the condition that the system must follow for invariant response according to Wei and Kuo (1969). (It might be noted that about the same time, systems mathematicians and econometricians derived the same result in their areas.)

We will see that this leads to some very useful properties of this lumping approach, and we will define it as "Wei–Kuo Exact Lumping" (WKEL). We will also see that there are other mathematically rigorous lumping schemes that do not have these features, and so it is useful to specifically distinguish this special case.

Two extensions useful at this point were given by Coxson and Bischoff (1987a,b). One shows how the time-continuous results are also true for the equivalent discrete-time approach, which will be useful in what follows (and is what is usually actually measured). For a time increment, τ, using the formal solution for the continuous system

$$x(t) = e^{Kt}x_0$$

nd

$$x((h+1)\tau) = e^{K\tau}x(h\tau)$$
$$\equiv G\,x(h\tau)$$

r

$$x(h) = G^h x_0 \tag{9–5}$$

here G is the $n \times n$ system matrix. The other useful result is that Eq. –4) can be solved for $\hat{K}$ by the expression

$$\hat{K} = LKL^{+}$$
$$= LKL^{T} \cdot \mathrm{Diag}[1/n_1, 1/n_2, \ldots, 1/n],$$

here L^{+} is a so-called generalized inverse (remember L is not square) nd the diagonal elements are the number of species in each lump 1, 2, ... his seems simpler than the original Wei–Kuo (1969) method, and can be seful in defining a reduced-order description of a complex kinetic scheme or purposes, e.g., of on-line automatic control algorithms.

OME EXTENSIONS

Iere we will first discuss a useful result of WKEL in the important area of ranslation of a lumping scheme from one reactor type to another. Then an xample of a mathematically precise, but not WKEL, scheme will be escribed that has different results in the above important area. Next the ssues in the practical area of approximate Wei–Kuo lumping, including hoice of lumps will be described. Finally, nonlinear kinetics will be briefly onsidered.

For monomolecular kinetic systems, a general way to consider different eactor types is through residence time distribution theory (RTD). It is well nown that for first-order kinetics, the output of a steady-state flow reactor independent of the state of micromixing, and so the general "segregated ow" solution is always valid:

$$x = \int_0^\infty E(t)e^{Kt}x_0\,dt,$$

/here $E(t)$ is the RTD of the reactor, and $e^{Kt}x_0$ is the batch (or ideal plug ow) reactor result. Then, Bischoff and Coxson (1987a) showed that for the monomolecular lumped system

$$\hat{x} = Lx = \int_0^\infty E(t)Le^{Kt}\, dt.$$

It has here been assumed that $E(t)$ is a scalar, independent of species and s unaffected by L; this would be true of the common turbulent flow, but n for situations where the RTD depends on species diffusivities, for exampl

Now, following a method of Wei and Kuo (1969):

$$Le^{Kt} = L + LKt + (1/2!)LK^2t^2 \cdots.$$

Using the condition for WKEL, Eq. (9–4),

$$LK = \hat{K}L,$$

$$LK^2 = (LK)K = \hat{K}LK = \hat{K}(\hat{K}L).$$

Thus

$$Le^{Kt} = L + \hat{K}L + (1/2!)\hat{K}^2t^2L \cdots$$

$$= e^{\hat{K}t}L$$

and so

$$\hat{x} = \int_0^\infty E(t)e^{\hat{K}t}Lx_0\, dt$$

$$= \int_0^\infty E(t)e^{\hat{K}t}\hat{x}_0\, dt. \qquad (9–$$

Therefore, the batch WKEL scheme carries through for any RTD and thus valid for *any* reactor type.

These same types of relations are often also used for heterogeneous an catalytic systems if the kinetics can be approximated by first-order kinetic However, here the results can also depend on the flow patterns of both phas and may not uniquely depend on the RTD of the primary flowing phase; general analysis is by Shinnar and Rumschitzki (1989).

Other cases where Eq. (9–6) is not rigorous/unique will be for non-WKE schemes (and, of course, nonlinear rate forms). A useful but simple exampl was provided by Aris (1968) for total lumping of very many (infinite) parall reactions, with practical relevance to hydrodesulfurization of many sulfu ontaining compounds, for example. Often here, even though model com-ound studies of individual species are first order, the desired model in terms f total sulfur has a higher order. The qualitative reasons for this effect are lear, considering the varying reactivities, but Aris provided the quantitative nalysis. Suppose each species is identified by its characteristic rate constant nore general approaches are discussed by Aris and by Astarita in this olume) so that each species has a composition

$$x(t, k) = x(0, k)e^{-kt}.$$

he lumping here will give

$$\hat{x} = \int_0^\infty x(t, k)\, dk$$

$$= \int_0^\infty x(0, k)e^{-kt}\, dk.$$

hus the lumped kinetics can be found if the initial species distribution given. For a gamma distribution with parameter α ($\alpha = 1$: exponential istribution; $\alpha = \infty$: impulse function), Aris gave the result

$$\frac{\hat{x}(t)}{\hat{x}(0)} = (1 + \bar{k}t/\alpha)^{-\alpha}$$

here $\bar{k}$ is the mean value of the rate constants. Alternately, the lumped ate equation is

$$\frac{d\hat{x}}{dt} = (k/\hat{x}_0^{1/\alpha})(\hat{x})^{1+1/\alpha}.$$

hus the lumped order is $1 + 1/\alpha > 1$ as qualitatively discussed above. In articular, for $\alpha = 1$, the lumped order is 2, as is commonly observed with eal HDS processes. One can argue that this initial distribution is rather trange, since it has $k = 0$ as the most prevalent value; physically, it epresents a large amount of very refractory feed. Complications for non-rst-order original rates are discussed by Aris and by Astarita, as above.

Note that the above lumping violates the criteria for WKEL—namely, nat the lumped kinetics be monomolecular as in the original system. ollowing Luss and Golikeri (1975), for $\alpha = 1$, if one solves each first-order eaction, and then lumps for the RTD for a perfectly mixed reactor, the following is obtained (late lumping)

$$\hat{x}(\tau) = \int_0^{\infty} \frac{x(0, h)}{1 + k\tau}\, dk = \frac{e^{1/\bar{k}\tau}}{\bar{k}\tau} E_1\left(\frac{1}{\bar{k}\tau}\right).$$

If one lumps first to obtain second-order kinetics (early lumping), an
then solves the usual PMR mass balance, the following results:

$$\hat{x} = -\frac{1}{2\bar{k}\tau} + \left[\left(\frac{1}{2\bar{k}\tau}\right)^2 + \frac{1}{\bar{k}\tau}\right]^{1/2};$$

clearly these two results are not identical, so here the lumping schen derived for a batch (or ideal PFR) *does not* carry over to a PMR reacto Therefore, one must be very careful in translating between reactor types fo non-WKEL schemes. Other situations where the attempt to lump paralle consecutive reactions cannot be even qualitatively correct are given b Golikeri and Luss (1974).

Next, the issue of choice of lump for WKEL and approximate suc schemes will be considered. Coxson and Bischoff (1987a,b) have describe an approach based on systems mathematics. It involves constructing th responses of the complex system to specific input feeds, and then analyzin these responses by a formal, objective statistical technique: cluster analysi This is only one approach, of course, but will be briefly described below.

One constructs a so-called "response matrix" whose columns consist the kinetic responses to each pure (unit composition) feed species (fee mixtures can also be handled); the kinetic responses are the compositions a function of discrete time increments. This response matrix can be con structed from only experimental data by following the time evolution of th reacting mixture in the usual way. Utilizing the above theory for discre monomolecular kinetic systems, which is described by the system matri $G = e^{K\tau}$, the response matrix mathematically is

$$R = \begin{bmatrix} L \\ LG \\ LG^2 \\ \vdots \\ LG^{h-1} \end{bmatrix}. \qquad (9\text{–}$$

Column j contains the lumped responses following an experiment wi pure unit composition species j, with the first block row being the initi lumped feed composition that should lead to invariant response, or appro imately so. By multiplying out the other block rows, it is clear that they represent the lumped compositions according to Eq. (9–5).

For exact lumping, the columns representing the feed species to be lumped should be identical. However, constructing the response matrix from real data with errors will never have truly identical columns, nor will approximate lumpings that may still be adequate in practice. A useful approach to test the similarity between columns is the statistical technique of cluster analysis, one version of which computes and compares the mean Euclidian difference (squares of differences) between each successive element of each pair of columns. The result is to determine the most similar columns (least difference "error"), which then empirically have the most similar kinetic patterns and are the best candidates to be lumped or clustered.

Each of the pairs of columns are so compared and clustered, down to the most lumped system that might be of interest. The formal output of the algorithm is a table of the successive clusters, with the increases in "error" (increases in dissimilarities) as the system is further lumped. Coxson and Bischoff (1987a) applied these techniques to the Mobil 10-Lump Model for Catalytic Cracking—a realistic published model based on actual data (Jacob et al. 1976). Without going into details (see the above reference), Table 9–1 illustrates the results.

If one peruses the lumping of the various species (actually, already partially lumped pseudospecies), the steps do not violate chemical intuition—well known after ~50 years of investigation. For new processes, this might not be so clear, and the cluster analysis results could lead to new chemical insight. Also note that the error increases are not uniform, but "jump" at certain lumpings, indicating that there are relatively sharp demarcations between "better" and "worse" lumpings. Using the 4-cluster

Table 9–1 Ten Original Species 3 Lumped Measurements (8-1-1) Clusters Within the Feedstock Lump

Cluster Pattern	No. Clusters	Error Incr.
1 2 3 4 5 2 7 8	7	0.01903
1 1 3 4 5 1 7 8	6	0.07766
1 1 3 4 5 1 5 8	5	0.13283
1 1 3 4 3 1 3 8	4	0.30812
1 1 1 4 1 1 1 8	3	1.43921
1 1 1 4 1 1 1 4	2	1.47388
1 1 1 1 1 1 1 1	1	7.81369

(4 + 2 lump) case, Coxson and Bischoff (1987a) showed that there was virtually no difference between the 6-lump versus the full 10-lump gasoline yields. In Bischoff and Coxson (1987a) it was shown that addition of ± 10–30% random error in the simulations and clusterings led to essentially the same lumping results—the approach seems to be very robust. However, there is still the need to trace through the algorithm for the exact relation between error in the data and the mean Euclidean errors of clustering. One would expect that when the errors of clustering are equal to the inherent data errors, that this is an optimum result.

All of the above has only been mathematically proven for monomolecular, first-order systems, and an important issue is how much is true for nonlinear kinetics. To illustrate that rather precise lumping is possible, consider two parallel Michaelis–Menten or LHHW reactions:

$$r_1 = \frac{v_{m1}C}{K_{m1} + C}, \quad r_2 = \frac{v_{m2}C}{K_{m2} + C}.$$

We wish to lump these into a single rate expression of the same form:

$$r = \frac{v_m C}{K_m + C}.$$

If one exactly matches the high-C and the low-C asymptotes, it is found that

$$r \sim v_m = v_{m1} + v_{m2}$$

and

$$r \sim \frac{v_m}{K_m} C = \frac{v_{m1}}{K_{m1}} C + \frac{v_{m2}}{K_{m2}} C.$$

Thus the lumped parameters are

$$v_m = \sum v_{mi},$$

$$\frac{1}{K_m} = \frac{1}{\sum v_{mi}} \sum \frac{v_{mi}}{K_{mi}}.$$

In simulations, the lumped versus sum of individual rates are virtually identical over the entire concentration range. Astarita and Ocone (1988) have generalized this idea to reactions in continuous mixtures.

For more general nonlinear reaction networks, Li (1984) and Li and Rabitz (1989) have followed the same type of derivation as Wei and Kuo (1969) to derive conditions for exact lumping. The results are rather complicated and cannot be given here. However, there would seem to be some unresolved issues. For example, it is now well known that with nonlinear dynamical systems with at least three dependent variables, such strange behavior as chaos can occur, but if these were lumped to two variables, this would presumably not occur. That this is still a puzzle was agreed by Rabitz (1990, personal communication), and remains to be clarified.

SOME NEW RESULTS

Considering the unresolved issues raised above, there are a few specific examples that can be given. One is based on the translation of WKEL systems from one steady state reactor to another, which was shown above to preserve the identical lumping scheme. For a transient PMR, this also seems to be true, as shown below. The mass balance is

$$\begin{aligned}\frac{dx}{dt} &= \frac{1}{\tau}(x_F - x) + Kx, \; x(0) = x_0 \\ &= \left(K - \frac{I}{\tau}\right)x + \frac{1}{\tau}x_F.\end{aligned}$$

Then,

$$\frac{d\hat{x}}{dt} = L\frac{dx}{dt} = L\left(K - \frac{I}{\tau}\right)x + \frac{L}{\tau}x_F.$$

Using the condition for WKEL, and also the fact that $LI = \hat{I}L$,

$$\begin{aligned}\frac{d\hat{x}}{dt} &= \left(\hat{K} - \frac{\hat{I}}{\tau}\right)Lx + \frac{1}{\tau}\hat{x}_F \\ &= \left(\hat{K} - \frac{\hat{I}}{\tau}\right)\hat{x} + \frac{1}{\tau}\hat{x}_F, \quad \hat{x}(0) = Lx_0.\end{aligned}$$

Thus, once again, the WKEL system will give invariant response. This result could be generalized using Theorem 2 of Wei and Kuo (1969) for analytic functions.

Finally, an example will be given of exact lumping of a second-order kinetic system of Li (1984), worked out by Nigam (1989, personal communication). The full model is given by

$$A_1 + A_3 \underset{1}{\overset{2}{\Leftrightarrow}} A_6 + A_8,$$

$$A_1 + A_4 \underset{1}{\overset{2}{\Leftrightarrow}} A_5 + A_8,$$

$$A_2 + A_3 \underset{1}{\overset{2}{\Leftrightarrow}} A_6 + A_7,$$

$$A_2 + A_4 \underset{1}{\overset{2}{\Leftrightarrow}} A_5 + A_7.$$

This complicated overall second-order system of reactions can be shown to be exactly lumpable as the equation

$$\hat{A}_1 + \hat{A}_2 \underset{1}{\overset{2}{\Leftrightarrow}} \hat{A}_3 + \hat{A}_4.$$

The lumped species are defined by

$$\hat{A}_1 = (A_1 + A_2),$$

$$\hat{A}_2 = (A_3 + A_4),$$

$$\hat{A}_3 = (A_5 + A_6),$$

$$\hat{A}_4 = (A_7 + A_8).$$

For second-order kinetics, of course, the response matrix cannot be determined by only feeding single species, since there would be no reaction occurring. Nigam decided that a reasonable alternative would be to have a basis of unit composition of all species, and then perturb the initial states by another unit composition of each feed species. Then, if this system is exactly lumpable, the dynamics of the lumped species remains unchanged between any pair of the initial conditions

$$(2,1,1,1,1,1,1,1)^T \quad \text{or} \quad (1,2,1,1,1,1,1,1)^T$$
$$(1,1,2,1,1,1,1,1)^T \quad \text{or} \quad (1,1,1,2,1,1,1,1)^T$$
$$(1,1,1,1,2,1,1,1)^T \quad \text{or} \quad (1,1,1,1,1,2,1,1)^T$$
$$(1,1,1,1,1,1,2,1)^T \quad \text{or} \quad (1,1,1,1,1,1,1,2)^T.$$

Nigam computed the simulations of the time changes of the reacting mixture, constructed the response matrix, which was then clustered, with the following results:

Cluster Pattern	Increase	Total ESS
1 1 3 4 5 6 7 8	0.00000	0.00000
1 1 3 3 5 6 7 8	0.00000	0.00000
1 1 3 3 5 5 7 8	0.00000	0.00000
1 1 3 3 5 5 7 7	0.00000	0.00000
1 1 3 3 1 1 7 7	12.27390	12.27390
1 1 3 3 1 1 3 3	12.27390	24.54781
1 1 1 1 1 1 1 1	22.00000	46.54781

It is seen that exact lumping is indeed achieved with the proposed lumping, or any other consistent less lumped scheme, with large increases in clustering error with any further lumping beyond that proposed. This verifies that some nonlinear systems are exactly lumpable. For this system, one can also solve the mass balances for a PMR, with the results shown as

Cluster Pattern	Increase	Total ESS
1 1 3 4 5 6 7 8	0.00000	0.00000
1 1 3 3 5 6 7 8	0.00000	0.00000
1 1 3 3 5 5 7 8	0.00000	0.00000
1 1 3 3 5 5 7 7	0.00000	0.00000
1 1 3 3 5 5 1 1	11.26886	11.26896
1 1 3 3 3 3 1 1	11.40171	22.67057
1 1 1 1 1 1 1 1	20.17399	42.84457

Again, it is seen that the proposed lumping scheme provides exact lumping in this extremely different reactor flow pattern, as before with the batch or ideal-PFR.

A final simulation considers approximately exact lumping in this system for the different rate constants shown below:

$$A_1 + A_3 \underset{0.5}{\overset{1.5}{\Leftrightarrow}} A_6 + A_8,$$

$$A_1 + A_4 \underset{1.2}{\overset{1.8}{\Leftrightarrow}} A_5 + A_8,$$

$$A_2 + A_3 \underset{1.5}{\overset{2.5}{\Leftrightarrow}} A_6 + A_7,$$

$$A_2 + A_4 \underset{0.8}{\overset{2.2}{\Leftrightarrow}} A_5 + A_7.$$

Here the following clustering table shows that the proposed lumping is still an extremely good choice, at least relative to the clustering errors of further lumping.

Cluster Pattern	Increase	Total ESS
1 2 3 3 5 6 7 8	0.28875	0.28875
1 2 3 3 5 5 7 8	0.54689	0.83564
1 1 3 3 5 5 7 8	0.55692	1.39256
1 1 3 3 5 5 7 7	1.02021	2.41277
1 1 3 3 5 5 3 3	11.42281	13.83558
1 1 3 3 1 1 3 3	12.43014	26.26572
1 1 1 1 1 1 1 1	21.41090	47.67663

Note, however, that the order of clustering of the various species is not identical to the exactly lumpable system.

CONCLUSIONS

We have reviewed some classic mathematical analyses of lumping theory, providing both exact and approximate schemes. The special role of Wei–Kuo Exact Lumping (WKEL) has been emphasized, since this approach has many valuable corollaries, e.g, translation between reactor types, obviously very important in practice. We have also tried to point out many unresolved issues, but also illustrating some recent understanding of the problems.

A more comprehensive review of the mathematical aspects of lumping is given in Bischoff and Coxson (1987b). It is interesting that lumping practice has been utilized for many years, but the mathematical basis developed during 1963–1975 showed both advances and problems. Then there was a hiatus until 1987, after which a plethora of new results have appeared, for some of which we have given an outline.

A final comment seems appropriate with respect to how much lumping does one want to do. As exemplified by Allen, Froment, Klein, and Quann in this volume, current practice permits much more comprehensive chemical analytical data, so that kinetic analyses have a much richer database for analysis. In years past, with only PONA (paraffins, olefins, naphthalenes, aromatics) analysis available, only much simpler kinetic schemes were feasible. Those days are gone, and we must now deal with the greatly improved chemical analytical bases of kinetic networks, often able to be based on reasonably detailed fundamental chemistry. This is certainly the preferred approach, all the way to utilizing quantum chemistry to elucidate reaction pathways. However, the extremely complex mixtures of concern in industry, especially petroleum processing, will always require some aspects of lumping—hopefully helped by the above works.

REFERENCES

Aris, R. (1968) *Arch. Ratl. Mech. Anal.* **27**:356.

Astarita, G. and R. Ocone (1988) *A.I.Ch.E.J.* **34**:1299.

Bischoff, K. B. and P. G. Coxson (1987a) *Sadhana-Proc. Indian Acad. Sci.* **10**:299.

Bischoff, K. B. and P. G. Coxson (1987b) *Recent Trends in Chemical Reaction Engineering*, B. D. Kulkarni, R. A. Mashelkar, and M. M. Sharma (eds.) Wiley Eastern Ltd., New Delhi.

Coxson, P. G. and K. B. Bischoff (1987a) *Ind. Engin. Chem. Res.* **26**:1239.

Coxson, P. G. and K. B. Bischoff (1987b) *Ind. Engin. Chem. Res.* **26**:2151.

Golikeri, S. V. and D. Luss (1974) *Chem. Engin. Sci.* **29**:845.

Jacob, S. M., et al. (1976) *A.I.Ch.E.J.* **22**:701.

Li, G. (1984) *Chem. Engin. Sci.* **39**:1261.

Li, G. and H. Rabbitz (1989) *Chem. Engin. Sci.* **44**, 1413.

Luss, D. and S. V. Golikeri (1975) *A.I.Ch.E.J.* **21**, 865.

Ramage, M. P., K. R. Graziani, P. H. Schipper, F. J. Krambeck, and B. C. Choi (1987) *Adv. Chem. Engin.* (J. Wei, J. L. Anderson, K. B. Bischoff, M. M. Denn, and J. H. Seinfeld, eds.) **13**, 193.

Shinnar, R. and D. Rumschitzki (1989) *A.I.Ch.E.J.* **35**:1651.

Weekman, V. W. (1979) *A.I.Ch.E. Monogr. Ser.*, **75**:3.

Wei, J. and J. Kuo (1969) *Ind. Engin. Chem. Fund.* **8**:114.

Wei, J. and C. D. Prater (1962) *Adv. Catal.* **13**:203.

10

Applications of Chemical Reaction Network Theory in Heterogeneous Catalysis

MARTIN FEINBERG

I want to talk about the differential equations that govern complex chemical reactors. The focus here will be on complexity that arises from the chemistry itself rather than from thermal effects, and so I shall confine my remarks to isothermal reactors. Because much has been said about "lumping" at this conference, I should state at the outset that the differential equations under consideration here will be those that derive from the "true" network of elementary reactions (or from a network that is intended to approximate the true network). With this in mind, *I shall always take the kinetics to be mass action.* The resulting differential equations will, therefore, involve polynomials in the species concentrations.

In a conference sponsored by Mobil, where reactors involving thousands of species are an everyday concern, it is perhaps useful to keep in mind that, from a mathematical viewpoint, coupled systems of polynomial differential equations with only seven or eight dependent variables can already be horribly complex. It is interesting to note that much of the semipopular chaos literature has focused on the Lorenz equations, which amount to a coupled system of polynomial differential equations in only three dependent variables.

There is, then, a major dilemma for chemical reaction engineering: The fact is that real reactors generally involve at least a few coupled reactions among a variety of distinct species. Even when the complexity is not so severe as in Mobil's kilospecies reactors, the governing equations are already quite difficult to study, and the mathematics literature has little to say of a general nature. In fact, systems of polynomial differential equations in more than four or five variables remain poorly understood. (A major problem posed by David Hilbert at the turn of the century was about systems of two polynomial equations in two variables.)

Why not just "go to the computer"?

To do what? At least in the initial stages of reactor analysis, not much is known about what the species and reactions actually are, and so one can only make a mechanistic conjecture. Even then, not much will be known about rate constants. And so, while the shape of differential equations corresponding to the putative mechanism might be known, parameter values cannot be filled in with any degree of precision (if they can be filled in at all).

It seems to me that preliminary analysis must be largely *qualitative* in nature (or, as we shall see later, semiqualitative). That is, one needs to know which mechanistic proposals are at least qualitatively consistent with whatever gross phenomena might have been observed. More precisely, one might like to know which candidate mechanisms have the property that, for at least *some* rate constant values, the corresponding differential equations admit Phenomenon X. Here Phenomenon X might represent, for example, the presence of multiple steady states, of periodic solutions, of sustained aperiodic solutions, and so on.

It is the aim of *chemical reaction network theory* to draw connections between the structure of a reaction network and the capacity of the corresponding differential equations to admit qualitative phenomena of a specified kind. I said before that not much is known *in general* about large systems of polynomial differential equations, and so it is natural to wonder why there should be any hope for nontrivial advances toward the goal I indicated. The reason is that we are not interested in polynomial equations *in general*, only those that derive from reaction networks. That is, we are interested in studying a *special* class of polynomial equations—a class that turns out to have remarkably beautiful properties.

The philosophy and aims of reaction network theory are most easily discussed in the context of homogeneous continuous flow stirred tank reactors (CFSTRs), and so I'll begin with a very brief discussion of them. Then I'll turn to catalytic reactors. In another review (Feinberg 1991), I went into considerable detail on homogeneous CFSTRs, but I said relatively little about heterogeneous catalysis. Here I'll do just the opposite. In a meeting sponsored by a company that is heavily involved in catalysis, it is perhaps fitting that most of my discussion should be focused there.

MULTIPLE STEADY STATES IN COMPLEX HOMOGENEOUS CFSTRs

So that I can explain the philosophy and aims of reaction network theory in more concrete terms, I am going to begin by considering the relationship between the structure of a reaction network and its capacity to admit isothermal multiple steady states in a homogeneous CFSTR context. When I say that a reaction network has the capacity for multiple steady states, I shall mean that the corresponding CFSTR equations admit multiple steady states for at least one combination of rate constants, residence time and feed composition.

Consider, for example, an isothermal homogeneous CFSTR in which the chemistry is believed to be well-modeled by the reaction network shown in (10–1).

$$
\begin{gathered}
A + B \underset{2}{\overset{1}{\rightleftarrows}} D \underset{4}{\overset{3}{\rightleftarrows}} 2C \\
C \underset{6}{\overset{5}{\rightleftarrows}} E \\
A + F \underset{8}{\overset{7}{\rightleftarrows}} G \\
2B \underset{10}{\overset{9}{\rightleftarrows}} H \underset{12}{\overset{11}{\rightleftarrows}} I \\
I + J \underset{14}{\overset{13}{\rightleftarrows}} K
\end{gathered}
\tag{10–1}
$$

In (10–2) I have displayed the (isothermal) CFSTR equations that correspond to network (10–1). Here $c_A, \ldots, c_K$ are the species concentrations in the reactor (and in the exit stream), the superscript f indicates a feed stream concentration, $k_1, \ldots, k_{14}$ are rate constants, θ is the residence time, and the overdot indicates time differentiation.

$$\dot{c}_A = (1/\theta)(c_A{}^f - c_A) - k_1 c_A c_B + k_2 c_D - k_7 c_A c_F + k_8 c_G$$

$$\dot{c}_B = (1/\theta)(c_B{}^f - c_B) - k_1 c_A c_B + k_2 c_D - 2k_9 c_B{}^2 + 2k_{10} c_H$$

$$\dot{c}_C = (1/\theta)(c_C{}^f - c_C) + 2k_3 c_D - 2k_4 c_C{}^2 - k_5 c_C + k_6 c_E$$

$$\dot{c}_D = (1/\theta)(c_D{}^f - c_D) + k_1 c_A c_B + k_4 c_C{}^2 - (k_2 + k_3) c_D$$

$$\begin{aligned}
\dot{c}_E &= (1/\theta)(c_E{}^f - c_E) + k_5 c_C - k_6 c_E \\
\dot{c}_F &= (1/\theta)(c_F{}^f - c_F) - k_7 c_A c_F + k_8 c_G \\
\dot{c}_G &= (1/\theta)(c_G{}^f - c_G) + k_7 c_A c_F - k_8 c_G \\
\dot{c}_H &= (1/\theta)(c_H{}^f - c_H) + k_9 c_B{}^2 + k_{12} c_I - (k_{10} + k_{11}) c_H \\
\dot{c}_I &= (1/\theta)(c_I{}^f - c_I) + k_{11} c_H - k_{12} c_I - k_{13} c_I c_J + k_{14} c_K \\
\dot{c}_J &= (1/\theta)(c_J{}^f - c_J) - k_{13} c_I c_J + k_{14} c_K \\
\dot{c}_K &= (1/\theta)(c_K{}^f - c_K) + k_{13} c_I c_J - k_{14} c_K
\end{aligned} \tag{10–2}$$

We can now ask whether the network (10–2) has the capacity for isothermal multiple steady states in a homogeneous CFSTR context. That is, we can ask if there is at least one combination of rate constants, residence time, and feed concentrations such that (10–2) admits two or more steady states. The question is not an easy one.

And this is just one example. There are *thousands* of different reaction networks that might present themselves for study at one time or another, each giving rise to its own distinctive system of CFSTR equations. Some of these systems will be far more complex than (10–2).

At first glance, things would seem pretty grim. There is, however, a source of hope: Note that there is a precise scheme for passing from a reaction network such as (10–1) to the corresponding dynamical equations. (Undergraduates are supposed to know how to do this.) If the reaction network determines the equations (up to parameter values), then there must be a connection between reaction network structure and the capacity of the corresponding equations to admit multiple steady states (or, more generally, qualitative behavior of a specified kind).

It is the aim of reaction network theory to draw just such connections. A successful theory would presumably enable the user to determine directly from the structure of network (10–1) whether, for at least some combination of parameter values, the corresponding system (10–2) can admit multiple steady states.

But such a theory, if it is to be broadly applicable, must be very, very subtle. Here are some facts about homogeneous CFSTRs that a good theory must "fit": Networks (10–3) and (10–5) have the capacity for steady-state multiplicity in an isothermal homogeneous CFSTR context, but the "in between" network (10–4) does not! And consider this: Although (10–3) has the capacity for multiple steady states, network (10–6) does not despite the fact that it differs from (10–3) in only one stoichiometric coefficient. This last observation leads one to conjecture that isothermal multiple steady states require the presence of something other than "1" as a stoichiometric coefficient. In fact, the conjecture is false: Network (10–7) gives multiple steady states despite the fact that no stoichiometric coefficient is higher than 1. These examples suggest that the capacity for multiple steady states depends upon reaction network structure in a very delicate way.

$$\begin{aligned} A + B &\rightleftarrows P \\ B + C &\rightleftarrows Q \\ C &\rightleftarrows 2A \end{aligned} \tag{10–3}$$

$$\begin{aligned} A + B &\rightleftarrows P \\ B + C &\rightleftarrows Q \\ C + D &\rightleftarrows R \\ D &\rightleftarrows 2A \end{aligned} \tag{10–4}$$

$$\begin{aligned} A + B &\rightleftarrows P \\ B + C &\rightleftarrows Q \\ C + D &\rightleftarrows R \\ D + E &\rightleftarrows S \\ E &\rightleftarrows 2A \end{aligned} \tag{10–5}$$

$$\begin{aligned} A + B &\rightleftarrows P \\ B + C &\rightleftarrows Q \\ C &\rightleftarrows A \end{aligned} \tag{10–6}$$

$$\begin{aligned} A + B &\rightleftarrows F \\ A + C &\rightleftarrows G \\ C + D &\rightleftarrows B \\ C + E &\rightleftarrows D \end{aligned} \tag{10–7}$$

What I want to say now is that there is already a good theory of isothermal multiple steady states in homogeneous CFSTRs, one that has sufficient subtlety and range as to draw the correct distinctions among networks (10–3)–(10–7) and also to indicate—within about five minutes!—that the very complex network (10–1) does *not* have the capacity for multiple steady states.

The theory derives from joint work with Paul Schlosser (who was the real hero of the effort). Application of the theory requires no knowledge of advanced mathematics (or, for that matter, of anything but arithmetic). From a reaction diagram of the kind shown in (10–1), one constructs another diagram called the *SCL Graph*, and inspection of the SCL Graph gives information about the possibility of isothermal multiple steady states.

Just how this works is explained in (Schlosser and Feinberg 1987) and in (Feinberg 1991). Proofs and still more results are given in Paul Schlosser's PhD thesis (1988). I won't attempt still another exposition here. I do want to say, however, that the SCL Graph results represent the kind of product that reaction network theory seeks to deliver. They permit engineers and chemists (not necessarily those on friendly terms with modern mathematics) to draw deep connections between reaction network structure and qualitative properties of the corresponding differential equations (in this case the capacity of the equations for multiple steady states).

On the other hand, I should also say that the SCL Graph results are very specific to isothermal *homogeneous* CFSTRs. In its current incarnation, the SCL Graph theory does not give much information about isothermal CFSTRs involving heterogeneous catalysis. There is, however, another and (in some ways) deeper part of reaction network theory that does. This is a subject I'll turn to next.

MULTIPLE STEADY STATES AS A KEY TO MECHANISM DISCRIMINATION IN HETEROGENEOUS CATALYSIS

In Feinberg (1987, 1988), I surveyed a part of reaction network theory that is based upon classification of networks by means of an easily computed non-negative integer index called the *deficiency*. Thus, there are reaction networks of deficiency zero, of deficiency one, of deficiency two, and so on. [Fred Krambeck (1970) was, as far as I know, the first person to understand the importance of ideas related to what is now called the deficiency of a reaction network. His focus then was on issues somewhat different from those I am considering here. In fact, it is interesting to see how the deficiency pops up in seemingly disconnected parts of the reaction network theory landscape. See also (Feinberg 1989).]

In this survey, it won't be necessary for me to indicate how the deficiency of a reaction network is calculated, but I should tell you this: The deficiency is *not* a measure of the "size" of the network. There are networks of deficiency zero that have hundreds of species and hundreds of reactions.

A central result is the Deficiency Zero Theorem, which has its roots in work of Horn and Jackson (1972). In rough terms, the theorem says that the mass action differential equations that derive from deficiency zero networks (however complicated) can only give rise to "stable" behavior, no matter what (positive) values the rate constants take. For example, there is no possibility of multiple steady states, of an unstable steady state, or of sustained periodic solutions.

This is not true of networks having higher deficiency. In particular, there are networks of deficiency one which have the capacity for multiple steady states, for unstable steady states, and for periodic composition oscillations.

On the other hand, deficiency one networks, like deficiency zero networks, have very special properties. They are far more subtle than deficiency zero networks, but, at least with respect to questions about multiple steady states, deficiency one networks are now well understood.

It happens that catalytic reactors—in particular, those designed for mechanistic studies—very often turn out to be described by reaction networks of deficiency one, and so *deficiency one theory* has much to say about them. [Leib et al. (1988) explain just why deficiency one theory is well-suited to catalytic reactors and, also, why deficiency one theory is not very good for homogeneous CFSTRs.]

Deficiency one theory brings what I think are simple but powerful analytical resources to problems of mechanism discrimination. Without saying much about how and why the theory works, I want to describe the kind of information it gives.

By way of background, I should tell you that deficiency one theory provides the machinery to come to this conclusion: *Virtually all the classical catalytic mechanisms have the capacity for multiple steady states in an isothermal CFSTR setting.* That is, for virtually all the classical catalytic mechanisms, there are combinations of rate constants, residence time, and feed composition such that the resulting isothermal steady-state equations (derived from the *elementary* reactions) admit multiple solutions. This, of course, is not to say that every catalytic experiment in a CFSTR will yield multiple steady states, only that the observation of multiple states should not be regarded as remarkable. In fact, multiple steady states in isothermal catalytic CFSTRs are observed routinely.

Deficiency one theory indicates not only that multiple steady states are unexceptional in isothermal catalytic CFSTRs, but also that *the underlying catalytic mechanism leaves a distinct "signature" in the multiplicity data actually recorded in the laboratory.* Moreover, this signature can be evident even when the data are fragmentary.

I'll try to explain how this works by means of a simple "case study." The illustration is hypothetical, but I can assure you that, in its basic features, the example captures quite well some recent experience with ethylene hydrogenation on rhodium (Yue 1989).

The Problem

Consider an isothermal CFSTR involving heterogeneous catalysis. In the reactor chamber, a low-pressure gas phase is in contact with a catalytic surface that promotes the overall reaction

$$A + B \rightarrow C.$$

A feed stream carrying reactants A and B enters the reactor, and an effluent stream carrying A, B, and C is withdrawn. The two streams are composed largely of an inert carrier gas so that, at steady state, their volumetric flow rates are virtually identical. In suitably chosen units, the residence time of the reactor is 1.0, and the molar concentrations of A and B in the feed are $c_A{}^f = 1.0$ and $c_B{}^f = 0.8$.

Under identical operating conditions, two different steady states are observed (corresponding to different startup procedures). In each of the steady states, concentrations of A, B, and C are measured in the effluent stream. Moreover, a sensor gives qualitative information about the relative amounts of A on the surface in the two steady states. (In the ethylene hydrogenation experiments, Yue determined the relative amounts of surface hydrogen in the two steady states by means of electrochemical methods developed by my colleague, Howard Saltsburg.)

The data are summarized in Table 10–1.

Three mechanisms for the chemistry on the catalyst surface are proposed. These are displayed as (M1)–(M3):

$$\begin{gathered} A + S \rightleftarrows AS \\ B + S \rightleftarrows BS \\ AS + BS \rightarrow C + 2S \end{gathered} \tag{M1}$$

$$\begin{gathered} A + S \rightleftarrows AS \\ B + 2S \rightleftarrows BS_2 \\ AS + BS_2 \rightarrow C + 3S \end{gathered} \tag{M2}$$

$$\begin{gathered} A + S \rightleftarrows AS \\ B + S \rightleftarrows BS \\ A + BS \rightarrow C + S \end{gathered} \tag{M3}$$

Here S denotes a vacant active site. In (M1), for example, the first two lines represent (reversible) binding of A and B from the gas phase onto single vacant sites. The third line represents a surface reaction whereby the product C is formed and immediately enters the gas phase, leaving behind two vacant sites. Mechanisms (M1) and (M2) are of the Langmuir–Hinshelwood family, while (M3) is of the Eley–Rideal type, involving a gas-phase attack of A on surface-bound B.

Table 10-1 Two Steady States for a Catalytic Experiment

Gas Phase Data	
Steady State No. 1	Steady State No. 2
$c_A = 0.6$	$c_A = 0.3$
$c_B = 0.4$	$c_B = 0.1$
$c_C = 0.4$	$c_C = 0.7$
Surface Data	
More A on surface in Steady State No. 1 than in Steady State No. 2	

Each mechanism is presumed to be governed by mass action kinetics, and, in both steady states, concentrations of adsorbates and of vacant sites are presumed to be spatially uniform over the catalyst surface. Mass transfer resistance in the gas phase is assumed negligible.

For each mechanism, one can write (in terms of unknown rate constants) a system of six coupled polynomial equations that govern the steady-state concentrations of A, B, and C in the gas phase of vacant sites, adsorbed A and adsorbed B on the catalyst surface. Using values prescribed for the residence time and for the concentrations of A and B in the feed stream, one can write the six steady-state equations corresponding to mechanism (M1), for example, in the form

$$
\begin{aligned}
0 &= (1/1.0)\cdot(1.0 - c_A) - \alpha c_A c_S + \beta c_{AS} \\
0 &= (1/1.0)\cdot(0.8 - c_B) - \gamma c_B c_S + \varepsilon c_{BS} \\
0 &= -(1/1.0)\cdot c_C + \nu c_{AS} c_{BS} \\
0 &= -\alpha c_A c_S + \beta c_{AS} - \gamma c_B c_S + \varepsilon c_{BS} + 2\nu c_{AS} c_{BS} \\
0 &= \alpha c_A c_S - \beta c_{AS} - \nu c_{AS} c_{BS} \\
0 &= \gamma c_B c_S - \varepsilon c_{BS} - \nu c_{AS} c_{BS}.
\end{aligned}
\qquad (10\text{–}8)
$$

Here α, β, γ, ε, and ν are the (unknown) rate constants for the five reactions in (M1). One can write a similar system for each of the mechanisms (M2) and (M3).

We might hope that distinctions among mechanisms (M1)–(M3) could be drawn simply on the basis of their *qualitative* capacity to admit multiple steady states. For example, we might hope that the system (10–8) is incapable of admitting two solutions (having the same total site concentration), no matter what positive values the rate constants take. In this case we could deny the viability of (M1) on the grounds that the mechanism is *qualitatively* inconsistent with the steady-state multiplicity observed.

Deficiency one theory, however, indicates that (M1)–(M3) *all* have the capacity for multiple steady states in an isothermal CFSTR context: For each there are rate constants such that the corresponding steady-state equations have multiple solutions. Thus, *the mere qualitative capacity for multiple steady states cannot provide a basis for discriminating among mechanisms* (M1)–(M3).

But there is more to Table 10–1 than meets the eye. However fragmentary the data there might seem, Table 10–1 carries crucial *quantitative* information—information that permits the formulation of a sharpened question: For which of the mechanisms (M1)–(M3) are there rate constant values such that the corresponding steady–state equations admit two solutions *consistent with the entries in Table 10–1*? It turns out that this question provides the basis for acute distinctions among (M1)–(M3), but its resolution is not a simple matter.

It is here that the notion of *mechanistic signature* comes into play. In order that the idea might be suitably motivated, I want to digress a little by making an (imperfect) analogy with batch reactors.

Motivation: How Chemical Mechanisms Reveal Themselves in the Equilibria of Homogeneous Batch Reactors

There will be nothing new in this section. But, by way of preparation for the next section, it will be useful to recall the extent to which mechanisms (independent of rate constant values) shape—and reveal themselves in—the locus of equilibria for batch reactors.

Consider, then, an isothermal homogeneous (well-stirred) constant volume batch reactor in which there are four species, A, B, C, and D. The species interact through a network of elementary reactions that is to be determined. Suppose that the reactor is permitted to come to equilibrium and that measurements of the molar concentrations yield values c_A^*, c_B^*, c_C^*, and c_D^*. The reactor is then opened, more A is added, and another equilibrium is realized. At the second equilibrium, measurements yield new values for the species concentrations, c_A^{**}, c_B^{**}, c_C^{**}, and c_D^{**}.

We are interested in the extent to which the pair of measured equilibria give clues about the underlying network of elementary reactions. More precisely, we shall want to know how to tell which candidate mechanisms (taken with mass action kinetics) are compatible with the two measurements taken.

Consider the following candidate mechanism (with rate constants indicated by Greek letters alongside the reaction arrows):

$$\mathrm{A} + \mathrm{B} \underset{\beta}{\overset{\alpha}{\rightleftarrows}} \mathrm{C} \underset{\delta}{\overset{\gamma}{\rightleftarrows}} 2\mathrm{D}. \tag{10–9}$$

The differential equations for the species concentrations are

$$\begin{aligned} \dot{c}_{\mathrm{A}} &= -\alpha c_{\mathrm{A}} c_{\mathrm{B}} + \beta c_{\mathrm{C}} \\ \dot{c}_{\mathrm{B}} &= -\alpha c_{\mathrm{A}} c_{\mathrm{B}} + \beta c_{\mathrm{C}} \\ \dot{c}_{\mathrm{C}} &= \alpha c_{\mathrm{A}} c_{\mathrm{B}} - (\beta + \gamma) c_{\mathrm{C}} + \delta c_{\mathrm{D}}^2 \\ \dot{c}_{\mathrm{D}} &= 2\gamma c_{\mathrm{C}} - 2\delta c_{\mathrm{D}}^2. \end{aligned} \tag{10–10}$$

From these it can be determined easily that equilibrium concentrations are characterized by the equations

$$(\alpha/\beta) c_{\mathrm{A}} c_{\mathrm{B}} = c_{\mathrm{C}} = (\delta/\gamma) c_{\mathrm{D}}^2. \tag{10–11}$$

Now let the numbers μ_{A}, μ_{B}, μ_{C}, and μ_{D} be constructed from the two measured equilibria according to the prescription

$$\mu_\theta := \ln(c_\theta{}^{**}/c_\theta{}^{*}), \quad \theta = \mathrm{A, B, C, D}. \tag{10–12}$$

From (10–11) it is easily determined that, if the mechanism is to be viable, these numbers must satisfy the relations

$$\mu_{\mathrm{A}} + \mu_{\mathrm{B}} = \mu_{\mathrm{C}} = 2\mu_{\mathrm{D}}. \tag{10–13}$$

In other words, the candidate mechanism shown in (10–9) is compatible with the pair of measured equilibria only if the numbers μ_{A}, μ_{B}, μ_{C}, and μ_{D} are consistent with the stringent conditions set by (10–13).

Note that (10–13) derived from the reaction diagram (10–9) *but makes no explicit reference to rate constants associated with the various reactions.* This is to say that (10–13) is an attribute *of the mechanism itself*—a "*signature*" of the mechanism that must be respected by all pairs of (mass action) equilibria consistent with it.

It will be useful to view the passage from the mechanism (10–9) to its signature relations (10–13) in a certain way: In language of Horn and Jackson (1972), the *complexes* of a reaction network are the objects that appear before and after the reaction arrows. Thus, the complexes in the simple network (10–9) are

$$\{A + B, C, 2D\}.$$

Now with the species A, B, C, and D of the network we associate the variables μ_A, μ_B, μ_C, and μ_D, and with the complexes we associate in an obvious way the linear forms

$$\{\mu_A + \mu_B, \mu_C, 2\mu_D\}.$$

The passage from mechanism (10–9) to its signature relations in (10–13) can be viewed as having resulted from a joining by *equal signs* of the linear forms corresponding to the complexes.

For catalytic CFSTRs, it will turn out that there is a similar passage from a mechanism to its signature, but the joining of linear forms corresponding to the complexes will, more often than not, be by way of *inequality signs*.

How Catalytic Mechanisms Reveal Themselves in Isothermal CFSTR Multiplicity Data

Deficiency one theory provides a natural and systematic way to carry the simple (but powerful) ideas of the last section from simple homogeneous batch reactors to reactors of greater complexity, including catalytic CFSTRs. More detail is given in (Feinberg 1987, 1988).

In order that mechanistic signatures for catalytic CFSTRs might be seen as natural extensions of those for batch reactors, it is important to understand how CFSTRs are described in reaction network terms. [The procedure and rationale are described in (Feinberg 1987).] Here it suffices to say that the catalytic CFSTR under consideration is given a network description by augmenting the (presumed) chemical mechanism with "reactions" $0 \rightarrow A$ and $0 \rightarrow B$ to account for the presence of A and B in the feed and by adding "reactions" $A \rightarrow 0$, $B \rightarrow 0$, and $C \rightarrow 0$ to account for the presence of A, B, and C in the effluent.[1] Thus, if the operative chemical mechanism is (M1),

[1] The reaction $0 \rightarrow A$ models a zeroth order (constant) addition of A to the reactor chamber via the feed stream, while $A \rightarrow 0$ models a first-order removal via the effluent stream.

then the CFSTR is encoded in the reaction network

$$
\begin{gathered}
A + S \rightleftarrows AS \\
B + S \rightleftarrows BS \\
AS + BS \rightarrow C + 2S \\
A \rightleftarrows 0 \rightleftarrows B \\
\uparrow \\
C.
\end{gathered}
\tag{M1'}
$$

In a similar fashion, one can formulate networks (M2′) and (M3′) that encode CFSTRs in which (M2) and (M3) are the operative chemical mechanisms.

In (M1′) the symbol "0" indicates the *zero complex*, a complex devoid of any species. Thus, the species in (M1′) are

$$\{A, B, C, S, AS, BS\}, \tag{10–14}$$

and, in the language of the section on Motivation, the complexes are

$$\{A + S, AS, B + S, BS, AS + BS, C + 2S, A, B, C, 0\}. \tag{10–15}$$

In order to describe the sense in which deficiency one theory gives rise to mechanistic signatures for catalytic CFSTRs, we shall assume, for the sake of concreteness, that the operative mechanism in the reactor under consideration is (M1) [so that the CFSTR is encoded by (M1′)]. In this case, we associate with the species set (10–14) a set of numbers

$$\{\mu_A, \mu_B, \mu_C, \mu_S, \mu_{AS}, \mu_{BS}\}, \tag{10–16}$$

where

$$\mu_\theta := \ln(c_\theta{}^{**}/c_\theta{}^{*}), \quad \theta = A, B, C, S, AS, BS. \tag{10–17}$$

Here $c_\theta{}^*$ is the (perhaps unknown) concentration of species θ in the first steady state, and $c_\theta{}^{**}$ is its (perhaps unknown) concentration in the second steady state. By way of analogy to the batch reactor case, we associate with the complexes (10–15) a set of linear forms

$$\{\mu_A + \mu_S, \mu_{AS}, \mu_B + \mu_S, \mu_{BS}, \mu_{AS} + \mu_{BS}, \mu_C + 2\mu_S, \mu_A, \mu_B, \mu_C, 0\}.$$

Deficiency one theory (Feinberg 1988) *indicates how these should be joined by signs in the set* $\{<, =, >\}$ *to give systems of relations that must be satisfied if* (M1′) *is to be compatible with whatever CFSTR multiplicity data are available.*

In fact, the theory gives *two* essentially distinct ways in which the network (M1′) can "sign its name" in CFSTR multiplicity data. *In order for the data to be consistent with* (M1′), *it is necessary that* $\mu_A, \mu_B, \mu_C, \mu_S, \mu_{AS}$, *and* μ_{BS} *be consistent with at least one of the following "signatures"*:

$$\begin{gathered}\{\mu_A, \mu_B\} < 0 < \mu_C = \mu_{AS} + \mu_{BS}\\ \mu_{AS} > \mu_A + \mu_S > \mu_{AS} + \mu_{BS} > \mu_B + \mu_S > \mu_{BS}\end{gathered} \tag{10–18a}$$

or

$$\begin{gathered}\{\mu_A, \mu_B\} < 0 < \mu_C = \mu_{AS} + \mu_{BS}\\ \mu_{AS} < \mu_A + \mu_S < \mu_{AS} + \mu_{BS} < \mu_B + \mu_S < \mu_{BS}.\end{gathered} \tag{10–18b}$$

Here $(\mu_A, \mu_B) < 0$ indicates that both μ_A and μ_B are to be less than zero.[2]

In a similar way, deficiency one theory gives signatures for the networks (M2′) and (M3′), the CFSTR counterparts of (M2) and (M3). [Unlike (M1′) and (M2′), network (M3′) has only one signature.] Signatures corresponding to the three networks are shown in Table 10–2.

The question now becomes this: Which (if any) of the signatures in Table 10–2 are consistent with the data shown in Table 10–1? By virtue of (10–17), we can calculate from the gas phase information in Table 10–1 that

$$\begin{aligned}\mu_A &= \ln(0.3/0.6) \approx -0.693\\ \mu_B &= \ln(0.1/0.4) \approx -1.386\\ \mu_C &= \ln(0.7/0.4) \approx 0.560.\end{aligned} \tag{10–19}$$

Moreover, from the surface information in Table 10–1, we can infer that

$$\mu_{AS} = \ln(c_{AS}^{**}/c_{AS}^{*}) < 0. \tag{10–20}$$

There is no similar information available about μ_S or μ_{BS} [μ_{BS2} in the case of (M2′)], but there is still something we can say: Since there is more A on the surface in the first steady state than in the second, site conservation

[2] The inequality systems that constitute the signature(s) are those that admit Step 8 solutions in the deficiency one algorithm (Feinberg 1988). Strictly speaking, there are two other signature systems for (M1′)—those obtained by reversing all the inequality signs in (10–18a) and (10–18b). However, the resulting signatures are essentially identical to those already indicated [up to an interchange of the roles of the two steady states in (10–17)].

Table 10–2 Signature Inequalities for Three Catalytic Mechanisms

Network	Signature(s)
(M1′)	
$A + S \rightleftarrows AS$	$\{\mu_A, \mu_B\} < 0 < \mu_C = \mu_{AS} + \mu_{BS}$
$B + S \rightleftarrows BS$	$\mu_{AS} > \mu_A + \mu_S > \mu_{AS} + \mu_{BS} > \mu_B + \mu_S > \mu_{BS}$
$AS + BS \rightarrow C + 2S$	or
$A \rightleftarrows 0 \rightleftarrows B$	
$\uparrow$	$\{\mu_A, \mu_B\} < 0 < \mu_C = \mu_{AS} + \mu_{BS}$
C	$\mu_{AS} < \mu_A + \mu_S < \mu_{AS} + \mu_{BS} < \mu_B + \mu_S < \mu_{Bs}$
(M2′)	
$A + S \rightleftarrows AS$	$\{\mu_A, \mu_B\} < 0 < \mu_C = \mu_{AS} + \mu_{BS2}$
$B + 2S \rightleftarrows BS_2$	$\mu_{AS} > \mu_A + \mu_S > \mu_{AS} + \mu_{BS2} > \mu_B + 2\mu_S > \mu_{BS2}$
$AS + BS_2 \rightarrow C + 3S$	or
$A \rightleftarrows 0 \rightleftarrows B$	
$\uparrow$	$\{\mu_A, \mu_B\} < 0 < \mu_C = \mu_{AS} + \mu_{BS2}$
C	$\mu_{As} < \mu_A + \mu_S < \mu_{AS} + \mu_{BS2} < \mu_B + 2\mu_S < \mu_{BS2}$
(M3′)	
$A + S \rightleftarrows AS$	
$B + S \rightleftarrows BS$	$\{\mu_A, \mu_B\} < 0 < \mu_C = \mu_A + \mu_{BS} < \mu_B + \mu_S < \mu_{BS}$
$A + BS \rightarrow C + S$	$\mu_A + \mu_S = \mu_{AS}$
$A \rightleftarrows 0 \rightleftarrows B$	
$\uparrow$	
C	

requires that there be more surface-adsorbed B or more vacant sites in the second steady state than in the first. From this and (10–17) it follows that

$$\textit{either } \mu_S \textit{ or } \mu_{BS}\ [\mu_{BS2}] \textit{ is positive.} \tag{10–21}$$

(Here, of course, it is not precluded that both μ_S and μ_{BS} [μ_{BS2} in the case of (M2′)] might be positive.)

It turns out that *information contained in* (10–19)–(10–21) *is not compatible with the signature for* (M3′) *shown in Table 10–2,*[3] *nor is it compatible with either of the signatures shown for* (M1′). That is, *there are no values for* μ_{AS}, μ_{BS}, *and* μ_S *which are consistent with* (10–20) *and* (10–21) *and which, when supplemented with values for* μ_A, μ_B, *and* μ_C *in* (10–19), *satisfy the indicated signature relations.*

In the case of network (M1′), for example, deficiency one theory then says this: *There is no set of (positive) rate constant values for which the balance equations* (10–14) *admit two steady states that are consistent with site conservation and with the data in Table 10–1. In particular, no matter how rate constants are assigned, there are no values of the (unmeasured) surface concentrations* c_{AS}, c_{BS}, *and* c_S *in the two steady states which are simultaneously consistent with site conservation, with Table 10–1, and with (10–14).* A similar statement can be made for network (M3′).

The situation for reaction network (M2′) is very different. The second of its signatures in Table 10–2 admits solutions consistent with (10–19)–(10–21), for example,

$$\begin{aligned} \mu_A &= -0.693 & \mu_S &= 1.066 \\ \mu_B &= -1.386 & \mu_{AS} &= -0.373 \\ \mu_C &= 0.560 & \mu_{BS2} &= 0.933 \end{aligned} \qquad (10\text{–}22)$$

In this case, deficiency one theory indicates that mechanism (M2′) is consistent with the data in Table 10–1 in the following sense: *There are values for the (unknown) rate constants in mechanism* (M2) *and values of the (unmeasured) surface concentrations* c_{AS}, c_{BS}, *and* c_S *in the two steady states such that*

(i) $c_{AS}{}^* > c_{AS}{}^{**}$,
(ii) *Site conservation is respected—that is,*

$$c_S{}^* + c_{AS}{}^* + 2c_{BS2}{}^* = c_S{}^{**} + c_{AS}{}^{**} + 2c_{BS2}{}^{**}.$$

(iii) *Taken with the gas phase data in Table 10–1, the balance equations for* (M2) [*analogous to* (10–8)] *are satisfied in both steady states.*

[3] It is interesting to note that, *even in the absence of the surface information in Table 10–1,* mechanism (M3′) is inconsistent with the remaining data. [In this case, (10–20) would be dropped, and (10–21) would be replaced by weaker requirement that μ_{AS}, μ_{BS}, and μ_S not all of the same sign.]

In fact, deficiency one theory gives sample values of the (unknown) rate constants and the (unmeasured) surface concentrations for which consistency with both Table 10–1 and the balance equations are ensured: The (computer-generated) rate constants

$$\begin{aligned} A + S &\underset{0.185516}{\overset{20.21862}{\rightleftarrows}} AS \\ B + 2S &\underset{5.861598}{\overset{871.00869}{\rightleftarrows}} BS_2 \\ AS + BS_2 &\overset{8.399984}{\rightarrow} C + 3S \end{aligned} \tag{10–23}$$

give rise to the following pair of steady states:

Steady State No. 1	Steady State No. 2
$c_A = 0.6$	$c_A = 0.3$
$c_B = 0.4$	$c_B = 0.1$
$c_C = 0.4$	$c_C = 0.7$
$c_S = 0.045833$	$c_S = 0.133116$
$c_{AS} = 0.840911$	$c_{AS} = 0.579062$
$c_{BS2} = 0.056628$	$c_{BS2} = 0.143911$

Note that these are consistent with Table 10–1 and respect site conservation. (In both steady states we have $c_S + c_{As} + 2c_{BS2} = 1.$)

Values shown for the rate constants and the surface concentrations are merely exemplary. Deficiency one theory cannot, of course, "know" these things [or even whether (M2) is, in fact, the operative catalytic mechanism]. But, at least in principle, it has the capacity to deliver all combinations rate constants for (M2) and surface concentrations in the two steady states that are compatible with Table 10–1.

In any case, *deficiency one theory can know only what it is told.* In this case it has been told only the information shown in Table 10–1. Were more detailed surface measurements in the two steady states made available, the theory would serve to narrow the range of rate constants for which (M2) is compatible with the augmented data. On the other hand, the theory might, on the basis of the new information, deny the viability of (M2) altogether [just as it denied the viability of (M1) and (M3) on the basis of Table 10–1 alone].

At this point, it is important to stand back and see what lessons the example teaches. All three of the catalytic mechanisms (M1)–(M3) are consistent with the overall reaction A + B → C, and all have the qualitative capacity to admit multiple steady states in an isothermal CFSTR setting. At first glance, then, it would seem that there is little in Table 10–1 that could serve to distinguish one mechanism from another. But, as we have seen, Table 10–1 is ripe with quantitative detail. The information there, however fragmentary, is inconsistent with the CFSTR signatures of either mechanism (M1) or (M3). To the extent that these mechanisms are correctly reflected in equations such as (10–8),[4] they cannot account for the quantitative observations. If there is a lesson here it is this: Taken with the idea of mechanistic signature, a little data carries a lot of information.

The example was a hypothetical one, but, as I said before, it reflects quite accurately M.-H. Yue's (1989) experience with ethylene hydrogenation on rhodium. If anything, deficiency one theory was more incisive in the real case than in the example. *None* of the classical ethylene hydrogenation candidate mechanisms had signatures consistent with the multiplicity data obtained, not even when gas-phase mass transfer resistance was taken into account. Consistency with the data could be realized only after construction of (speculative) models involving the formation of ethylidine islands on the catalyst surface (Yue, 1989).

CONCLUDING REMARKS

With respect to the design and operation of reactors as complex as Mobil's, I don't think that we in academic life can offer anything to replace the experience that Mobil has acquired over the years. Evolution in engineering, as in biology, can do wondrous things.

But evolution in reaction engineering plays itself out against a conceptual background. That background can, I think, be enriched substantially. In particular, I am convinced that there is still a great deal to be learned about the relationship between reactor behavior and the structure of the underlying network of chemical reactions. As I said at the outset, the differential equations that derive from chemical reaction networks have very special and often remarkable properties. Once those properties are better understood, I

[4] The passage from a mechanism such as (M1) to a system of equations such as (10–8) is based upon assumptions which should be regarded as an intrinsic part of the mechanistic picture. Among these is the presumption of steady states in which adsorbed species are distributed uniformly over the catalyst surface and in which mass transfer resistance from the gas phase to the catalyst surface is negligible. *There are ways to augment reaction networks such as those shown in Table 10–2 to incorporate departures from these conditions.* See, for example, Leib et al. (1988) and Yue (1989).

am certain that the conceptual background against which we study particular reactors will be far more tolerant of complexity, perhaps even of great complexity.

It is said that reaction engineering is a mature discipline. Yet, it was only thirty years ago that people such as Neal Amundson, Rutherford Aris, and James Wei began to put the subject on a firm mathematical footing. It seems to me that we are only getting started.

ACKNOWLEDGMENT

Work described here was supported by the National Science Foundation.

REFERENCES

Feinberg, M. (1987) Chemical reaction network structure and the stability of complex isothermal reactors: I. The deficiency zero and deficiency one theorems. *Chem. Engin. Sci.* **42**:2229.

Feinberg, M. (1988) Chemical reaction network structure and the stability of complex isothermal reactors: II. Multiple steady states for networks of deficiency one. *Chem. Engin. Sci.* **43**:1.

Feinberg, M. (1989) Necessary and sufficient conditions for detailed balancing in mass action systems of arbitrary complexity. *Chem. Engin. Sci.* **44**:1819.

Feinberg, M. (1991) Some recent results in chemical reaction network theory. In *Patterns and Dynamics in Reactive Media*, IMA Volumes on Mathematics and its Applications, Springer-Verlag, Berlin, Heidelberg, New York.

Horn, F. J. M. and Roy Jackson (1972) General mass action kinetics. *Arch. Rat. Mech. Anal.* **47**:81.

Krambeck, F. J. (1970) The mathematical structure of chemical kinetics in single phase systems. *Arch Rat. Mech. Anal.* **38**:317.

Leib, T. M., D. Rumschitzki, and M. Feinberg (1988) Multiple steady states in complex isothermal CFSTRs: I. General considerations. *Chem. Engin. Sci.* **43**:321.

Schlosser, P. (1988) A graphical determination of the possibility of multiple steady states in complex isothermal CFSTRs. PhD Thesis, Department of Chemical Engineering, University of Rochester.

Schlosser, P. and M. Feinberg (1987) A graphical determination of the possibility of multiple steady states in complex isothermal CFSTRs. In *Complex Chemical Reaction Systems*, J. Warnatz and W. Jager (eds.). Springer-Verlag, Berlin, Heidelberg, New York, pp. 102–115.

Yue, M.-H. (1989) Isothermal multiple steady states in ethylene hydrogenation over rhodium: Mechanistic screening and experimental studies. PhD Thesis, Department of Chemical Engineering, University of Rochester.

11

Development of Chemical Reaction Models

MICHAEL FRENKLACH

The current quest for "high technology," like in the areas of environment control and intensified search for new materials, places a demand for rapid and more sophisticated approaches to technology development. A common practice of "rules of thumb" is being replaced by the development of computer-aided-design (CAD) codes. In the fields associated with chemical transformations, the process description based on global modeling, when only a few overall reactions are considered, is not adequate and it is increasingly being sought in the form of detailed reaction mechanisms. Not only is this latter approach physically more realistic, but also the models it offers possess a much wider range of applicability, i.e., these models can be *extrapolated* to a variety of conditions, not necessarily those on which premises the model was developed initially.

The development of chemical reaction models involves a number of steps: compilation of a specific reaction network, collection and estimation of reaction rate parameters, analysis and verification of the composed reaction mechanism, and reduction of the model to the form and size suitable for practical applications. The first two parts rely on chemical insights and intuition of the modeler. The other two require rigorous mathematical methods. Although mostly heuristic in nature (Frenklach 1984), the development of reaction models can and should follow certain established procedures. In this chapter, an approach to reaction model development is presented and illustrated with examples taken primarily from the areas of hydrocarbon combustion. Several numerical techniques developed in our laboratory for application to model development and mechanism reduction will be discussed in more detail.

MODELING OBJECTIVES

The modeling should begin by clearly realizing what are the objectives the modeler is trying to achieve. The typical objectives are the following (Frenklach 1990).

Quantitative prediction of system responses. These are models developed for practical applications, such as computer-aided design of chemical reactors or forecast of pollution levels. The principal requirement for these models is a given prediction accuracy. The model itself can take any form as long as it delivers results with the required accuracy. The challenge is to make it least computer intensive so that it is suitable for computer-aided design applications.

Quantitative or qualitative analysis of certain experimental or theoretical relationships, dependencies, and trends with "established" models. Examples of these could be analysis of the chemical–kinetic—gas dynamic coupling in a reactive flow, simulation of flame stretching phenomena, testing a sooting correlation, or verification of a concept for modeling turbulence. The models used for these purposes have to contain a detailed description of the phenomenon in question, so that controlling factors can be identified. The numerical accuracy of model prediction is not necessarily at issue here. For instance, a model predicting the amounts of soot formed to within a factor of 2 will certainly be suitable for the determination of factors controlling the critical C-to-O ratios for the onset of sooting.

Parameter determination for well-established models. The most familiar example here is the determination of rate coefficients in chemical kinetic studies. It is not only necessary to establish the best-fit values of the "unknown" parameters, but also to estimate their uncertainties in rigorous statistical terms, i.e., to determine the joint confidence region, and to examine the correlations between the parameter values. The models in this category should include all *active* parameters, i.e., those rate coefficients, equilibrium constants, transport coefficients, etc., to which available experimental data exhibit significant sensitivity.

Exploratory modeling with "unknown" models. This could mean, for example, identification of possible reaction pathways leading to soot formation in hydrocarbon flames or nucleation of diamond in vapor-activated deposition processes. The focus here is on discovering the unknown by extrapolating the present knowledge. These models, therefore, should include

.s many details (i.e., chemical reactions and species) as possible in order not o miss important features.

Once the objective is set, the next step is to establish a strategy of achieving t. It is pertinent to mention here that the desire, for instance, to develop a mall-size model does not necessarily imply beginning with a small reaction nechanism; the strategy advocated here is to first develop a more compreensive reaction network, test its accuracy, and then reduce it to the size nd mathematical form suitable for practical applications.

:OMPOSITION OF A MECHANISM

:omposing the initial reaction mechanism, referred to as the *first-trial* nechanism, one begins by collecting information available: earlier reaction nechanisms proposed in the literature, related reaction mechanisms, and ompilations of pertinent reactions. A number of them in the areas of igh-temperature hydrocarbon chemistry have appeared in recent years (e.g., Ianson and Salimian 1984; Steinfeld et al. 1989; Tsang and Hampson 1986; 'zang 1987, 1988, 1990; Warnatz 1984; Westbrook and Dryer 1984; Westley t al. 1987). However, putting it all together may not (and most likely will ot) produce good results immediately. There are several reasons for this.

First, combining the information from several different sources may be nconsistent. For example, consider that one source assigns reaction products f methyl oxidation as

$$CH_3 + O_2 \rightarrow CH_2O + OH$$

nd another as

$$CH_3 + O_2 \rightarrow CH_3O + O.$$

ncluding both reactions in the mechanism is equivalent to increasing he rate coefficient by a factor of 2. This reaction may indeed have the two hannels, but their rate coefficients should add up to a total (experimental) alue. Another common source of errors is the use of inconsistent thermoynamic and rate constants. It is advisable, therefore, that the reaction nechanism is constructed following the principle of detailed balancing. The hermodynamic data can be found in several sources (e.g., Benson 1986; 3urcat 1984; Alberty, this volume; Smith, this volume).

Second, the mechanism composed of only "known" reactions may be (and isually is) incomplete for a specific objective chosen. Obviously, only further inetic experiments can provide the required information. Nonetheless, hould one wait until the missing rate coefficients are measured? No, because the missing data may be safely estimated within an order of magnitude. Several techniques can be employed for this: Benson's rules (Benson 1976), quantum-chemical calculations (Steinfeld et al. 1989), or simply chemical analogy and collision theory. The latter approach is particularly beneficial when the reaction network includes repetitive sequences of reactions, such as thermal degradation of a large-size polymer (Froment, this volume) or the build-up of polycyclic aromatic hydrocarbons (Frenklach et al. 1985). In such cases, the reactions are grouped into chemically similar classes, such as a class of hydrogen abstractions from olefinic C–H bonds, and each class is assigned a single rate coefficient (the size dependence of the rate coefficient can be taken into account, if necessary). The grouping of reactions into classes also provide a vehicle to furnish the "missing" reactions. For instance, knowing that a free hydrogen atom can abstract a hydrogen atom from acetylene,

$$C_2H_2 + H \rightarrow C_2H + H_2,$$

it is reasonable to assume that a similar abstraction can take place with a larger molecule as well,

$$R{-}C{\equiv}C{-}H + H \rightarrow R{-}C{\equiv}C\cdot + H_2,$$

and with approximately the same rate. Here R represents a hydrocarbon radical.

Finally, the third reason rests in the interactions between assumed parameter values, kinetic and transport, and the correlation of their uncertainties. That is, being an experimentally determined property, each parameter value has associated with it an uncertainty. Combining reactions into a mechanism, these uncertainties then form a hypercube in the parameter space. The central point of the hypercube, whose coordinates represent the "best fit values" of the individual parameters, is not necessarily the best-fit-point for the complete mechanism (see, e.g., Gardiner et al. 1987). Each point of the hypercube is equiprobable, and for a high dimension of the hypercube the probability that a randomly selected point represents the "right" combination of parameter values should be extremely low. That is, the composed mechanism must be optimized, or "tuned," to reproduce a selected set of experimental data within the ranges of parameter uncertainties. The latter are usually misconstrued to be merely percentage-magnitude "random errors" of individual measurements instead of being correctly ascribed to systematic errors, where the disagreement is typically between a factor of 2 and 10 for various determinations of the same parameter. We are facing what is generally referred to as the inverse problem.

Thus, the mechanism combined initially of "known" reactions should be carefully screened for consistency and omissions. Thermodynamic and rate parameters for the "missing reactions" should be estimated to within order of magnitude; this should be sufficient at the initial stage of model development, because many of the parameters may appear later to have no measurable effects on the modeling responses. Those parameters identified in a subsequent analysis to have high sensitivities should be subjected to a more scrupulous estimation. The composed reaction mechanism may contain "bugs"; for instance, a wrongly assigned rate coefficient may be too large, causing severe numerical difficulties. These, however, can be usually found by several random tests and cursory sampling of numerical results. We are interested in a systematic way of mechanism analysis and reduction, and this is the subject of the following discussion.

DETAILED MODEL REDUCTION

Not all of the reactions in the composed first-trial mechanism contribute significantly to specific responses of interest in a given modeling study and many can be safely removed. The question is how to identify these noncontributing reactions. Removing one reaction at a time is tedious and expensive for large reaction mechanisms. It is often advocated that formal sensitivity analysis be used for identification of "unimportant" reactions. However, it would be impractical to start analysis of the first-trial mechanism by a full-scale sensitivity analysis regardless of the numerical efficiency of the technique. More importantly, such an approach may not provide the required information. For instance, a high sensitivity of the concentration of a minor side product with respect to the rate coefficient of the reaction producing it does not mean that this reaction is contributing to the formation rate of a major product. On the other hand, a zero sensitivity does not necessarily identify a noncontributing reaction—the low value may indicate that the reaction is fast and not rate limiting for the response in question. Removing this reaction, based on its zero sensitivity, will send the modeling efforts into entirely wrong direction.

A traditional approach to mechanism development consists in analysis of reaction rates: given two parallel reactions with significantly different rates, the slower one can be neglected. However, it would be practically impossible to apply this simple rule to a large reaction mechanism with its complex network of reactions. Experience shows that an effective reduction strategy is to identify noncontributing reactions by comparing the individual reaction rates with the rate of a chosen reference reaction (Frenklach 1987, 1990; Frenklach et al. 1986). The reference reaction could be the rate-limiting step or the fastest reaction. It was further proposed (Frenklach et al. 1986) that the computations used to obtain the required rate data do not have to be performed with the actual full fluid-dynamic/chemical-kinetic model—the reduction can be achieved with a much simpler computation using a similar but greatly simplified geometry and conditions.

These ideas were demonstrated using convectively perturbed methane ignition (Frenklach et al. 1986). The reduction was based on the results computed with a simple constant-density model using a full detailed chemical reaction mechanism. The following criteria were applied to identify the noncontributing reactions

$$|R_i| < \varepsilon_1 |R_{\mathrm{rls}}|, \quad (11\text{–}1)$$

$$|R_i \Delta H_i| < \varepsilon_2 \dot{Q}_{\max}, \quad (11\text{–}2)$$

where R_i is the rate of reaction i, R_{rls} is the rate of the rate-limiting step (which for the most part was the reaction $H + O_2 \rightarrow OH + O$), ΔH_i is the heat of reaction i, $\dot{Q}_{\max}$ is the maximum value among all the terms $|R_i \Delta H_i|$ and ε_1 and ε_2 are chosen parameters considerably smaller than unity.

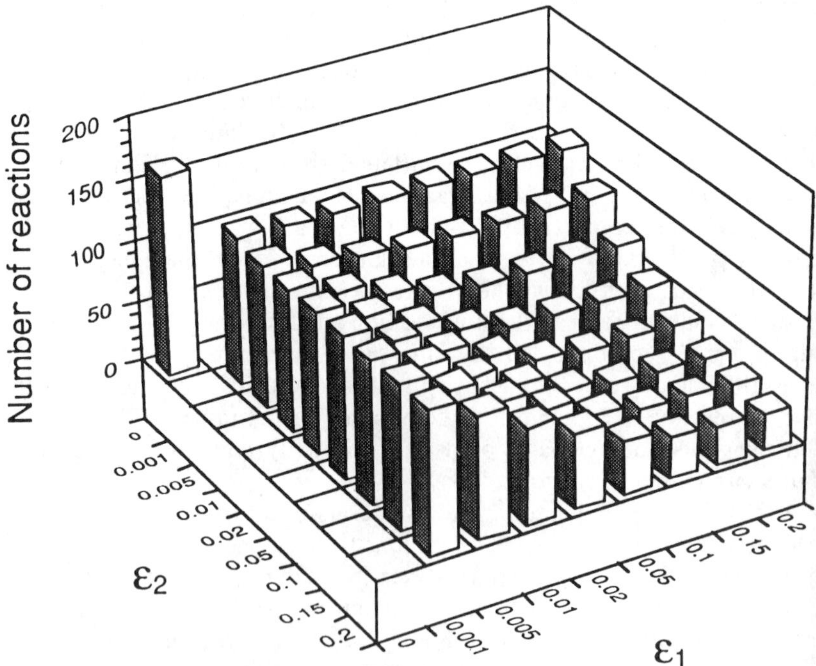

Figure 11–1 Number of reactions as a results of the detailed reduction with different values of ε_1 and ε_2 for a stoichiometric methane–air laminar premixed flame. (Preliminary results from Wang and Frenklach, 1990.)

Reactions whose corresponding rates satisfied inequalities (11–1) and (11–2) were removed from the mechanism. This procedure eliminates a specific species when all reactions of such a species happen to be removed. Three reduced reaction mechanisms were developed based on several assumed values of ε_1 and ε_2. The larger these values are, the smaller and less accurate the reduced mechanism becomes. The reduced reaction mechanisms were then used in a series of reactive flow simulations with and without a large-amplitude sinusoidal perturbation applied to a system that is initially quiescent and whose temperature is high enough to start ignition processes. The results showed that in every case the trends computed with the perturbation mimic those without them and the faithfulness of the numerical description with the reduced mechanisms increases with a decrease in the values of ε_1 and ε_2.

This method was recently applied to mechanism reduction of a premixed flame (Wang and Frenklach, 1990). The same two inequalities, (11–1) and (11–2), were applied to a zero-dimensional simulation describing the chemistry of an atmospheric methane–air flame. Preliminary results of this study are presented in Figures 11–1 and 11–2. Figure 11–1 depicts the reduction

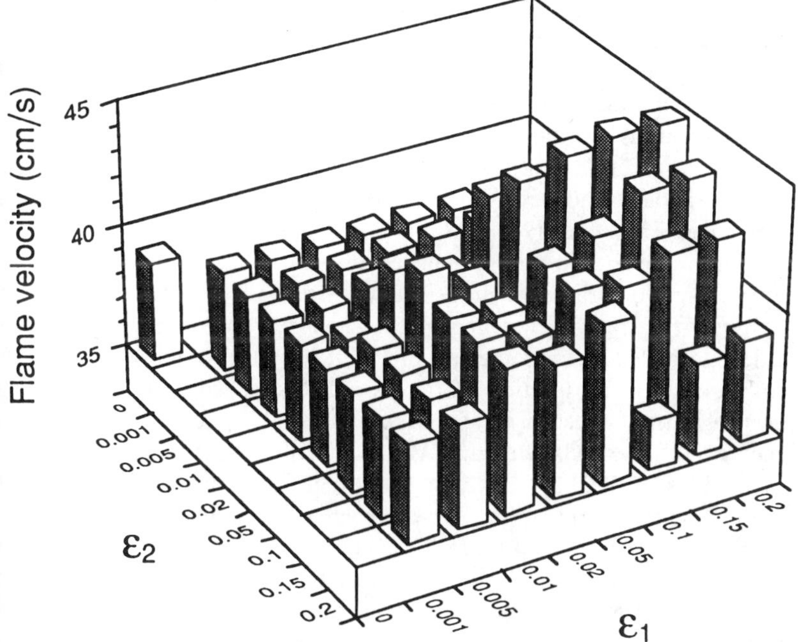

Figure 11–2 Computed velocity of a stoichiometric methane–air laminar premixed flame with different reduced reaction mechanisms. (Preliminary results from Wang and Frenklach, 1990.)

in size of the reaction mechanism with the increased values of ε_1 and ε_2, and Figure 11–2 shows the accuracy of the one-dimensional fluid-dynamic simulations of this flame performed using the Sandia laminar burner code (Kee et al. 1985) with the obtained reduced mechanisms.

RESPONSE MODELING OF A MODEL

After the noncontributing reactions are removed, and before the mechanism is used for various applications, it needs to be tested and optimized against the available experimental data. In other words, we are facing the problem of determining the optimal point of a hypercube in the parameter space which has been referred to as mechanism calibration or, more generally, a the inverse problem. The mechanism reduced in the manner described above still contains many chemical reactions and species, i.e., is still described by a large number of differential equations. The inverse problem for a dynamic model is not a trivial task. Compounding the difficulties are the large dimensionality of the hypercube and the vast span of the individual dimensions. For these reasons and the inherent nonlinearity in chemical reaction systems, the optimization objective function is highly structured, having multiple ridges and valleys (Frenklach 1984). Various algorithms based on linearization of the objective function (Bock 1981; Deuflhard and Nowak 1986; Milstein 1981; Rosenbrook and Story 1966) become ill-conditioned for such functions; optimization methods using linear sensitivity coefficients for gradient search (e.g., Rabitz et al. 1983) fall into the same category. A method capable of addressing the stated problem, referred to as *solution mapping* (see Frenklach 1984, 1987, 1990), is discussed below.

We first note that in considering reaction mechanism optimization, it is not necessary to include *all* model parameters (i.e., not all the rate coefficients). For the specific modeling objective chosen, the model responses do not depend on all the parameters. In fact it has been noted by many that usually only a small fraction of the parameters, called *active parameters*, show a significant effect on measured responses. This phenomenon has been termed *effect sparsity* by Box and Meyer (1985). The active parameters are immediately identified by a screening sensitivity analysis when the sensitivities are ranked according to their absolute values. In doing so, the parameters naturally separate into two groups: the active parameters, whose effects on the response(s) are above the experimental noise level, and those below it. Figure 11–3 presents an example of such ranking. Only active parameters need to be considered for optimization. Inclusion of parameters with noise level sensitivity into optimization only worsen the character of the objective function surface, necessarily increasing the dimensionality of the valley structure. If these parameters are to be the subject of optimization, then

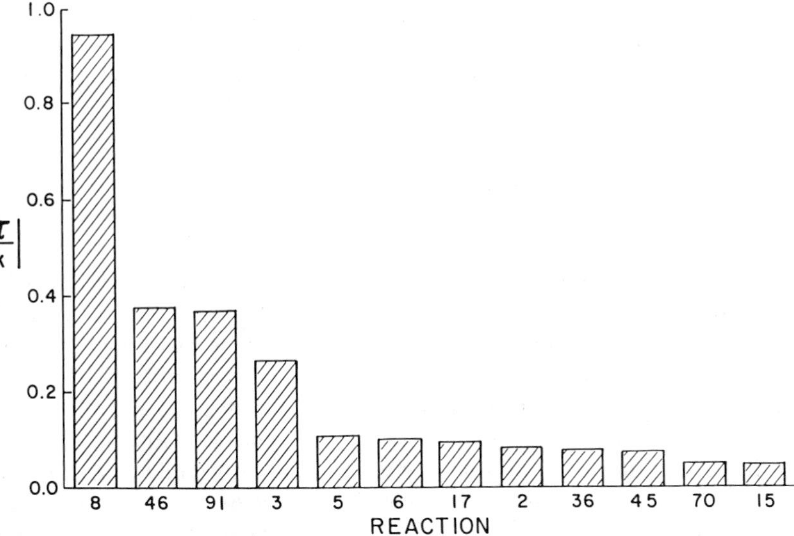

igure 11–3 Ranking of absolute sensitivities of induction time with respect o rate coefficients for methane oxidation (from Frenklach and Bornside 1984).

conditions have to be found where they become active. One does not have o search for a single condition in which all the parameters of interest are active. Instead, a strategy of performing experiments under conditions where different subsets of parameters are active, and combining the results into a joint optimization can be adopted.

The problem is then stated as follows. Given a system of (in this example, ordinary) differential equations,

$$\frac{d\boldsymbol{\eta}}{dt} = \mathbf{g}(t, \boldsymbol{\vartheta}), \qquad (11\text{–}3)$$

where $\boldsymbol{\eta}$ is the vector of state variables (species concentrations, temperature, etc.), t is the reaction time, $\boldsymbol{\vartheta}$ is the vector of active parameters (e.g., rate constants to be determined), and $\mathbf{g}$ is the vector of functions (expressing the mass-action law, energy balance, etc.), for a given set of initial conditions $\boldsymbol{\xi} = \boldsymbol{\eta}_{t=0}$, one is interested in determining the values of parameters $\boldsymbol{\vartheta}$ such that the solution of Eq. (11–3)

$$\boldsymbol{\eta} = \mathbf{f}(\boldsymbol{\vartheta}, \boldsymbol{\xi}, t),$$

matches the experimental observations, $\boldsymbol{\eta}^{obs}$. The mathematical statemen of this condition is minimization of an *objective function*, Φ, with respect tc parameters $\boldsymbol{\vartheta}$. The objective function is usually specified as a weighted leas squares,

$$\Phi(\boldsymbol{\vartheta}) = \sum_{i=1}^{r} \omega_i[\eta_i^{\text{obs}} - \eta_i(\boldsymbol{\vartheta})]^2, \quad (11\text{–}4$$

where r is the number of responses and ω_i is the statistical weight of the ith response.

A straightforward optimization of Eq. (11–4) can be envisioned as a search algorithm, within the multidimensional parameter space $\boldsymbol{\vartheta}$, with calls to a differential equation solver each time a new value of $\boldsymbol{\eta}(\boldsymbol{\vartheta})$ is required Although this approach is suitable for low-dimensionality systems with well-behaved functional dependence of Eq. (11–4), it would be hardly applic-able to problems of practical importance. In chemical kinetics, there are usually a large number of parameters with significant uncertainty, resulting in a large dimensionality and span of the $\boldsymbol{\vartheta}$ space. The surface of $\boldsymbol{\Phi}$ in the $\boldsymbol{\vartheta}$ space is typically highly structured—narrow, sinuous, gently sloping valleys exhibiting multiple local minima. Under these conditions, using even highly efficient search strategies, a very large number of function evaluations would be required to find a global minimum. When functions are evaluated by numerical integration, the search would place prohibitively large demands on computer time.

The essence of the solution mapping technique is approximation of responses $\boldsymbol{\eta}$ by simple algebraic expressions,

$$\mathbf{f}_t,\ \xi(\boldsymbol{\vartheta}) \approx \boldsymbol{\Psi}(\boldsymbol{\vartheta}), \quad (11\text{–}5)$$

within parameter space $\boldsymbol{\vartheta}$. The functions $\boldsymbol{\psi}(\boldsymbol{\vartheta})$ are developed by *computer experiments*. The objective function (11–4) is then reduced to

$$\Phi(\boldsymbol{\vartheta}) = \sum_{i=1}^{r} \omega_i[\eta_i^{\text{obs}} - \Psi_i(\boldsymbol{\vartheta})]^2,$$

that is, optimization no longer requires costly numerical integration of differential equations, but rather the simple evaluation of algebraic expres-sions $\boldsymbol{\psi}(\boldsymbol{\vartheta})$. There is, in principle, no restriction on the mathematical form of $\boldsymbol{\psi}(\boldsymbol{\vartheta})$. In our work we have used polynomials in $\boldsymbol{\vartheta}$ whose coefficients have been determined via computer experiments arranged in orthogonal factorial designs (Box and Draper 1987).

An example of a design matrix—a second-order orthogonal composite design based on a 2^{7-1} fractional factorial design—is shown in Table 11–1 (from Frenklach and Rabinowitz 1988). In this table the variables **X** are coded factorial variables, defined as

$$X = \frac{\ln(k/k')}{\ln f},$$

where for each subscript i, indicating the corresponding reaction, k is the rate coefficient, k' is the rate coefficient at the center point of the computer-experiment design, and f is the span of k. Responses in Table 11–1—$\tau_{\xi = \alpha}$,

le 11–1 A Design Matrix

mputer eriment No.	Variables[a]							Responses		
	X_1	X_2	X_3	X_4	X_5	X_6	X_7	τ_α (μs)	τ_β (μs)	$[CH_3]_\gamma$ (mol/m^3)
1	+1	+1	+1	+1	+1	+1	+1	54	703	0.785
2	−1	+1	+1	+1	+1	+1	−1	151	882	0.682
3	+1	−1	+1	+1	+1	+1	−1	42	605	0.921
4	−1	−1	+1	+1	+1	+1	+1	53	676	0.798
5	+1	+1	−1	+1	+1	+1	−1	150	1760	0.960
6	−1	+1	−1	+1	+1	+1	+1	174	2207	0.842
7	+1	−1	−1	+1	+1	+1	+1	31	1512	1.246
⋮	⋮	⋮	⋮	⋮	⋮	⋮	⋮	⋮	⋮	⋮
62	−1	+1	−1	−1	−1	−1	−1	347	3321	0.585
63	+1	−1	−1	−1	−1	−1	−1	140	1614	1.084
64	−1	−1	−1	−1	−1	−1	+1	149	1710	0.943
65	0	0	0	0	0	0	0	118	1403	0.812
66	+α	0	0	0	0	0	0	89	1245	9.272
67	−α	0	0	0	0	0	0	158	1555	7.411
68	0	+α	0	0	0	0	0	141	1504	0.763
69	0	−α	0	0	0	0	0	96	1288	0.861
⋮	⋮	⋮	⋮	⋮	⋮	⋮	⋮	⋮	⋮	⋮
78	0	0	0	0	0	0	+α	103	1332	0.832
79	0	0	0	0	0	0	−α	134	1478	0.816

ce: Frenklach and Rabinowitz (1988).
1.885.

$\tau_{\xi = \beta}$, and $[CH_3]_{\xi = \gamma}$—are the induction times and methyl concentration computed at different sets of initial conditions: α, β, and γ. These computations were performed with a 121-reaction, 29-species mechanism of Frenklach and Rabinowitz (1988) developed for methane combustion, each run with seven active rate coefficients preset according to the **X** values of the corresponding row of the design matrix.

The computed in such a manner responses can be expressed as (polynomial) functions of rate variables **X**. These functions can then be used in various applications (Frenklach 1987): parametric sensitivity analysis, model optimization, statistically rigorous parameter estimation, storage and transfer of modeling information, and representation of chemistry in large-scale modeling. For example, in a recent modeling study (Frenklach, Wang, and Rabinowitz, in press), a large-size chemical kinetic model of methane combustion was optimized in thirteen active variables against experimental data collected under different experiments: ignition delays, methyl radical concentration profiles, and velocities of laminar premixed flames. A five-dimensional cross section of the parameter space is shown in Figure 11–4. The shaded areas in this figure are the zeros of the objective function; that is, the amount and quality of the experimental information considered in this optimization is not sufficient to determine the values of all the active parameters—the number of degrees of freedom is negative. To resolve this uncertainty one needs either to increase the number of responses or to make assumptions about several parameters. No mathematical technique can compensate for insufficient experimental data; however, quantifying the available information in rigorous mathematical terms, the objective achieved with the solution mapping methodology, helps to identify the future strategy for solving the problem. When the number of degrees of freedom is positive, the solution mapping technique can be used for statistically rigorous determination of parameter values and associated confidence intervals (Frenklach and Miller 1985).

The same method of solution mapping, only now replacing the roles of $\boldsymbol{\vartheta}$ and $\boldsymbol{\xi}$ in computer experiments, that is, developing the approximations to $\mathbf{f}$ in terms of $\boldsymbol{\xi}$ while keeping $\boldsymbol{\vartheta}$ constant, i.e. [cf. Eq. (11–5)],

$$\mathbf{f}_t,\ \boldsymbol{\vartheta}(\boldsymbol{\xi}) \approx \boldsymbol{\Psi}(\boldsymbol{\xi}),$$

can be used to "replace" the detailed description of chemical kinetics by simple algebraic relationships in complex fluid-dynamic simulations of reactive flows (Frenklach 1987, 1990). The method was recently demonstrated using an example of a three-dimensional simulation of the ozone level in the Los Angeles Air Basin (Marsden et al. 1987). A chemical reaction mechanism composed of 52 reactions and 24 species was adopted to describe

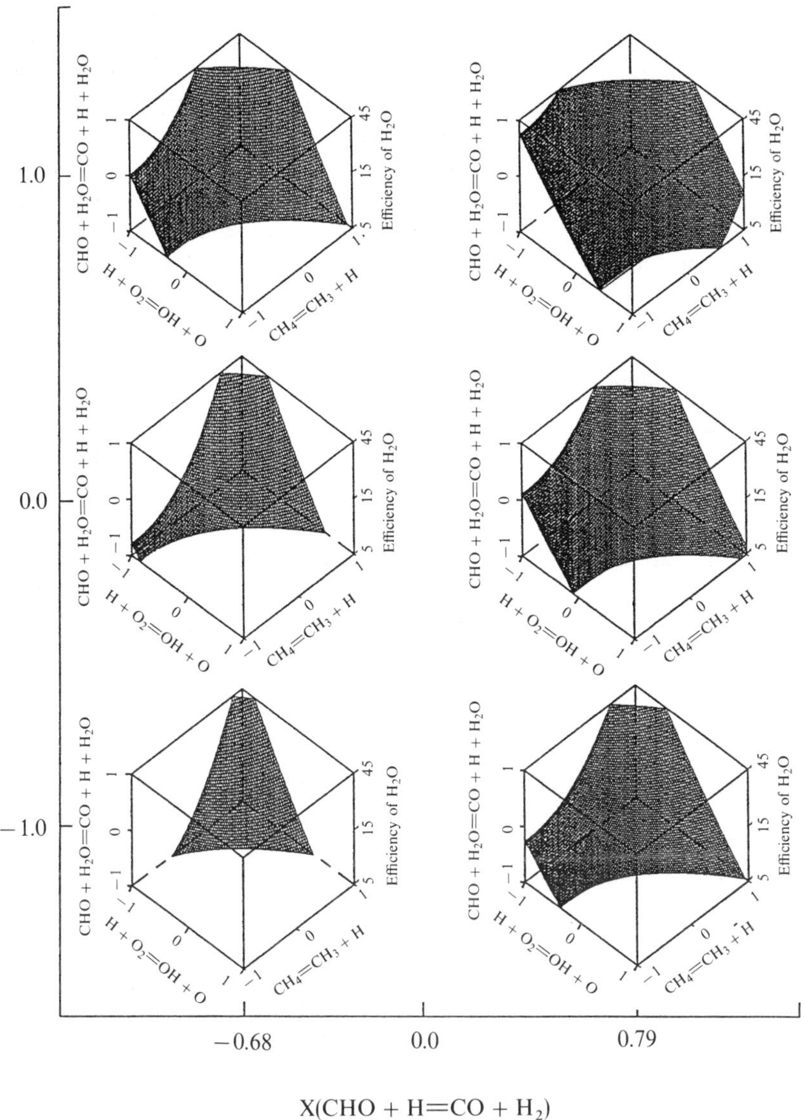

Figure 11–4 A five-dimensional cross section of the objective function in an optimization study. (Preliminary results from Frenklach, Wang, and Rabinowitz, in press.)

the chemical kinetics of ozone production. The concentrations of 15 of the chemical species at a particular reaction time, t, were chosen as model responses and the concentrations of these chosen species at $t = 0$ were assumed to be model variables. The concentrations of the remaining nine species were computed based on the pseudo-steady-state approximations. Using a factorial design for computer runs with different sets of initial concentration values, response surfaces for each of the 15 chosen species concentrations were developed in the following polynomial form:

$$\ln C_t = a_0 + \sum_{i=1}^{15} a_i \ln C_{i,0} + \sum_{ij=1}^{15} a_{ij} \ln C_{i,0} \ln C_{j,0}, \tag{11–6}$$

where $C_{i,0}$ is the initial concentration of species i and the a's are fitted coefficients. These runs were performed for isothermal isobaric conditions; otherwise the initial temperature and pressure could also be included as model variables. The developed polynomials, Eqs. (11–6), served then as an input for gas-dynamic simulations, which were performed employing the operator splitting method. The computed ozone concentration profiles were found to be in a close agreement with the solution obtained using the detailed-chemistry approach. The advantage of the solution mapping approach is a substantial reduction in the computational time without loss of detailed chemical kinetic information.

CHEMICAL LUMPING

The reaction mechanisms possessing certain topologies can be substantially reduced by lumping methods (see, for example, Astarita; Aris; Bischoff; this volume). One particular topology, which encounters in various reaction systems, is polymerization-type kinetics, where, beginning with a certain polymer size, the chemical reactions describing the polymer growth remain basically the same and the associated thermochemical and rate parameters remain essentially constant or exhibit only a weak dependence on the polymer size. An example can be provided by the mechanism of the formation and growth of polycyclic aromatic hydrocarbons (PAH) in high-temperature pyrolysis and flame environment, elucidated in detailed kinetic simulations with mechanisms containing more than 600 elementary reversible chemical reactions of 100–200 chemical species (Frenklach and Warnatz 1987; Frenklach et al. 1985, 1988). In this reaction system, the PAH growth beyond a certain size was identified to proceed by a replication-type reaction sequence (Figure 11–5). It appeared, however, that reversibility of chemical reactions

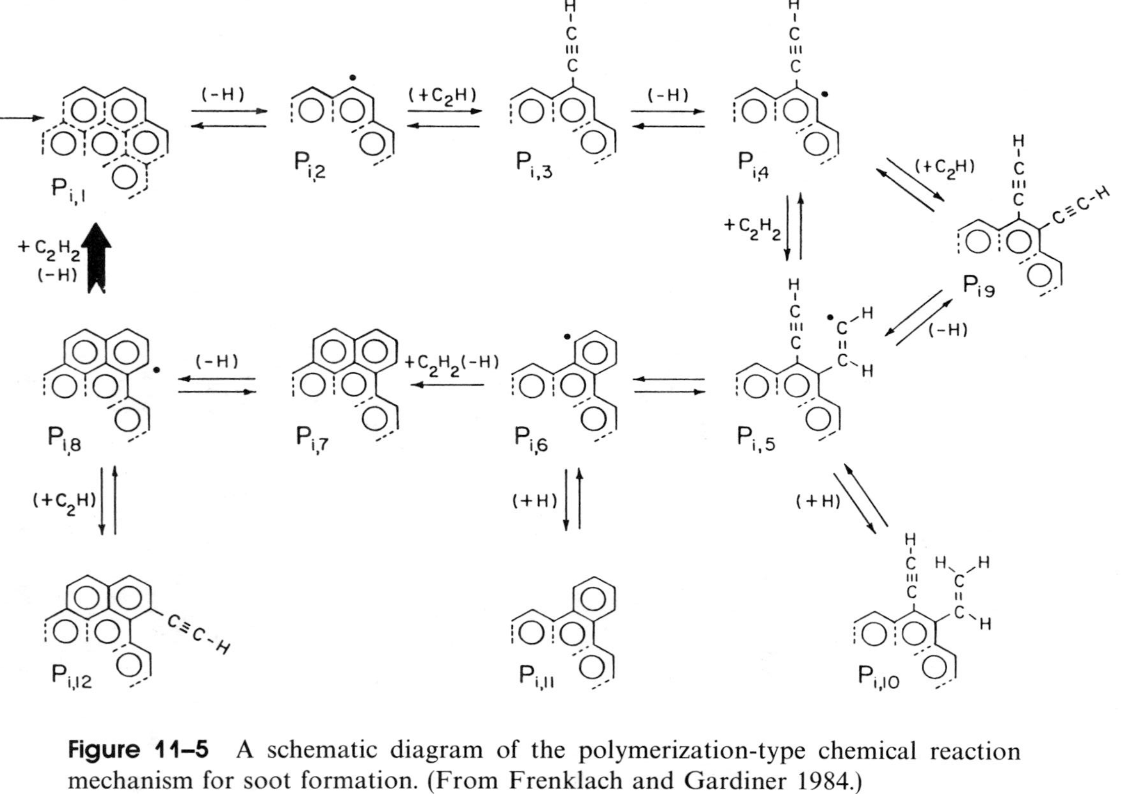

Figure 11–5 A schematic diagram of the polymerization-type chemical reaction mechanism for soot formation. (From Frenklach and Gardiner 1984.)

in the replicating reaction sequence is critical for accurate modeling of the process and, therefore, none of the classic methods developed for free-radical polymerization kinetics could be applied for they all assume irreversible growth steps.

Although easily conceived by the human mind, a detailed description of such a polymerization-type process requires a very large, practically infinite number of reaction steps. Lumping therefore is necessary. The approach outlined below (Frenklach 1985) consists in the development of a finite number of differential equations for the moments of the polymer distribution function. The underlying principle of the method of moments is that the properties one considers in most practical applications are determined by average measures (moments) and the history of individual species can be ignored; and what is lost in the resolution is compensated by a significant increase in computational speed and a dramatic reduction in computer memory requirements.

In the method of chemical lumping described here, the lumping procedure is guided by similarity in chemical structure or chemical reactivity of the reacting species in the replicating reaction sequence. A simple example (Frenklach 1985) is used here to illustrate the underlying ideas involved. Let us consider reaction system

$$\xrightarrow{r_0} A_1 \rightleftarrows B_1 \rightleftarrows C_1 \rightleftarrows D_1 \rightarrow A_2 \rightleftarrows B_2 \rightleftarrows C_2 \rightleftarrows D_2 \rightarrow \cdots \qquad (11\text{–}7)$$

which represents an infinite reaction sequence continuously building up molecular size. Reactions $A_i \rightleftarrows B_i \rightleftarrows C_i \rightleftarrows D_i$ constitute a single polymerization step. There are no restrictions on reactions within the individual polymerization step: they can be reversible or irreversible, consecutive or parallel, of any reaction order, with temperature-dependent or, more generally, time-dependent rate coefficients, etc. It is essential, however, that the reactions connecting the polymerization steps, and there can be a number of them, be irreversible.

It is further assumed that reaction sequence (11–7) is a replication-type reaction. That is, species $A_1, A_2, \ldots$ are similar to each other in the sense that their reactions $A_1 \rightarrow B_1$, $A_2 \rightarrow B_2, \ldots$ have the same kinetic parameters. Let us for simplicity in presentation assume that all the reactions in Eq. (11–7) are first order. We also assume that $B \rightarrow C$ and $D \rightarrow A$ are the monomer addition steps, each adding one monomer unit, the number of monomer units in A_1 is m_0, and r_0 is the rate of formation of A_1 from initiation reactions.

The differential equations describing the population of polymer species specified by reaction sequence (11–7) are

$$\frac{dA_1}{dt} = r_0 - k_{ab}A_1 + k_{ba}B_1, \tag{11–8A1}$$

$$\frac{dB_1}{dt} = k_{ab}A_1 - k_{ba}B_1 - k_{bc}B_1 + k_{cb}C_1, \tag{11–8B1}$$

$$\frac{dC_1}{dt} = k_{bc}B_1 - k_{cb}C_1 - k_{cd}C_1 + k_{dc}D_1, \tag{11–8C1}$$

$$\frac{dD_1}{dt} = k_{cd}C_1 - k_{dc}D_1 - k_{da}D_1, \tag{11–8D1}$$

$$\frac{dA_2}{dt} = k_{da}D_1 - k_{ab}A_2 + k_{ba}B_2, \tag{11–8A2}$$

$$\frac{dB_2}{dt} = k_{ab}A_2 - k_{ba}B_2 - k_{bc}B_2 + k_{cb}C_2, \tag{11–8B2}$$

$$\vdots \tag{11–8–∞}$$

where t is the reaction time, A_i, B_i, C_i, D_i are the concentrations of species $\mathrm{A_i}$, $\mathrm{B_i}$, $\mathrm{C_i}$, $\mathrm{D_i}$, respectively, and $k_{\alpha\beta}$ is the rate coefficient of reaction $\alpha \to \beta$ assumed to have the same value in each cycle i. Summing Eqs. (11–8) we obtain

$$\left(\frac{dA_1}{dt} + \frac{dB_1}{dt} + \frac{dC_1}{dt} + \frac{dD_1}{dt} + \frac{dA_2}{dt} + \frac{dB_2}{dt} + \cdots\right)$$
$$= \frac{d}{dt}(A_1 + B_1 + C_1 + D_1 + A_2 + B_2 + \cdots) = r_0,$$

or

$$\frac{dM_0}{dt} = r_0, \tag{11–9}$$

where M_0 is the zeroth concentration moment,

$$M_0 = A_1 + B_1 + C_1 + D_1 + A_2 + B_2 + \cdots.$$

Equation (11–9) describes the evolution of the total concentration, M_0, defined above.

Multiplying Eqs. (11–8) by the corresponding molecular mass, that is,

$$m_0 \times \frac{dA_1}{dt} = r_0 - k_{ab}A_1 + k_{ba}B_1,$$

$$m_0 \times \frac{dB_1}{dt} = k_{ab}A_1 - k_{ba}B_1 - k_{bc}B_1 + k_{cb}C_1,$$

$$(m_0 + 1) \times \frac{dC_1}{dt} = k_{bc}B_1 - k_{cb}C_1 - k_{cd}C_1 + k_{dc}D_1,$$

$$(m_0 + 1) \times \frac{dD_1}{dt} = k_{cd}C_1 - k_{dc}D_1 - k_{da}D_1,$$

$$(m_0 + 2) \times \frac{dA_2}{dt} = k_{da}D_1 - k_{ab}A_2 + k_{ba}B_2,$$

$$(m_0 + 2) \times \frac{dB_2}{dt} = k_{ab}A_2 - k_{ba}B_2 - k_{bc}B_2 + k_{cb}C_2,$$

$$\vdots$$

and summing up the obtained equations, we obtain

$$\left(m_0 \frac{dA_1}{dt} + m_0 \frac{dB_1}{dt} + (m_0 + 1)\frac{dC_1}{dt} + (m_0 + 1)\frac{dD_1}{dt} + (m_0 + 2)\frac{dA_2}{dt} + (m_0 + 2)\frac{dB_2}{dt} + \cdots\right)$$
$$= m_0 r_0 + (k_{bc}B_1 - k_{cb}C_1 + k_{da}D_1) + (k_{bc}B_2 - k_{cb}C_2 + k_{da}D_2) + \cdots$$

or

$$\frac{dM_1}{dt} = m_0 r_0 + k_{bc}\sum_{i=1}^{\infty} B_i - k_{cb}\sum_{i=1}^{\infty} C_i + k_{da}\sum_{i=1}^{\infty} D_i \tag{11–10}$$

where

$$M_1 = m_0 A_1 + m_0 B_1 + (m_0 + 1)C_1 + (m_0 + 1)D_1 + (m_0 + 2)A_2 + (m_0 + 2)B_2 + \cdots$$

is the first concentration moment of the polymer distribution function—the total number of monomer units accumulated in all polymer species.

The terms $\sum_{i=1}^{\infty} B_i$, $\sum_{i=1}^{\infty} C_i$, and $\sum_{i=1}^{\infty} D_i$ appearing in Eq. (11–10) are the lumped concentrations. Let us define them as b, c, and d, respectively, along with $a = \sum_{i=1}^{\infty} A_i$. These lumped concentrations are determined by solving differential equations

$$\frac{da}{dt} = r_0 - k_{ab}a + k_{ba}b + k_{da}d, \qquad (11\text{–}11a)$$

$$\frac{db}{dt} = k_{ab}a - k_{ba}b - k_{bc}b + k_{cb}c, \qquad (11\text{–}16b)$$

$$\frac{dc}{dt} = k_{bc}b - k_{cb}c - k_{cd}c + k_{dc}d, \qquad (11\text{–}11c)$$

$$\frac{dd}{dt} = k_{cd}c - k_{dc}d - k_{da}d \qquad (11\text{–}11d)$$

which are obtained by summing up the equations in (11–8) for species A_i, $i = 1, 2, \ldots$ [i.e., Eq. (11–8A1) + Eq. (11–8A2) + $\cdots$ = Eq. (11–11.a)], for species B_i, $i = 1, 2, \ldots, \infty$, etc.

Integration of Eqs. (11–9), (11–10), and (11–11), combined with the differential equations that describe the kinetics of the initiation reactions defining r_0, determines M_0 and M_1 at any given reaction time. The ratio

$$\mu = \frac{M_1}{M_0}$$

determines the average number of monomer units in the polymer, or in other words, the first moment of the polymer size distribution function. In a similar manner, equations for higher concentration moments are derived and higher moments of the polymer size are computed, thus specifying such statistical properties as the width, skewness, etc. (Johnson and Leone 1977) of the size distribution function. As an example, the first three moments of the PAH distribution function, expressed in units of carbon atoms, are shown in Figure 11–6.

The lumping described by Eqs. (11–11) can be expressed in the matrix from

$$\begin{pmatrix} a \\ b \\ c \\ d \end{pmatrix} = \begin{pmatrix} 1 & 0 & 0 & 0 & 1 & 0 & 0 & 0 & \ldots \\ 0 & 1 & 0 & 0 & 0 & 1 & 0 & 0 & \ldots \\ 0 & 0 & 1 & 0 & 0 & 0 & 1 & 0 & \ldots \\ 0 & 0 & 0 & 1 & 0 & 0 & 0 & 1 & \ldots \end{pmatrix} \begin{pmatrix} A_1 \\ B_1 \\ C_1 \\ D_1 \\ A_2 \\ B_2 \\ \vdots \end{pmatrix} \qquad (11\text{–}12)$$

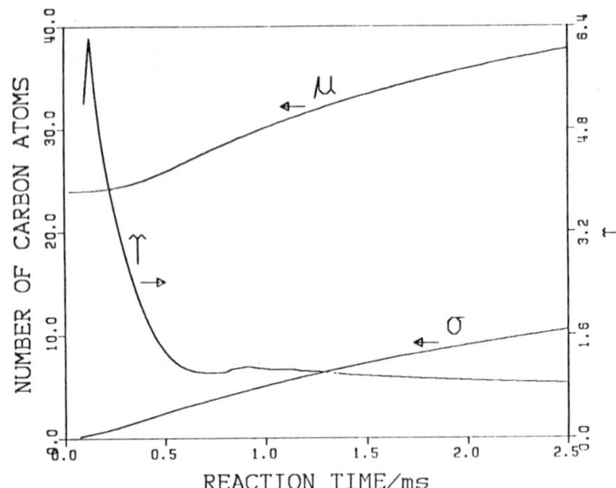

Figure 11–6 First three moments of the PAH size distribution: μ, mean; σ, standard deviation; and γ, skewness; μ and σ are in the units of carbon atoms. The computations were performed for a constant-pressure, constant-temperature pyrolysis of acetylene at 1,600 K and initial concentration of acetylene 4×10^{-7} mol/cm^3. (From Frenklach 1985.)

It can be shown by a straightforward matrix multiplication that the lumping matrix in (11–12), converting species concentrations A_i–D_i to lumped concentration variables a–d, satisfies the condition of exactness of Wei and Kuo (1969). The lumping expressed by Eq. (11–12) can schematically be represented as

$$\xrightarrow{r_0} a \rightleftarrows b \rightleftarrows c \rightleftarrows d \rightarrow a,$$

indicating the simplicity of computer implementation of this algorithm.

The presented chemical lumping technique can be applied to modeling surface deposition processes (Frenklach and Wang 1990). Also, if polymer–polymer interactions become of interest, the moment equations developed for chemical lumping can be directly combined with the moment equations describing the polymer–polymer reactions. This approach was recently demonstrated with a detailed kinetic modeling of soot particle nucleation and growth in laminar premixed flames (Frenklach and Wang 1991). The model begins with fuel pyrolysis, followed by the formation of polycyclic aromatic hydrocarbons, their planar growth and coagulation into spherical particles, and finally, surface growth and oxidation of the particles. The formation and growth of PAHs were modeled using the described here

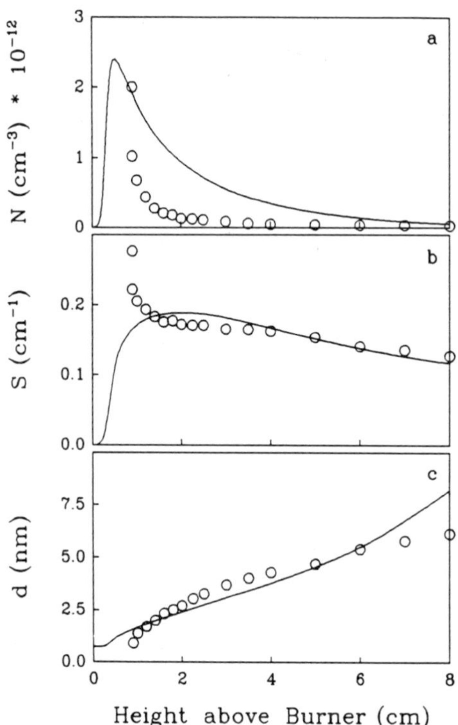

Figure 11–7 Comparison of model prediction and experimental data for soot particle formation in a 25.4% C_2H_2–19.6% O_2–Ar laminar premixed flame at pressure 90 torr; (*a*) soot particle number density; (*b*) specific surface area of soot particle; (*c*) average soot particle diameter. The experimental data (circles) are of Wieschnowsky et al. (1989); the calculations (lines) are from Frenklach and Wang (1991).

technique of chemical lumping. The PAH nucleation, coagulation, and surface reactions—with a technique of nonlinear lumping (Frenklach and Harris 1987). Figure 11–7 demonstrates the agreement obtained between the computational results and experimental measurements.

CONCLUSION

Chemical reaction models suitable for practical applications are sought to be sufficiently "compact," so that they can fit a desktop computer or to be used in conjunction with a sophisticated fluid dynamic code. It is advocated here that the development of such models should begin not from postulating a small-size model a priori; but rather from establishing first a full-size detailed reaction mechanism, verifying and optimizing its adequacy, and then reducing its mathematical form to a required size. Various lumping techniques can and should be used at different stages of model development.

ACKNOWLEDGMENTS

The methods and ideas presented in this manuscript were developed while working over the years on several projects supported by: NASA—Lewis Research Center, Contract No. NAS 3-23542 and Grants NAG 3-477, NAG 3-668, NAG 3-991; the Gas Research Institute, Contact 5086-260-1320; and the Air Force Office of Scientific Research, Grant 88-0072.

NOMENCLATURE

a_i	coefficients of Eq. (11–6)
a	lumped species concentration—sum of concentrations A_i for $i = 1, 2, \ldots, \infty$
A_i	concentration of species A_i
b	lumped species concentration—sum of concentrations B_i for $i = 1, 2, \ldots, \infty$
B_i	concentration of species B_i
c	lumped species concentration—sum of concentrations C_i for $i = 1, 2, \ldots, \infty$
C_i	concentration of species C_i
C_t	concentration of a given species at a reaction time t
$C_{i,0}$	the initial concentration of species i
d	lumped species concentration—sum of concentrations D_i for $i = 1, 2, \ldots, \infty$
D_i	concentration of species D_i
$\mathbf{f}$	range of variation for rate coefficient k
f	exact solution of the differential equations describing kinetic evolution of chemical species
$\mathbf{g}$	vector of functions expressing the mass-action law
ΔH_i	standard enthalpy change of reaction i
k_i	rate coefficient of reaction i
k'	center point for the rate coefficient in computer-experiment design
$k_{\alpha\beta}$	rate coefficient of reaction $\alpha \rightarrow \beta$
m_0	the number of monomer units in species A_1
M_i	ith concentration moment
$\dot{Q}_{max}$	maximum rate of heat generation
r	number of responses

r_0 rate of formation of species A_1 from initiation reactions
R_i rate of reaction i
r_{rls} rate of the rate-limiting reaction
t reaction time
X design matrix

Greek Letters

ε reduction-level parameter
$\boldsymbol{\eta}$ vector of computed state variables
$\boldsymbol{\eta}^{obs}$ vector of experimentally determined state variables
$\boldsymbol{\vartheta}$ vector of active parameters
μ first moment of the polymer size distribution function
$\boldsymbol{\xi}$ vector of initial conditions
τ induction time
$\boldsymbol{\Phi}$ objective function for optimization
$\boldsymbol{\Psi}$ vector of approximating function developed by computer experiments
ω_i statistical weight of the ith response

REFERENCES

Benson, S. W. (1976) *Thermochemical Kinetics.* John Wiley, New York.

Bock, H. G. (1981) Numerical treatment of inverse problems in chemical reaction kinetics. In *Modelling of Chemical Reaction Systems* (K. H. Ebert, P. Deuflhard, and W. Jäer, Eds.), p. 102. Springer-Verlag, Berlin.

Box, G. E. P. and N. R. Draper (1987) *Empirical Model-Building and Response Surfaces.* John Wiley, New York.

Box, G. E. P. and R. D. Meyer (1985) Some new ideas in the analysis of screening designs. *J. Res. Natl. Bur. Stand.* **90**:495.

Burcat, A. (1984) Thermochemical data for combustion calculations. In *Combustion Chemistry* (W. C. Gardiner, Jr., Ed.), Chap. 8. Springer-Verlag, New York.

Deuflhard, P. and U. Nowak (1986) Efficient numerical simulation and identification of large chemical reaction systems. *Ber. Bunsenges. Phys. Chem.* **90**:940.

Frenklach, M. (1984) Modeling. In *Combustion Chemistry* (W. C. Gardiner, Jr., Ed.), Chap. 7. Springer-Verlag, New York.

Frenklach, M. (1985) Computer modeling of infinite reactions sequences: a chemical lumping. *Chem. Engin. Sci.*, **40**:1843.

Frenklach, M. (1987) Modeling of large reaction systems. In *Complex Chemical Reaction Systems, Mathematical Modelling and Simulation* (Warnatz, J. and W. Jäer, Eds.), p. 2. Springer-Verlag, Berlin.

Frenklach, M. (1990) Reduction of Chemical Reaction Models. In *Numerical Approaches to Combustion Modeling* (Oran, E. S., and J. P. Boris, Eds.). American Institute of Aeronautics and Astronautics: Washington, D.C.

Frenklach, M. and D. L. Bornside (1984) Shock-initiated ignition in methane–propane mixtures. *Combust. Flame* **56**:1.

Frenklach, M. and W. C. Gardiner, Jr. (1984) Representation of homogeneous polymerization in detailed computer modeling of chemical kinetics. *J. Phys. Chem.* **88**:6263.

Frenklach, M. and S. J. Harris (1987) Aerosol dynamics modeling using the method of moments. *J. Colloid Interface Sci.* **118**:252.

Frenklach, M. and D. L. Miller (1985) Statistically rigorous parameter estimation in dynamic modeling using approximate empirical models. *A.I.Ch.E.J.* **31**:498.

Frenklach, M. and M. J. Rabinowitz (1988) Optimization of large reaction systems. *Proceedings of 12th IMACS World Congress on Scientific Computation, Vol. 3* (Vichnevetsky, R., P. Borne, and J. Vignes, Eds.), p. 602. Gerfidn, France.

Frenklach, M. and H. Wang (1991) Detailed surface and gas-phase chemical kinetics of diamond deposition. *Phys. Rev.* B. **43**.

Frenklach, M. and H. Wang (in press) Detailed modeling of soot particle nucleation and growth. In *Twenty-Third Symposium (International) on Combustion*, The Combustion Institute, Pittsburgh.

Frenklach, M., H. Wang, and M. Rabinowitz (in press) Optimization and analysis of large chemical reaction mechanisms using the solution mapping method—Combustion of methane. *Prog. Energy Combust. Sci.*

Frenklach, M. and J. Warnatz (1987) Detailed modeling of PAH profiles in a sooting low-pressure acetylene flame. *Combust. Sci. Technol.* **51**:265.

Frenklach, M., D. W. Clary, W. C. Gardiner, Jr., and S. E. Stein (1985) Detailed kinetic modeling of soot formation on shock-tube pyrolysis of acetylene. In *Twentieth Symposium (International) on Combustion*, p. 887. The Combustion Institute, Pittsburgh.

Frenklach, M., K. Kailasanath, and E. S. Oran (1986) Systematic development of reduced reaction mechanisms for dynamic modeling. *Prog. Astron. Aeron.* **105**:365.

Frenklach, M., T. Yuan, and M. K. Ramachandra (1988) Soot formation in binary hydrocarbon mixtures. *Energy Fuels* **2**:462.

Gardiner, W. C., Jr., S. M. Hwang, and M. J. Rabinowitz (1987) Shock tube and modeling study of methyl radical in methane oxidation. *Energy Fuels* **1**:545.

Hanson, R. K. and S. Salimian (1984) Survey of rate coefficients in the N/H/O system. In *Combustion Chemistry* (Gardiner, W. C., Jr., Ed.), Chap. 6. Springer-Verlag, New York.

Johnson, N. L. and F. C. Leone (1977) *Statistics and Experimental Design in Engineering and the Physical Sciences, Vol. 1*, Chap. 3. John Wiley, New York.

Kee, R. J., J. F. Grgar, M. D. Smooke, and J. A. Miller (1985) A Fortran program for modeling steady laminar one-dimensional premixed flames. Report No. SAND85-8240, Sandia, Livermore, California.

Marsden, A. R., Jr., M. Frenklach, and D. D. Reible (1987) Increasing the computational feasibility of urban air quality models that employ complex chemical mechanisms. *J. Air Pollut. Contr. Assoc.* **37**:370.

Milstein, J. (1981) The inverse problem: estimation of kinetic parameters. In *Modelling of Chemical Reaction Systems* (Ebert, K. H., P. Deuflhard, and W. Jäger, Eds.), p. 92. Springer-Verlag, Berlin.

Rabitz, H., M. Kramer, and D. Dacol (1983) Sensitivity analysis in chemical kinetics. *Annu. Rev. Phys. Chem.* **34**:419.

Rosenbrook, H. H. and C. Storey (1966) *Computational Techniques for Chemical Engineers.* Pergamon, Oxford.

Steinfeld, J. I., J. S. Francisco, and W. L. Hase (1989) *Chemical Kinetics and Dynamics*, p. 535. Prentice Hall, Englewood Cliffs, New Jersey.

Tsang, W. (1987) Chemical kinetic data base for combustion chemistry. Part 2. Methanol. *J. Phys. Chem. Ref. Data* **16**:471.

Tsang, W. (1988) Chemical kinetic data base for combustion chemistry. Part 3. Propane. *J. Phys. Chem. Ref. Data* **17**:887.

Tsang, W. (1990) Chemical kinetic data base for combustion chemistry. Part 4. Isobutane. *J. Phys. Chem. Ref. Data* **19**:1.

Tsang, W. and R. F. Hampson (1986) Chemical kinetic data base for combustion chemistry. Part 1. Methane and related compounds. *J. Phys. Chem. Ref. Data* **15**:1087.

Wang, H. and M. Frenklach (1990) Detailed mechanism reduction for flame modeling. In *Chemical and Physical Processes in Combustion*, 23rd Fall Technical Meeting of the Eastern Section of the Combustion Institute. The Combustion Institute, Pittsburgh.

Warnatz, J. (1984) Rate coefficients in the C/H/O system. In *Combustion Chemistry* (Gardiner, W. C., Jr., Ed.), Chap. 5. Springer-Verlag, New York.

Wei, J. and J. C. W. Kuo (1969) A lumping analysis in monomolecular reaction systems. Analysis of exactly lumpable system. *Ind. Engin. Chem. Fundam.* **8**:114.

Westbrook, C. K. and F. L. Dryer (1984) Chemical kinetic modeling of hydrocarbon combustion. *Prog. Energy Combust. Sci.* **10**:1.

Westley, F., J. F. Herron, and R. J. Cvetanovic (1987) Compilation of chemical kinetic data for combustion chemistry. I. Non-aromatic C, H, O, N and S containing compounds. *Natl. Stand. Ref. Data Ser.*, NBS 73.

Wieschnowsky, U., H. Bockhorn, and F. Fetting (1989) Some new observations concerning the mass growth of soot in premixed hydrocarbon–oxygen flames. In *Twenty-Second Symposium (International) on Combustion*, p. 343. The Combustion Institute, Pittsburgh, Pennsylvania.

12

Kinetic Modeling at Mobil: An Historical Perspective

A. V. SAPRE

Over the years, Mobil Research and Development Corporation has been active in developing kinetic modeling technology to enhance operation, optimization, and design of several processes of commercial importance in the petroleum refining industry. Although similar efforts are quite common in large industrial organizations in the petroleum and petrochemical field, the unique character of the Mobil organization has always been to provide a timely feedback to the scientific community on the state of the art and science of methodologies being used through the years. This long-standing open communication policy to enhance basic understanding, raise awareness of the problems facing the industry, and pose challenging key research objectives for the future, has been the hallmark of the Mobil organization. The steady flow of articles in the scientific and trade journals and joint research programs with the universities, government laboratories, research institutes, industry groups, and other companies over years is a good measure of this commitment.

A recent literature search of the American Petroleum Institute (API) data base is summarized in Table 12–1. We have included three major processes—catalytic cracking, reforming, and hydroprocessing—in this survey. A schematic of process units in a typical modern refinery is shown in Figure 12–1, and these key processes are highlighted. We chose these processes as they primarily involve handling of complex feedstocks with a large number

Table 12–1 API Literature Data Base on Process Kinetics[a]

Process	Total Number of Papers	Other Oil Companies	Mobil	Mobil % of Oil Industry
Fluid catalytic cracking	149	14	11	44
Reforming	737	21	9	30
Hydroprocessing	2,491	143	29	17

[a] Search conducted in February 1990.

of components in the feed and the product. Although a large number of articles appear in the open literature, the contribution of the oil industry is relatively small. However, Mobil's contribution stands out at least in terms of numbers. We hope it is the quality, not just the number of these contributions, that reflect our commitment to advancing the science of process modeling, and sharing our problems with the scientific community. A summary of key articles from Mobil in the process kinetics area is included in Appendix A. Important contributions by others to this field of kinetic and thermodynamic modeling of complex reaction mixtures are summarized in Appendix B. This is not an exhaustive survey by any means, however, it is a good start to a new researcher in this field.

An historical perspective on Mobil's kinetic modeling philosophy is highlighted in this brief review. The principal motivation for developing modeling technology at Mobil has been that models should be a carrier of one's understanding of the process technology and include as many fundamental principles as possible to be able to quantify phenomena in their greatest detail. The material presented here is primarily based on published information and is testimony to the above ideal. In particular, Mobil has published quite extensively on our Fluid Catalytic Cracking (FCC) and Reforming Kinetic models. These models represent typical examples of complexity involved with petroleum processes handling feedstocks containing innumerable individual components. We believe these examples focus on problems in translating laboratory data into commercial operations and characterizing these operations for process optimization. The published work may not have included the full details on the rate constants or complete identification of the chemistry. This is understandable in view of the usefulness of these models in commercial applications. The models, in general though, have benefited significantly from Mobil's extensive data base of precision commercial test runs. The broad acceptance of the use of models in Mobil has proven that reaction kinetics can contribute strongly to the development as well as to subsequent optimization of the process technology.

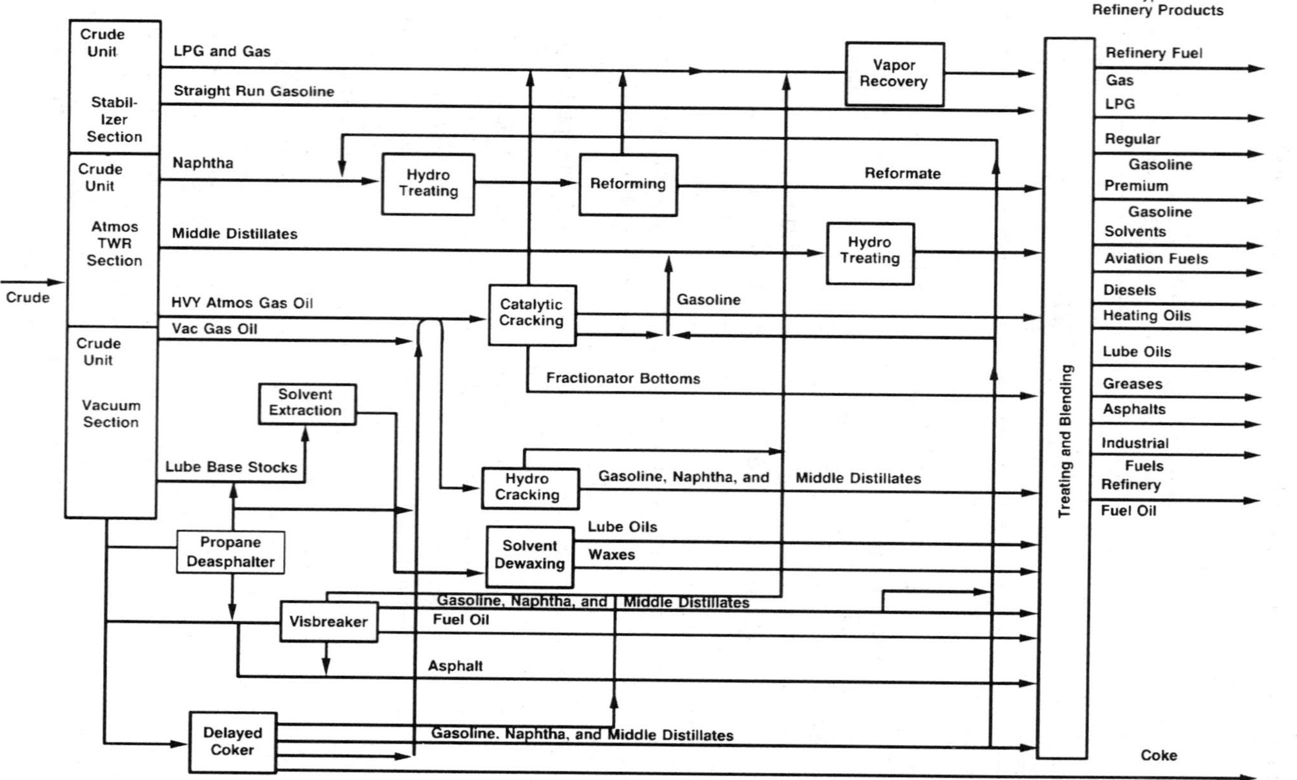

Figure 12–1 Flow diagram of a typical modern refinery.

The academic contributions undoubtedly provided the theoretical basis to much of the work in the industry, and it is complementary to our efforts. This workshop emphasizes the new horizons for the kinetics and thermodynamics of complex mixtures with a large number of components.

HISTORY

The tremendous growth of the petroleum industry in the 20th century is closely tied to the increasing use of transportation vehicles, in particular automobiles and airplanes. The requirements of high octane aviation gasolines during World War II brought about the wide use of catalytic processes with complex reactor technologies for petroleum refining processes such as catalytic cracking, polymerization, alkylation, and reforming. But perhaps a more important impact on the industry after the war was the availability of dedicated, trained professionals. During the war, academic scientists were given the primary responsibility to develop technologies, such as the atomic bomb. One of the biggest benefits of this effort was that people acquired skills to bring fundamental scientific principles to practical applied problems, and these people were now available to the industrial community. Around that period, the Research Department of Socony Vacuum at Paulsboro, which later became Mobil, decided to expand the research labs. A physics section parallel to the then existing chemistry section was started. Several physicists, biophysicists to be exact, including Dwight Prater and Paul Weisz, joined this group and went into the oil business and became "engineers by osmosis." At that time the emphasis of the chemistry section was somewhat different from that of the physics section, although the work in the two groups was complementary. The newly formed physics group concentrated more on fundamental problems rather than day-to-day firefighting. The group flourished rapidly and led to several innovative ideas that made a positive impact on the oil business. Diffusion theories in catalyst particles originated in this group. This work by Weisz and Prater is now a part of all chemical reaction engineering textbooks. Mathematical modeling activities were aggressively pursued.

The most widely used industrial process models in the late 1940s and early 1950s were typically graphical representation of plant data in easy to use nomographs. One such model, a runaround chart for commercial TCC units, is summarized in Figure 12–2. These nomographs were purely correlative with no fundamental basis built into them. As a result, extrapolation beyond the range of data used for the development of such charts was quite hazardous. The task at hand for the physics group was to change this state of the art. The pioneering work by P. B. Weisz and co-workers on kinetics of coke burning in the mid-1950s was the first successful kinetic model in

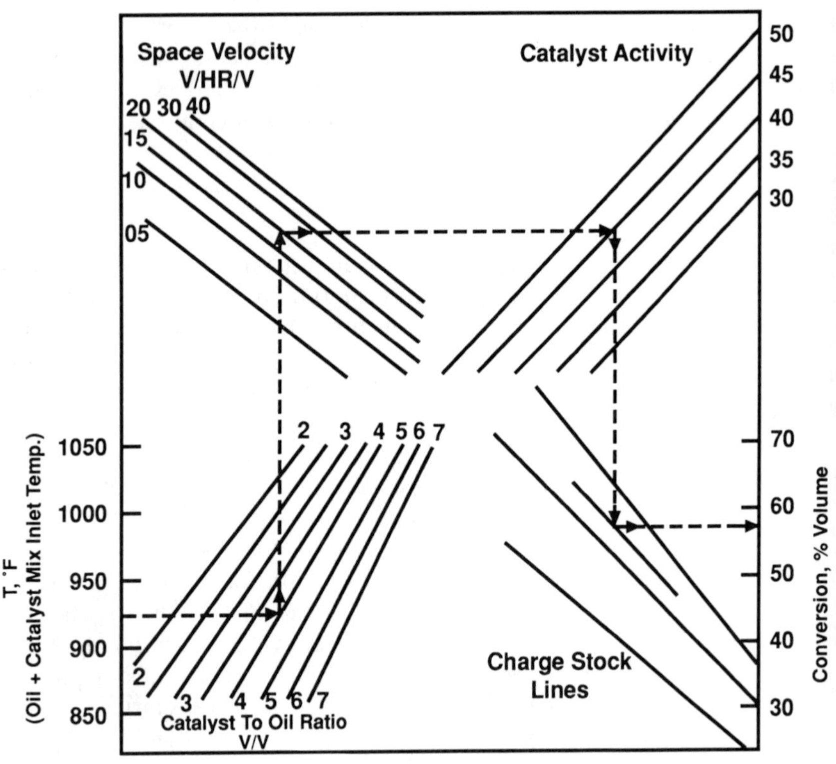

Figure 12–2 Runaround chart for commercial TCC unit.

Mobil. This model included such effects as fast and slow coke, intrinsic carbon burning kinetics and its relation to oxygen utilization, and impact of diffusion on coke burning in TCC beads. Quantitative predictions of resulting "fish eye" phenomena observed commercially in TCC beads and intrinsic kinetics are well documented in a series of articles by Weisz and Goodwin (1963, 1966). The first successful so-called "cold turkey model" was the TCC kiln model which has been reviewed in detail by Prater et al. (1983). This fundamental approach convinced refinery and research engineers of the importance of rigorous modeling on improving commercial processes. With management's blessing, the group focused on basic research. An academic environment in the industrial setting was fostered. Several members of the group even took sabbatical leaves to pursue interest in academia; one-year assignments outside the Mobil laboratories was a norm of the day. James Wei, for example, actively pursued research interests at Cal Tech and Princeton universities. The primary work of Wei and Prater on reversible monomolecular reaction systems was summarized in 1962 in Advances in

Catalysis, under the title, "The structure and analysis of complex reaction systems." This work again is widely quoted today in chemical kinetics textbooks. The physics group laid the foundation for research philosophy in kinetic modeling, the legacy of which is seen even today through the works of Mobil researchers. Both Prater and Weisz were later elected to membership in the National Academy of Engineering. Election to the Academy is the highest professional distinction that can be conferred upon an engineer and honors those who have made important contributions to engineering theory and practice.

Professor Shinnar of City College of New York, a long-time Mobil consultant, has also helped to shape our thinking in industrial process modeling. According to Professor Shinnar, "Reaction models are rather imperfect approximations of the real system." He classifies all models into the following three classes:

1. *Correlative Models*: Organize a large amount of data for practical applications.
2. *Learning Models*: An iterative process that helps us understand structural relationships, the modeling process itself. These models provide guidance for efficient experimentation which finally leads to good predictive models.
3. *Design Models*: Predict system properties based on whatever information is available. Truly predictive models are generally excellent design models.

Process models developed over the years at Mobil encompass the full range of this classification.

Kinetic models are the heart of the complete process models. For a process model to be successful, the following elements are essential:

- Impact of operating variables on product yields and properties
- Impact of different feedstocks on product yields and properties
- Impact of catalyst type on product yields and properties
- Impact of reactor hydrodynamics and scale-up nonidealities
- Impact of plant constraints on reactor optimization .

Furthermore, for the models to be useful in the plant setting, they must have easy interface with plant data and be user-friendly to novice users. Models that encompass fundamental physicochemical phenomena, assimilate information from independent experiments describing some parts of the problem and coupling of smaller independent events are generally successful. With complex systems, however, it is not always practical to describe each effect in its most fundamental form, and calibration factors are sometimes

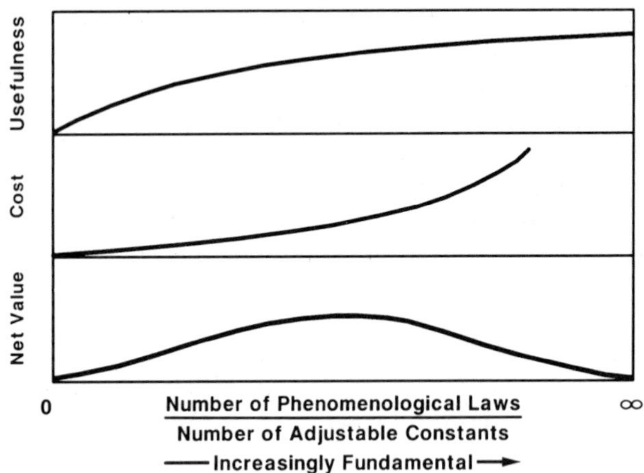

Figure 12–3 Prater's principle of optimum sloppiness.

needed. We at Mobil apply Prater's optimum sloppiness principle, described schematically in Figure 12–3, to develop our modeling technology. In Figure 12–3, we span the spectrum of purely correlative models to the left to completely fundamental models at the right. Achieving the optimum on the net value basis is the goal.

The mid and late 1950s was also the period when Mobil scientists recognized the potential of zeolites for catalytic reactions. The work of P. B. Weisz and V. J. Frilette (1960) laid the foundation for zeolite research at Mobil. In the early 1960s, Plank and Rosinski's work on Re_2O_3 stabilized Y-zeolite in catalytic cracking revolutionized the petroleum industry. By the early 1960s, Mobil laboratories were involved in the transition from the amorphous catalysts to the new zeolite catalysts. In the middle 1960s, we were considering building commercial fluid-bed units to take full advantage of the new zeolites rather than our older moving-bed TCC units. The history of cracking technology has recently been reviewed by Avidan et al. (1990). Vern Weekman (1979) has given an excellent summary of the kinetic modeling aspects of the catalytic cracking and Mobil's successful 10-lump cracking model. Here we will briefly discuss some additional facets of the cracking modeling technology. Some of the progress in the last ten years and future challenges ahead of us will be addressed.

CATALYTIC CRACKING MODEL

The first kinetic models for cracking treated the entire gasoil feedstock as one lump (Blanding 1953), and the overall conversion to gasoline and lighter products was described by a second-order reaction. This model was refined

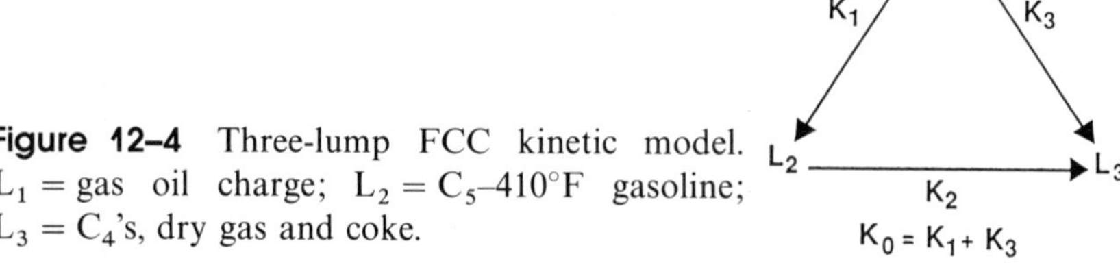

Figure 12–4 Three-lump FCC kinetic model. L_1 = gas oil charge; $L_2 = C_5$–410°F gasoline; $L_3 = C_4$'s, dry gas and coke.

to a three-lump model by Weekman and co-workers. Figure 12–4 shows the three-lump model. Since the catalyst decays with coke laydown, it is assumed that activity declines in direct proportion to coke formation. The coke formation can be reliably estimated from the Voorhies (1945) time-on-stream theory. The three-lump model with the above deactivation law could reliably predict the performance of various types of reactors such as fixed-bed, moving-bed, fixed-fluid bed, and riser reactors for a given feedstock and catalyst. But unfortunately the rate parameters are strongly dependent on the feedstock. However, the models were able to sort out some apparently anomalous effects of time-averaged yields on the performance of older silica–alumina catalyst and the newer zeolites.

Advantages of zeolites were almost missed initially. Mobil almost gave away the zeolite cracking technology due to the apparent poor performance in the catalyst screening test. In the standard laboratory long-time screening tests, where product yields are time-averaged over the catalyst decay cycle, the high-activity zeolites gave poor performance. Typical data are summarized in Figure 12–5. These results for the transient laboratory tests can be explained on the basis of high deactivation rate associated with the higher activity zeolite catalysts and potential lower time-averaged conversion at long contact times. In the short contact time riser reactor, however, the zeolite performance was far superior than the low-activity slower-decaying silica-alumina catalysts. In fact, just the reverse problem arose 3 to 4 years ago when catalyst vendors started to market lower-activity ultra stable *Y* (USY) catalysts in the market to satisfy the increasing octane demand. In the transient laboratory tests, such as Mobil's FFB or industry standard Micro Activity Test (MAT) tests, these USY catalysts gave phenomenally superior performance relative to the much optimized Rare Earth Y (REY) catalysts. Of course, this advantage is significantly reduced in a steady-state riser. A typical set of data are summarized in Table 12–2. The simple second-order reaction kinetics model could easily sort out these potential pitfalls in erroneous ranking of catalysts. For catalyst screening, therefore, this simple model is adequate. We use this approach to translate catalyst performance among different reactor types, as summarized by Sapre and

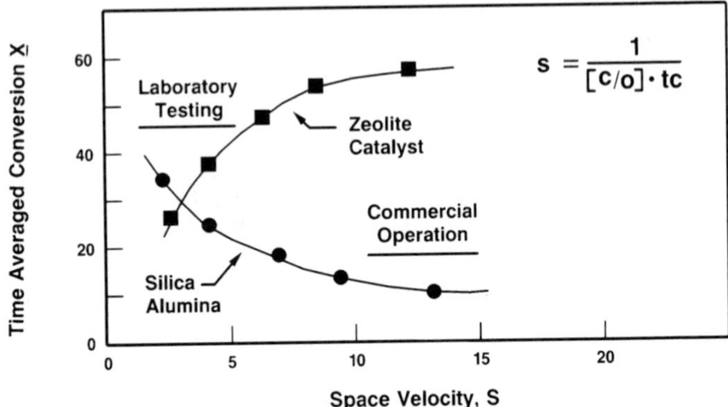

Figure 12–5 Transient laboratory catalyst characterization tests can rank catalysts erroneously. (From V. W. Weekman, Jr., *A.I.Ch.E.* Monograph Series No. 11, 1979.)

Leib (1990). Experimental data and model predictions for the MAT and fixed fluid bed reactor types for USY and REY catalysts are summarized in Figure 12–6. The agreement between model predictions and experimental data is excellent.

For a heat-balanced FCC operation, coke-conversion selectivity, or the dynamic activity as it is termed in the petroleum industry, is the key catalyst characterization parameter. For an overall second-order reaction, it is defined as

$$K_c = \frac{\text{coke}}{C_r},$$

where

$$C_r = \frac{x}{1 - x}, \quad (12\text{–}1)$$

and x is conversion. Weekman's earlier articles (1968, 1970, 1979) discussed the impact of the transient nature of catalyst screening tests on measured catalyst activity or conversion. More recent papers from Mobil (Chin et al. 1989; Krambeck's chapter in this volume; Sapre and Leib 1990) discuss in detail the coke-conversion selectivity, or the K_c parameter, which is perhaps the most critical parameter for catalyst screening. As Table 12–2 points out, ranking by catalyst activity could be different from that based on the coke-conversion selectivity.

The second-order kinetics approach has been extended to the entire boiling range as discussed by F. J. Krambeck. For a given feedstock, this

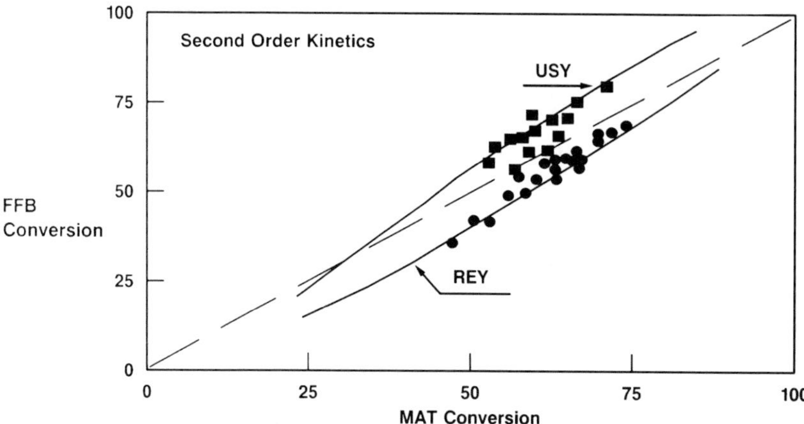

Figure 12–6 Second-order kinetics can accurately translate test results between different catalyst characterization tests. Lines represent model predictions.

approach is excellent. However, the resulting kinetic parameters are feedstock dependent. The inherent difficulty with this continuum approach is that not enough chemistry is included into the kinetic description. An attempt to include some of the important reaction steps in the kinetic analysis is the recent work of Allen and his co-workers (1989). Allen's chapter in this volume summarizes the key concepts and extends the approach to accurately estimate product properties such as gasoline octanes.

Table 12–2 Catalyst Ranking Based on Experimental Data

	FFB Test		MAT Test		Riser Results	
	Conv., %	K_c	Conv., %	K_c	Conv. %	K_c
Catalyst [A]	63	0.53	66	1.39	70	1.47
Catalyst [B]	65	0.38	62	0.95	64	1.20
Catalyst [C]	68	0.42	67	1.04	67	1.33
	Ranking for Activity			Ranking for Coke Selectivity		
FFB Ranking	C	B	A	B	C	A
MAT Ranking	C	A	B	B	C	A
Riser Ranking	A	C	B	B	C	A

Catalyst A, REY; catalyst B, USY; catalyst C, RE-USY.

Table 12–3 Carbon Number, Atmospheric Equivalent Boiling Point (AEBP), and the Number of Acyclic Alkane Isomers

Carbon No.	*n*-Alkane AEBP, °F	No. of Isomers	Examples of Petroleum Distillation Cuts
5	97	3	Gasoline (C5–C12)
8	258	18	
10	345	75	
12	421	355	
15	519	4,347	Diesel and jet fuels, middle distillates (C15–C25)
20	651	36.6×10^4	
25	755	36.7×10^6	Atmospheric residue (C25 and above)
30	840	41.1×10^8	Vacuum gas oil (C30–C45)
35	912	49.3×10^{10}	
40	972	62.4×10^{12}	
45	1,022	82.2×10^{14}	Vacuum residue, asphalt (C45 and above)
60	1,139	221.5×10^{20}	
80	1,242	$1{,}056.4 \times 10^{28}$	"Nondistillable" residue (C80 and above)
100	1,306	$5{,}920 \times 10^{36}$	

A complete description of the system in terms of individual compounds is impractical for complex feedstocks typically processed in modern FCCs. The immense complexity of petroleum can be illustrated by using acyclic alkanes as an example. Table 12–3 shows the number of possible acyclic alkane isomers at a given carbon number or boiling point. The numbers become astronomical as the carbon number increases. Although alkanes are not a major component of heavy petroleum, a large percentage of the carbon atoms in other types of molecules is represented by paraffinic side chains. The molecules having naphthenic and aromatic rings also have a large number of possible structural arrangements. Thus, it is impossible to characterize and describe kinetics at a molecular level. Thus lumping many individual compounds into a smaller number of groups is necessary. Lumping of similar chemical species greatly reduces the complexity of kinetic formulation. Considerable use is made of this technique in commercial reactor modeling. The first such successful lumped model is the 10-lump Mobil FCC model (Jacob et al. 1976).

The theoretical background for lumped monomolecular reactions was provided by Wei and Kuo (1969). This work showed that it is possible to lump a number of species together and still have the lumped kinetics accurately describe the overall reaction behavior of the system. The 10-lump model is called LINK-1 in Mobil, which stands for *L*umped *I*nvariant

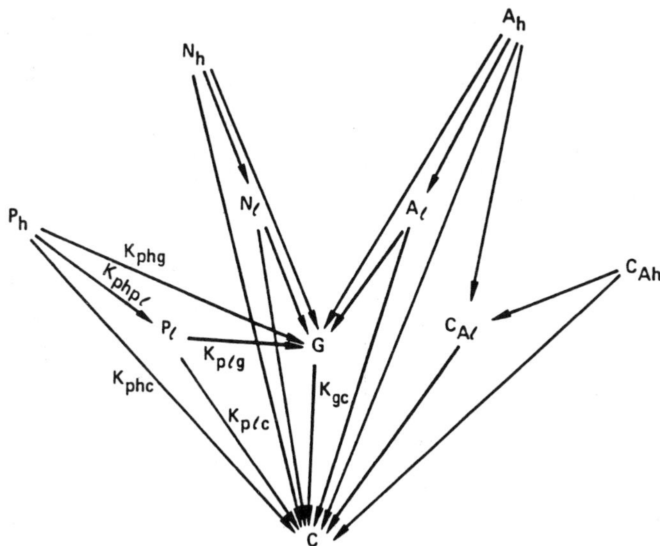

Figure 12–7 Ten-lump FCC kinetic model. P_l = Wt.% paraffinic molecules, (mass spec analysis), 430–650°F; N_l = Wt.% naphthenic molecules, (mass spec analysis), 430–650°F; C_{Al} = Wt.% carbon atoms among aromatic rings, (n-d-M method), 430–650°F; A_l = Wt.% aromatic substituent groups (430–650°F); P_h = Wt.% paraffinic molecules, (mass spec analysis), 650°F$^+$; N_h = Wt.% naphthenic molecules, (mass sec analysis), 650°F$^+$; C_{Ah} = Wt.% carbon atoms among aromatic rings, n-d-M method, 650°F$^+$; A_h = Wt.% aromatic substituent groups (650°F); G = G lump (C_5–430°F); C = C lump (C_1 to C_4 + coke); $C_{Al} + P_l + N_l + A_l$ = LFO (430–650°F); $C_{Ah} + P_h + N_h + A_h$ HFO (650°F). Adapted nomenclature for rate constants is detailed in the above figure for the paraffinic molecules. Similar rules apply for the other reaction steps.

Kinetics. The structure of the model is summarized in Figure 12–7. The rate constants in this model were invariant with respect to the original crude source of the lumps. LINK-1 did an excellent job for vacuum gas oils typically processed in Mobil at the time of its development. Improvements to the model have been continuous since then. Significant additional delumping was necessary to represent a large variety of stocks being processed in our refineries today. The feeds vary from hydrotreated to atmospheric resids, including significant portions of nonvirgin stocks such as lube extracts, coker gas oils, and other cracked stocks. The current version of the model, LINK7, is the seventh generation refinement to the original 10-lump model. The latest model attempts to recognize the variation in feedstock composition as a function of boiling point. Some key components are summarized in Figure 12–8.

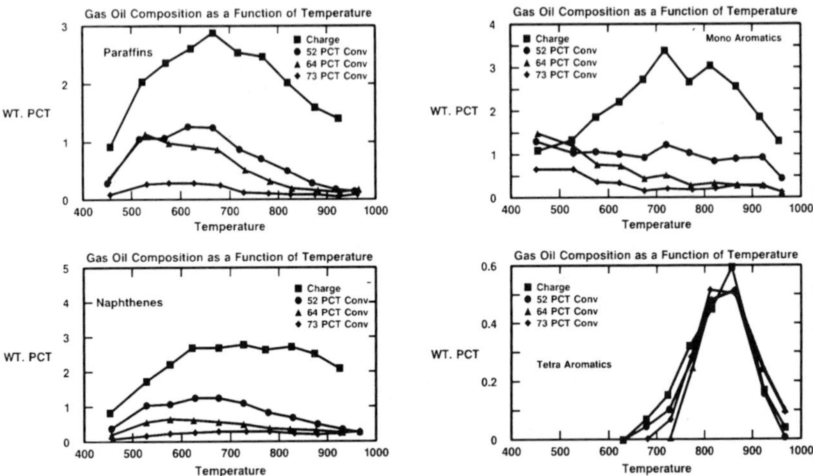

Figure 12–8 Composition shifts as a function of boiling point and conversion of a typical FCC feed.

One of the challenges in the Research Department is to package the best available modeling technology to our refineries, such that novice users can effectively use them. This technology transfer aspect imposes a certain degree of robustness in the modeling technology. More accurate fundamental kinetic models definitely make this task easier, and so the models get widely used.

The FCC kinetic model forms the heart of the entire process model. In the process simulator, all aspects of the FCC process including reactor regenerator, main column, and unsaturated gas plant are included. The FCC complex is summarized in Figure 12–9. An optimization package is integrated with the process model to determine the optimum operating conditions to maximize profits within the plant constraints such as wet gas and air blower capacity constraints. In addition, the model predicts product qualities such as gasoline octanes, light fuel oil sulfur, bottoms viscosity, etc. Product qualities are as important as yields. The objective function for the optimizer can be maximizing total profit or gasoline octane barrels, etc.

For a widespread refinery use these complex process models are integrated with the refinery management information systems and are accessed by menu-driven PC-based programs. The process simulators are customized for each refinery to reflect the actual equipment in the plant. Accuracy of the model to reflect a commercial plant performance is shown in Figure 12–10. Both conversion and gasoline yield are predicted quite accurately. This advanced modeling technology has helped our commercial plants to maintain a competitive advantage in the marketplace. Such models for each major

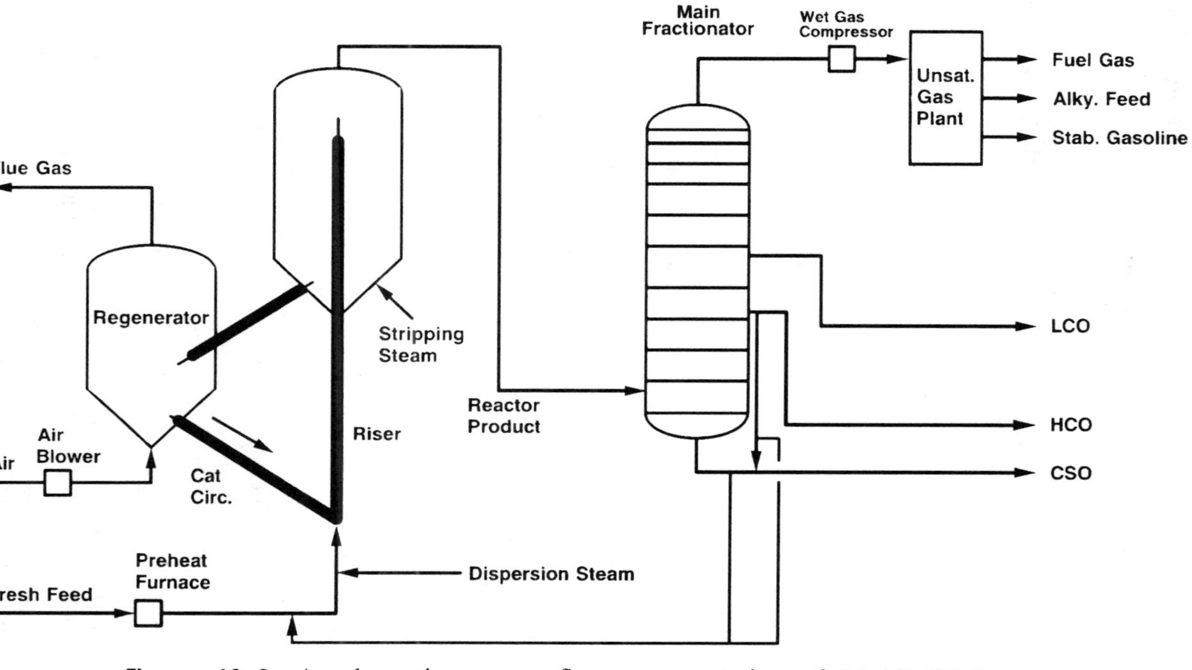

Figure 12–9 A schematic process flow representation of Mobil FCC process modeling technology.

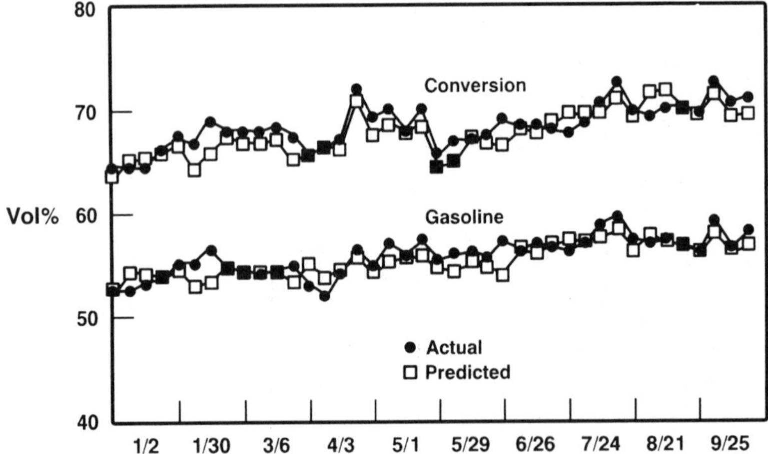

Figure 12–10 Mobil FCC model accurately tracks commercial performance.

process units are integrated in the refinery environment under a common advanced monitoring system umbrella. The key features of this Mobil advanced monitoring and optimization system are schematically presented in Figure 12–11. Apart from process monitoring and optimization, they serve a useful function as excellent training tools, local data bases, and an effective communication between the research department and operating divisions. These process optimization models integrate with the overall refinery optimization programs, LP models, for example. Also, these process models provide set point guidance to the advanced control systems.

REFORMING MODEL

The catalyst deactivation aspect, although important in FCC, does not significantly affect product selectivities, at least not as much as the catalyst selection itself. However, in other processes such as reforming, the catalyst deactivation significantly impacts product distribution and qualities. Therefore, a fundamental catalyst deactivation model is an integral part of the reforming model.

The Mobil reforming model was originally developed by a team led by Krambeck in 1971. Details of the reformer model are given in an excellent summary by Ramage et al. (1987). Some of the earliest work on dual functional catalysis typical of reforming is summarized by Weisz and Prater (1957). This fundamental work on how reforming catalysts function formed a basis of later kinetic modeling effort. In this reforming model, KINPtR reaction rates for the start-of-cycle reforming system are described by 1.

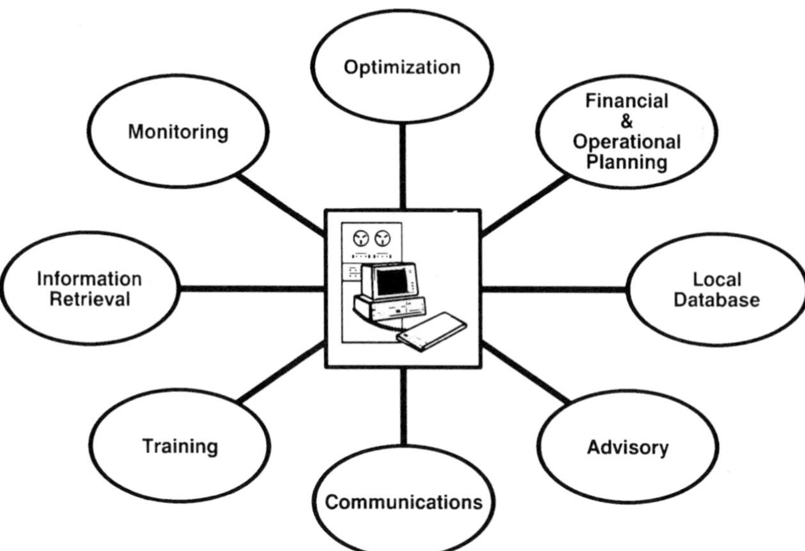

igure 12–11 A schematic representation of advanced monitoring system.

pseudomonomolecular kinetic lumps. On the other hand, there are 22 leactivation kinetic lumps to describe the aging phenomena. Therefore, while he 13 hydrocarbon lumps represent the hydrocarbon conversion kinetics, hey are delumped for deactivation kinetics. Additional delumping is necesary to estimate many of the product properties such as octanes and RVP. There are 34 lumps to completely characterize the product properties. The lifferent lumps in the reforming model are summarized in Figure 12–12 and Tables 12–4 and 12–5. Paraffins, five and six ring naphthenes, and aromatics

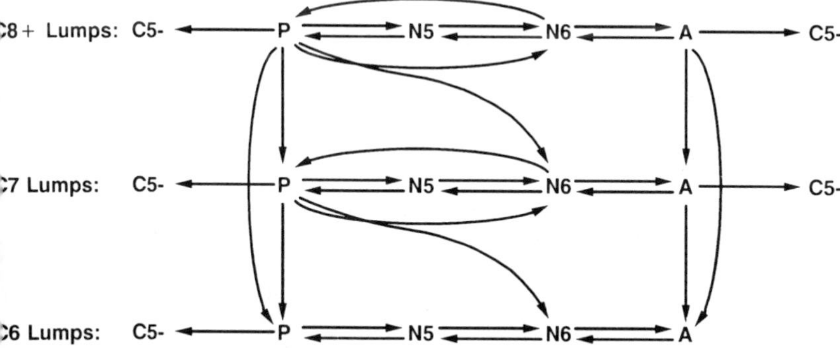

igure 12–12 Reforming lump reaction network. N, cyclopentane and cyclohexane naphthenes; P, C_{6+} paraffins; A, aromatics; C_{5-}, pentane and lighter.

Table 12–4 Components of Aging

Component Number	Component Name	Component Abbreviation
22	Hexanes	C_6P
21	Methylcyclopentane	C_6N_5
20	Cyclohexane	C_6N_6
19	Benzene	C_6A
18	C_7 Paraffins	C_7P
17	C_7 Cyclopentanes	C_7N_5
16	C_7 Cyclohexanes	C_7N_6
15	Toluene	C_7A
14	C_8 Paraffins	C_8P
13	C_8 Cyclopentanes	C_8N_5
12	C_8 Cyclohexanes	C_8N_6
11	C_8 Aromatics	C_8A
10	C_9 Paraffins	C_9P
9	C_9 Cyclopentanes	C_9N_5
8	C_9 Cyclohexanes	C_9N_6
7	C_9 Aromatics	C_9A
6	C_{10} Paraffins	$C_{10}P$
5	C_{10} Naphthenes	$C_{10}N$
4	C_{10} Aromatics	$C_{10}A$
3	C_{11} Paraffins	$C_{11}P$
2	C_{11} Napththenes	$C_{11}N$
1	C_{11} Aromatics	$C_{11}A$

Table 12–5 Components of Product Properties Delumped Components

Component Number	Component Lump
1	Hydrogen
2	Methane
3	Ethane
4	Propane
5	Isobutane

Table 12–5 *(continued)*

Component Number	Component Lump
6	*n*-Butane
7	Isopentane
8	*n*-Pentane
9	Isohexane
10	*n*-Hexane
11	Methylcyclo C_5
12	Cyclohexane
13	Benzene
14	Isoheptane
15	*n*-Heptane
16	C_7 Cyclo-C_5
17	Methylcyclo C_6
18	Toluene
19	Isooctane
20	*n*-Octane
21	C_8 Cyclo-C_5
22	C_8 Cyclo-C_6
23	C_8 Aromatic
24	Isononane
25	*n*-Nonane
26	C_9 Cyclo-C_5
27	C_9 Cyclo C_6
28	C_9 Aromatic
29	C_{10} Paraffin
30	C_{10} Naphthene
31	C_{10} Aromatic
32	C_{11} Paraffin
33	C_{11} Naphthene
34	C_{11} Aromatic
	2-Methylpentane
	3-Methylpentane
	Dimethylbutane
	2-Methylhexane
	3-Methylhexane
	Dimethylpentane
	Monomethyl C_{7+}
	Dimethyl C_{6+}
	Normal C_{8+}

for each carbon number are the critical lumps to describe the yield patterns as summarized in Figure 12–12. The four dominant reaction types are dehydrogenation, isomerization, ring closure, and cracking to C_{5-} lighter products. In general, the lumping scheme and reaction network of this model are considerably less complicated than that of Kmak (1971) and Froment (1987).

Catalyst aging is a strong function of feedstock and increases with increasing amounts of C_{8+} species in the feed. Therefore, there are 22 deactivation kinetic lumps, summarized in Table 12–4. Furthermore, catalyst aging affects the four reaction types differently. Therefore, there are four deactivation rate parameters, each corresponding to dehydrogenation, isomerization, ring closure, and cracking reactions. However, these are independent of carbon number, but are a function of local hydrocarbon composition, temperature and time.

Additional delumping is necessary to estimate many of the product properties and process conditions important to an effective process model. The C_{5-} kinetic lump is delumped into C_1 to C_5 light gas components; the paraffin kinetic lumps into isoparaffin and *n*-paraffin components; and the C_{8+} kinetic lumps into C_8, C_9, C_{10}, and C_{11} components by molecular type as summarized in Table 12–5. The paraffin distribution is constrained by known equilibrium.

The reliability of our reformer model to predict pilot plant and commercial data is illustrated in Figure 12–13. Gasoline yield versus octanes for a variety of feedstocks, catalysts, and operating conditions are shown in Figure 12–13A. This large variation in the gasoline yield at a given octane, as much as 25%, results from the wide range of process conditions and feed quality. In Figure 12–13B the same data are normalized using the KINPtR model. Over a wide range of octanes, KINPtR predictions are quite accurate. The predictions of cycle lengths also closely match the plant performance as summarized in Figure 12–14 and these predictions demonstrate the strength of the aging kinetics. The first version of the KINPtR model was developed for fixed-bed semiregenerative reformers. The model has now been extended to the latest generation of continuous reformers as well. Having a fundamental kinetic basis made this task simple.

The gasoline octane predictions from the KINPtR model are summarized in Figure 12–15. This is one of the earliest composition-based octane model developed to our knowledge. More than 6,000 runs went into validating the model. A small fraction of the data are presented in Figure 12–15 along with the form of the octane correlation. In this model, ON_i is pure component octane number and v_i is volume fraction of that component in the gasoline. This model recognizes nonlinear blending of octanes and blending octane

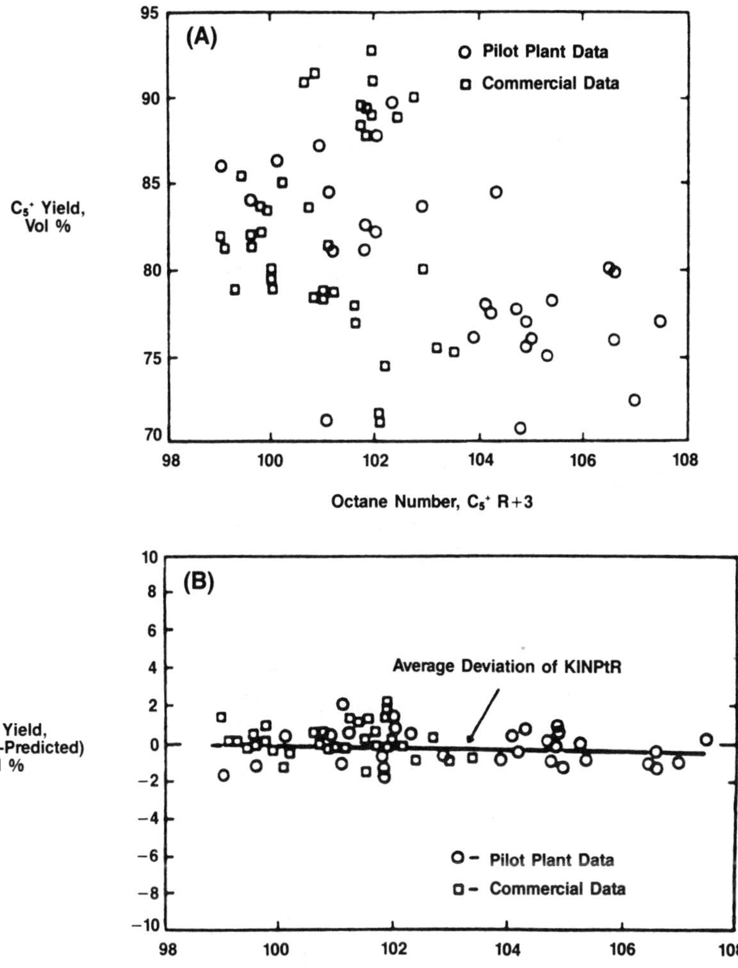

Figure 12–13 (*A*) Actual reformate yield-octane data; (*B*) KINPtR yield comparisons.

numbers are a function of the absolute octane level. The standard error of predictions is ±0.7 octane number.

These two examples, FCC and reforming, bring out some of the complexities of the industrial kinetic models. The complementary aspects of kinetic modeling such as feed/product characterization and impact of advances in computer technology are discussed next.

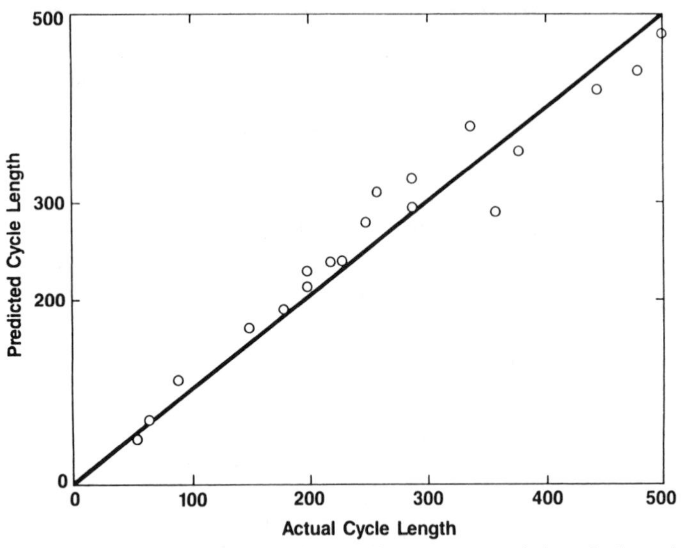

Figure 12–14 KINPtR accurately predicts commercial cycle lengths.

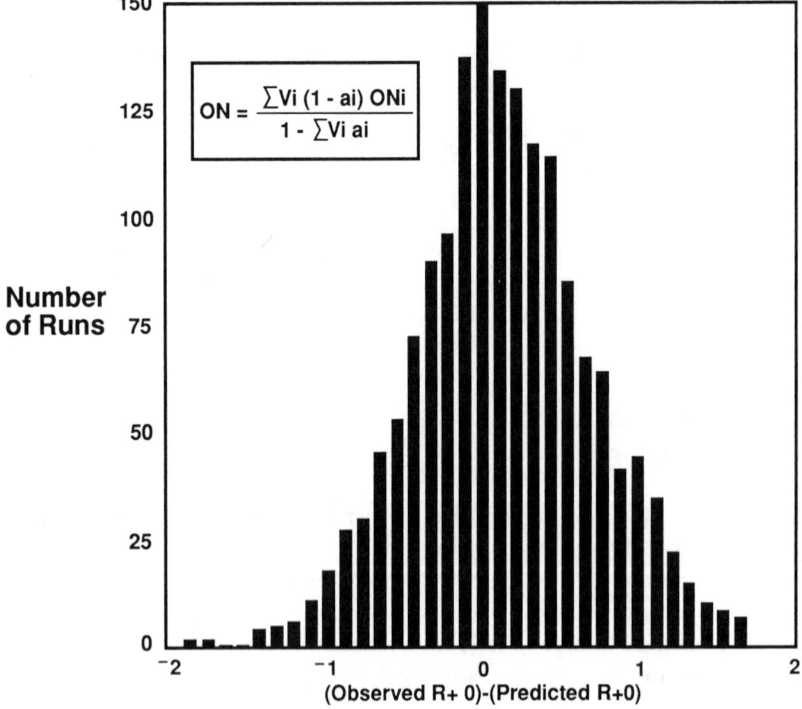

Figure 12–15 Accuracy of composition-based octane number model.

NALYTICAL TECHNIQUES

he progress of kinetic modeling is closely tied to the extent of available edstock characterization. Advancement in analytical techniques such as IPLC, GC, GC-MS, NMR, IR and other methods has greatly improved ur understanding of basic chemistry and therefore its representation in the inetic analysis. Recent advances in analytical chemistry of heavy oils and esids are summarized in a proceedings of ACS symposium held in Dallas, exas, April 1989. This symposium covers the state of the art in analytical hemistry as applied to petroleum fractions. The use of HPLC, GC-MS, IMS, and other techniques has allowed a significant progress in feedstock haracterization. Typical data for a Kern River Petroleum (Boduszynski 988) are summarized in Figures 12–16 and 12–17. This type of detailed verage molecular characterization was obviously not available in the late 960s and early 1970s when the first generation models were developed in ne industry. Making full use of this analytical information for efficient kinetic nodeling is the current challenge. Future progress in the analytical tools uch as planar wave guide spectroscopy and GC on a chip would certainly llow rapid accurate in-line analysis in the field and improve the process ontrol technology. A trend toward availability of advanced analytical echniques in the refinery laboratories, exclusively used in the research boratories in the past, has also been beneficial. This activity to upgrade efinery capabilities definitely provides a competitive advantage for the etailed kinetic models to optimize the unit performance.

Product properties are usually the most difficult to quantify in terms of asic chemical or physical phenomena. Some of the tests, such as octane umber for example, are measured on a standard engine following a set ecipe. Although several compositional based models for predicting prop-rties are now appearing in the literature and appear to be reasonably ccurate, their use is still limited. One such model for the reformate octane rediction was described previously. The kinetic models which can predict ufficient compositional details would definitely allow accurate predictions f product qualities. Process optimization invariably involves improved roduct quality as well as enhanced yields of the valuable products.

OMPUTER TECHNOLOGY

he increasing sophistication of the process modeling technology is also nked to the progress in computer hardware and software technology. We ave come a long way from the days of the two-dimensional slide rule used n the TCC kiln calculations to the use of supercomputers today. Mobil was ne of the first corporations to get dedicated computers to the research epartment in mid-50s. We were also the leaders in laboratory automation

Compound-Class Distributions in Kern River Petroleum

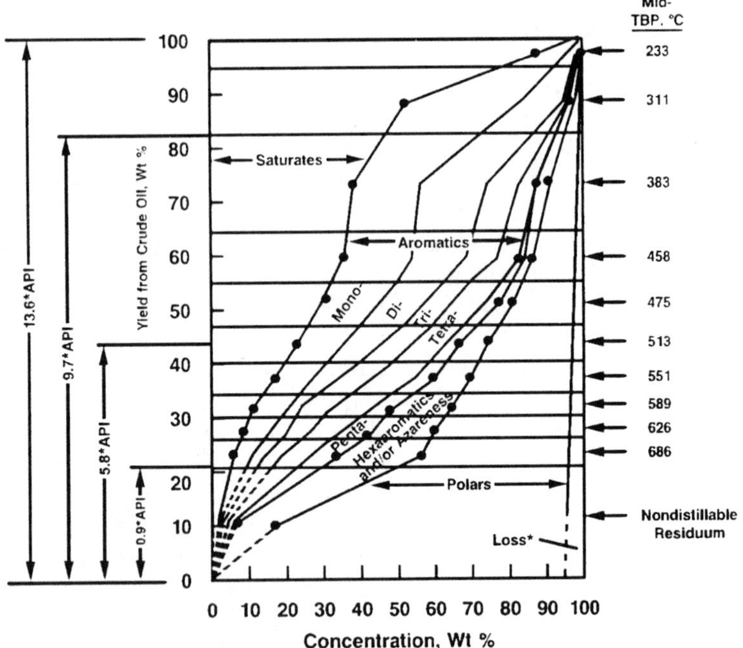

Distributions of Alkane Homologous Series in Kern River Petroleum

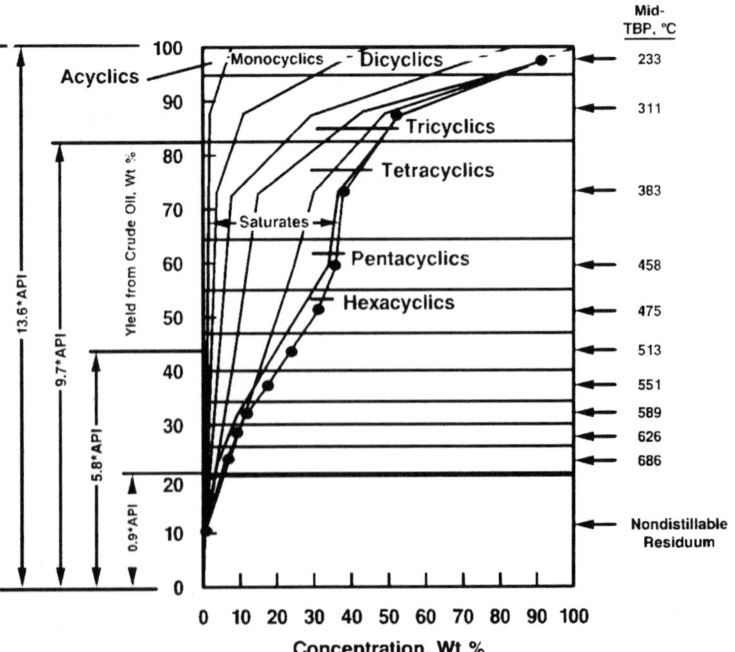

Figure 12–16 Typical advanced feedstock characterization.

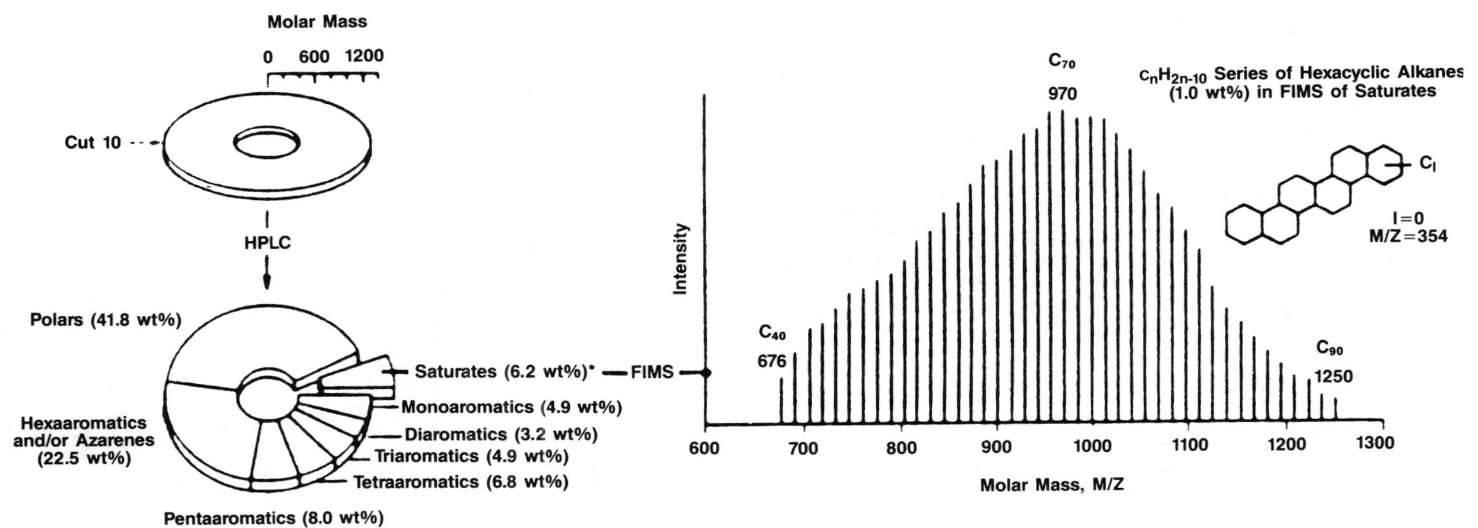

*All wt% Figures are on the Cut 10 Basis

Figure 12–17 Schematic diagram of HPLC-FIMS analysis.

and data acquisition technologies. Wide-spread use of advanced computer languages, such as APL, allow efficient data manipulation, interactive data analysis and timely model building.

Advances in speed, memory size, vectorizing and parallel processing are more than doubling the effectiveness of the computer each year. Advances in analytical chemistry now allow a significantly improved characterization of complex petroleum mixtures. This analytical information is now being used to build advanced kinetic models. Understanding of the reaction chemistry can be reflected in generating rule-based reaction networks. Allen's work is one such example from academia. Advanced computer software generating rule-based reaction pathways, including graphical displays portraying extent of reaction, are already being used in the industry. These rule-based methodologies, along with underlying fundamental reaction rate laws, clearly reflect a new generation of kinetic models. To transform these complex models into useful refinery optimization tools would require the full use of the advances in computing technology.

The advent of super computers has made it possible to do calculations that were not even imaginable previously. Increasing the use of these supercomputer capabilities in handling complex kinetics will definitely make the last decade in this century and the coming of a new millennium very exciting.

CONCLUSION

In the midst of the pursuit of progress in a given research field, it is sometimes fruitful to pause occasionally and take an historical stock of what has been accomplished so far. Although the focus here was restricted mainly to the Mobil work, it gives a flavor of what we believe to be a realistic picture of the industry as a whole. Recent research efforts by the industry and our university colleagues definitely are very important to the continued vitality of this field. The future for developing more accurate models for complex systems is even more challenging as exemplified by the chapters in this volume.

ACKNOWLEDGMENT

Useful discussions with C. D. Prater are sincerely appreciated.

REFERENCES

Avidan, A. A., M. Edwards, and H. Owen (1990) Innovative improvements highlight FCC's past and future. *O.G.J.* **1**:33.

Blanding, F. H. (1953) Reaction rates in catalytic cracking of petroleum. *I.E.C.* **45**:1186.

Boduszynski, M. M. (1988) Composition of heavy petroleums 2. Molecular characterization. *Energy Fuels* **2**:597.

Chin, A. A., J. E. Child, G. Hall, L. J. Altman, P. H. Schipper, and A. V. Sapre (1989) FCC cracking of coker gas oils. *A.I.Ch.E. Annual Meeting Preprint*, San Francisco.

Froment, G. F. (1987) The kinetics of complex catalytic reactions. *Chem. Engin. Sci.* **42**:1073.

Jacob, S. M., B. Gross, S. E. Voltz, and V. W. Weekman, Jr. (1976) A lumping and reaction scheme for catalytic cracking. *A.I.Ch.E.J.* **22**:701.

Kmak, W. S. (1971) Reforming kinetics. *A.I.Ch.E. Annual Meeting Preprint*, Houston.

Liguras, D. K. and D. T. Allen (1989a) Structural models for catalytic cracking 1. Model compound reactions. *Ind. Engin. Chem. Res.* **28**:665.

Liguras, D. K. and D. T. Allen (1989b) Structural models for catalytic cracking 2. Reactions of simulated oil mixtures. *Ind. Engin. Chem. Res.* **28**:674.

Prater, C. D., J. Wei, V. W. Weekman, Jr., and B. Gross (1983) A reaction engineering case history: coke burning in thermofor catalytic cracking regenerators. Academic Press, Orlando, Florida. Edited by J. Wei. In *Advances in Chemical Engineering, Vol. 12, 1.*

Ramage, M. P., K. R. Graziani, P. H. Schipper, F. J. Krambeck, and B. C. Choi (1987) KINPtR (Mobil's Kinetic Reforming Model): a review of Mobil's industrial process modeling philosophy. *Adv. Chem. Engin.* **13**:193.

Sapre, A. V. and T. M. Leib (1990) Translation of laboratory FCC catalyst characterization tests to riser reactors. *ACS Symposium* Series 452, Fluid Catalytic Cracking II, Concepts in Catalyst Design. Edited by M. L. Occelli, 144.

Voorhies, A. (1945) "Carbon formation in Catalytic Cracking" *Ind. & Engin. Chem.* **37**:318.

Weekman, V. W. Jr. (1968) A model of catalytic cracking conversion in fixed, moving and fluid-bed reactors. *I.E.C. Proc. Des. Dev.* **7**:90.

Weekman, V. W. Jr. (1979) Lumps, models and kinetics in practice. *A.I.Ch.E. Monogr. Ser.* **75**:3.

Weekman, V. W. Jr. and D. M. Nace (1970) Kinetics of catalytic cracking selectivity in fixed, moving and fluid-bed reactors. *A.I.Ch.E.J.* **16**:397.

Wei, J. and C. D. Prater (1962) The structure and analysis of complex reaction systems. *Advances in Catalysis, Vol. 13*, p. 203. Academic Press, Orlando, Florida. Edited by Eley, D. D., P. W. Selwood, and P. B. Weisz.

Wei, J. and J. C. W. Kuo (1969) A lumping analysis in monomolecular reaction systems-analysis of the exactly lumpable system. *Ind. Engin. Chem. Fund.* **8**:115.

Weisz, P. B. (1964) Polyfunctional heterogeneous catalysis. *Adv. Catal.* **13**:1.

Weisz, P. B. (1966) Combustion of carbonaceous deposits within porous catalyst particles. III. The CO_2/CO product ratio. *J. Catal.* **6**:425.

Weisz, P. B. and V. J. Frilette (1960) Intracrystalline and molecular shape-selective catalysis by zeolite salts. *J. Phys. Chem.* **64**:382.

Weisz, P. B. and R. D. Goodwin (1963) Combustion of carbonaceous deposits within porous catalyst particles. I. Diffusion-controlled kinetics. *J. Catal.* **2**:397.

Weisz, P. B. and R. D. Goodwin (1966) Combustion of carbonaceous deposits within porous catalyst particles. II. Intrinsic burning rate. *J. Catal.* **6**:227.

Weisz, P. B. and C. D. Prater (1957) Basic activity—properties for Pt-type reforming catalysts. *Adv. Catal.* **9**:583.

APPENDIX A KEY MOBIL ARTICLES

General

Wei, J. and C. D. Prater (1962) The structure and analysis of complex reaction systems. *Adv. Catalysis* **13**:203.

Wei, J. (1962) Axiomatic treatment of chemical reaction systems. *J. Chem. Phys.* **36**:1578.

Wei, J. and C. D. Prater (1963) A new approach to first-order chemical reaction systems. *A.I.Ch.E.J.* **9**:77.

Wei, J. (1965) Structure of complex chemical reaction systems. *Ind. Engin. Chem. Fund.* **4**:161.

Prater, C. D., A. J. Silvestri, and J. Wei (1967) On the structure and analysis of complex systems of first-order chemical reactions containing irreversible steps—I. General properties. *Chem. Engin. Sci.* **22**:1587.

Silvestri, A. J., C. D. Prater, and J. Wei (1968) On the structure and analysis of complex systems of first-order chemical reactions containing irreversible steps—II. Projection properties of the characteristic reactors. *Chem. Engin. Sci.* **23**:1191.

Weekman, V. W., Jr. (1969) Industrial process-oriented reaction engineering. *Ind. Engin. Chem.* **61**:53.

Weekman, V. W., Jr. (1970) Chemical reaction engineering. *Ind. Engin. Chem.* **62**:52.

Silvestri, A. J., C. D. Prater, and J. Wei (1970) On the structure and analysis of complex systems of first-order chemical reactions containing irreversible steps—III. Determination of the rate constants. *Chem. Engin. Sci.* **25**:407.

Wei, J. and J. C. W. Kuo (1969) A lumping analysis in monomolecular reaction systems—Analysis of the exactly lumpable system. *Ind. Engin. Chem. Fund.* **8**:115.

Kuo, J. C. W. and J. Wei (1969) A lumping analysis in monomolecular reaction systems—Analysis of approximately lumpable system. *Ind. Engin. Chem. Fund.* **8**:124.

Krambeck, F. J. (1970) The mathematical structure of chemical kinetics in homogeneous single-phase systems. *Arch. Ratl. Mech. Anal.* **38**:317.

Ozawa, Y. (1973) The structure of a lumpable monomolecular system for reversible chemical reactions. *Ind. Engin. Chem. Fund.* **12**:191.

Ozawa, Y. and J. Wei (1975) Perturbation on the rate constant matrix of a first-order chemical reaction. *Chem. Engin. Sci.* **30**:653.

Weekman, V. W., Jr. (1975) Industrial process models—state of the art. *Adv. Chem. Ser.* **148**:98.

Weekman, V. W., Jr. (1979) Lumps, models, and kinetics in practice. *A.I.Ch.E. Monogr. Ser.* **75**:3.

Weisz, P. B., A. B. Schwartz, and V. W. Weekman, Jr. (1980) Kinetics and catalysis in petroleum processing. *Proc. World Pet. Congr.* **10**:325.

Prater, C. D., J. Wei, V. W. Weekman, Jr., and B. Gross (1983) A reaction engineering case history: coke burning in thermofor catalytic cracking regenerators. *Adv. Chem. Engin.* **12**:1.

Krambeck, F. J. (1984) Computers and modern analysis in reactor design. I. *Chem. Symp. Ser.* **87**, ISCRE-8.

Krambeck, F. J. (1984) Accessible composition domains for monomolecular systems. *Chem. Engin. Sci.* **39**:1181.

Tabak, S. A., F. J. Krambeck, and W. E. Garwood (1986) Conversion of propylene and butylene over ZSM-5 catalyst. *A.I.Ch.E.J.* **32**:1526.

Krambeck, F. J., A. A. Avidan, C. K. Lee, and M. N. Lo (1987) Predicting fluid-bed reactor efficiency using adsorbing gas tracers. *A.I.Ch.E.J.* **33**:1727.

Quann, R. S., L. A. Green, S. A. Tabak, and F. J. Krambeck (1988) Chemistry of olefin oligomerization over ZSM-5 catalyst. *Ind. Engin. Chem. Res.* **27**:565.

Krambeck, F. J. (1988) Letters to the Editor. *A.I.Ch.E.J.* **34**:877.

Catalytic Cracking

Weekman, V. W., Jr., M. D. Harter, and G. R. Marr (1967) Hybrid computer simulation of a moving bed catalyst regenerator. *Ind. Engin. Chem.* **59**:84.

Weekman, V. W., Jr. (1968) A model of catalytic cracking conversion in fixed-moving and fluid-bed reactors. *Ind. Engin. Chem. Proc. Des. Dev.* **7**:90.

Weekman, V. W., Jr. (1968) Optimum operation-regeneration cycles for fixed-bed catalytic cracking. *Ind. Engin. Chem. Proc. Des. Dev.* **7**:252.

Weekman, V. W., Jr. (1969) Kinetics and dynamics of catalytic cracking selectivity in fixed-bed reactors. *Ind. Engin. Chem. Proc. Des. Dev.* **8**:305.

Weekman, V. W., Jr. and D. M. Nace (1970) Kinetics of catalytic cracking selectivity in fixed-, moving-, and fluid-bed reactors. *A.I.Ch.E.J.*, **16**:397.

Nace, D. M., S. E. Voltz, and V. W. Weekman, Jr. (1971) Application of a kinetic model for catalytic cracking—effects of charge stocks. *Ind. Engin. Chem. Proc. Des. Dev.* **10**:530.

Voltz, S. E., D. M. Nace, and V. W. Weekman, Jr. (1971) Application of a kinetic model for catalytic cracking—some correlations of rate constants. *Ind. Engin. Chem. Proc. Des. Dev.* **10**:538.

Jacob, S. M., V. W. Weekman, Jr., S. E. Voltz, and D. M. Nace (1972) Application of a kinetic model for catalytic cracking—3. Some effects of nitrogen poisoning and recycle. *Ind. Engin. Chem. Proc. Des. Dev.* **11**:261.

Gross, B., D. M. Nace, and S. E. Voltz (1974) Application of a kinetic model for comparison of catalytic cracking in a fixed-bed microreactor and a fluidized dense bed. *Ind. Engin. Chem. Proc. Des. Dev.* **13**:199.

Lee, W. and V. W. Weekman, Jr. (1976) Advanced control practice in the chemical process industry: a view from industry. *A.I.Ch.E.J.*, **22**:27.

Jacob, S. M., B. Gross, S. E. Voltz, and V. W. Weekman, Jr. (1976) A lumping and reaction scheme for catalytic cracking. *A.I.Ch.E.J.* **22**:701.

Venuto, P. B. and E. T. Habib, Jr. (1979) *Fluid Catalytic Cracking with Zeolite Catalysts.* Marcel Dekker, New York.

Chin, A. A., J. E. Child, G. Hall, L. J. Altman, P. H. Schipper, and A. V. Sapre (1989) FCC cracking of coker gas oils. *A.I.Ch.E. Annual Meeting*, San Francisco, November.

Reforming

Graziani, K. R. and M. P. Ramage (1978) Development of Mobil's Kinetic Reforming Model. *ACS Symp. Ser.* **65**:282.

Ramage, M. P., K. R. Graziani, and F. J. Krambeck (1980) Development of Mobil's Kinetic Reforming Model. *Chem. Engin. Sci.* **35**:41.

Schipper, P. H., K. R. Graziani, B. C. Choi, and M. P. Ramage (1984) The extension of Mobil's Kinetic Reforming Model to include catalyst deactivation. *Int. Chem. Eng. Symp. ISCRE-8, Ser.* **87**:33.

Ramage, M. P., K. R. Graziani, P. H. Schipper, F. J. Krambeck, and B. C. Choi (1987) KINPtR (Mobil's Kinetic Reforming Model): a review of Mobil's industrial process modeling philosophy. *Adv. Chem. Engin.* **13**:193.

Hydroprocessing

Jaffe, S. B. (1974) Kinetics of heat release in petroleum hydrogenation. *Ind. Engin. Chem. Proc. Des. Dev.* **1**:34.

Weekman, V. W., Jr. (1976) Hydroprocessing reaction engineering. *Chem. React. Eng. Proc. Int. Symp.* **4**:615.

Heck, R. H. and T. R. Stein (1977) Kinetics of hydroprocessing distillate coal liquids *ACS Div. Pet. Chem. Prep.* **22**:948.

Farcasiu, M., T. O. Mitchell, and D. D. Whitehurst (1976) On the kinetics and mechanisms of solvent refined coal. *Prep. Coal Chem. Workshop*, 101.

Jaffe, S. B. (1976) Hot spot simulation in commercial hydrogenation processes. *Ind. Engin. Chem. Proc. Des. Dev.* **15**:410.

Heck, R. H., M. J. Dabkowski, T. R. Stein, and C. A. Simpson (1979) Catalytic hydroprocessing of heavy coal liquids. *87th A.I.Ch.E. National Meeting, Boston, Prepr. 2B.*

Jaffe, S. B (1981) Catalysts for hydrotreating residual petroleum oil. U.S. Patent 4,267,071.

Shih, S. S. and P. J. Angevine (1985) Kinetic studies of coal liquids upgrading. *A.I.Ch.E. National Meeting*, Houston, Prepr. No. 9F.

Child, J. E., Q. N. Le, S. S. Shih, and A. V. Sapre (1986) Dearsenation of shale oil with low hydrogen consumption. *Energy Prog.* **6**:61.

Quann, R. J., R. A. Ware, C. W. Hung, and J. Wei (1989) Catalytic hydrodemetallation of petroleum. *Adv. Chem. Engin.* **14**:95.

APPENDIX B KEY NON-MOBIL CONTRIBUTIONS

Aris, R. and G. R. Gavalas (1966) Theory of reactions in continuous mixtures. *Philos. Trans. Roy. Soc.* **A260**:351.

Aris, R. (1968) Prolegomena to the rational analysis of systems of chemical reactions. II. Some addenda. *Arch. Ratl. Mech. Anal.* **27**:356.

Hutchinson, P. and D. Luss (1970) Lumping of mixtures with many parallel first-order reactions. *Chem. Engin. J.* **1**:129.

Luss, D. and P. Hutchinson (1971) Lumping of mixtures with many parallel *n*th order reactions. *Chem. Engin. J.* **2**:172.

Kmak, W. S. (1971) A kinetic simulation model of the power forming process. *A.I.Ch.E. National Meeting*, Houston.

Bailey, J. E. (1972) Lumping analysis of reactions in continuous mixtures. *Chem. Engin. J.* **3**:52.

Golikeri, S. V. and D. Luss (1972) Analysis of activation energy of grouped parallel reactions. *A.I.Ch.E.J.* **18**:277.

Jacquez, J. A. (1972) *Compartmental Analysis in Biology and Medicine.* Elsevier, New York.

Liu, Y. A. and L. Lapidus (1973) Observer theory for lumping analysis of monomolecular reaction systems. *A.I.Ch.E.J.* **19**:467.

Golikeri, S. V. and D. Luss (1974) Aggregation of many coupled consecutive first-order reactions. *Chem. Engin. Sci.* **29**:845.

Luss, D. and S. V. Golikeri (1975) Grouping of many species, each consumed by two parallel first-order reactions. *A.I.Ch.E.J.* **21**:865.

Benson, S. W. (1976) *Thermochemical kinetics.* John Wiley, New York.

Bischoff, K. B. (1977) Pharmacokinetic considerations in environmental toxicology, in *Environmental Health-Quantitative Methods.* (Whittmore, A., Ed.), SIAM, Philadelphia.

Lee, H. (1977) Kinetic behavior of mixtures with many first-order reactions. *A.I.Ch.E.J.* **23**:116.

Grinday, G. B., R. G. Moran, and R. Weokheiser (1975) *Med. Chem. Ser. Monogr. Drug Design*, **5**.

Lee, H. (1978) Synthesis of kinetic structure of reaction mixtures of irreversible first-order reactions. *A.I.Ch.E.J.* **24**:116.

Smith, W. R. and R. W. Smitten (1982) *Chemical Reaction Equilibrium Analysis*. John Wiley, New York.

McRae, G. J. and J. H. Seinfeld (1983) Development of second-generation mathematical model for urban air pollution—II. Evaluation of model performance. *Atmos. Environ.* **17**:501.

Godfrey, K. (1983) *Compartmental Models and Their Applications*. Academic Press, Orlando, Florida.

Li, G. (1984) A lumping analysis in mono- or/and bimolecular reaction systems. *Chem. Engin. Sci.* **29**:1261.

Alberty, R. A. and I. J. Oppenheimer (1984) A continuous thermodynamics approach to chemical equilibrium within an isomer group. *J. Chem. Phys.* **81**:4603.

Frenklach, M. (1985) Computer modeling of infinite reaction sequences: a chemical lumping. *Chem. Engin. Sci.* **40**:1843.

Prasad, G. N., J. B. Agnew, and T. Sridhar (1986) Continuous reaction mixture model for coal liquefaction theory. *A.I.Ch.E.J.* **32**:1277.

Prasad, G. N., J. B. Agnew, and T. Sridhar (1986) Continuous reaction mixture model for coal liquefaction—II. Comparison with experiments on catalyzed and uncatalyzed liquefaction of coals of different rank. *A.I.Ch.E.J.* **32**:1288.

McDermott, J. B. and M. T. Klein (1986) Chemical and probabilistic modeling of complex reactions: a lignin depolymerization example. *Chem. Engin. Sci.* **41**:1053.

Ho, T. C. and R. Aris (1987) On apparent second-order kinetics. *A.I.Ch.E.J.* **33**:1050.

Coxson, P. G. and K. B. Bischoff (1987) Lumping strategy—1. Introductory techniques and applications of cluster analysis. *Ind. Engin. Chem. Res.* **26**:1239.

Coxson, P. G. and K. B. Bischoff (1987) Lumping strategy—2. A system theoretic approach. *Ind. Engin. Chem. Res.* **26**:2151.

Chen, J. M. and H. D. Schindler (1987) A lumped kinetic model for hydroprocessing coal extract. *Ind. Engin. Chem. Res.* **26**:921.

Froment, G. F. (1987) The kinetics of complex catalytic reactions. *Chem. Engin. Sci.* **42**:1073.

Parnas, R. S. and D. T. Allen (1988) Compound class modeling of hydropyrolysis. *Chem. Engin. Sci.* **43**:1845.

Chou, M. Y. and T. C. Ho (1988) Continuum theory for lumping nonlinear reactions. *A.I.Ch.E.J.* **34**:1319.

Astarita, G. and R. Ocone (1988) Lumping nonlinear kinetics. *A.I.Ch.E.J.* **34**:1299.

Chou, M. Y. and T. C. Ho (1989) Lumping coupled nonlinear reactions in continuous mixtures. *A.I.Ch.E.J.* **35**:533.

Astarita, G. (1989) Lumping nonlinear kinetics: apparent overall order of reaction. *A.I.Ch.E.J.* **35**:529.

Aris, R. (1989) On reactions in continuous mixtures. *A.I.Ch.E.J.* **35**:539.

Liguras, D. K. and D. T. Allen (1989) Structural models for catalytic cracking 1. Model compound reactions. *Ind. Engin. Chem. Res.* **28**:665.

Liguras, D. K. and D. T. Allen (1989) Structural models for catalytic cracking 2. Reactions of simulated oil mixtures. *Ind. Engin. Chem. Res.* **28**:674.

Baltnas, M. A., V. Raemdonck, K. Kristiaan, G. F. Froment, and S. R. Moledas (1989) Fundamental kinetic modeling of hydroisomerization and hydrocracking on noble metal-loaded faujasites. 1. Rate parameters for hydroisomerization. *Ind. Engin. Chem. Res.* **28**:899.

PART 4

THERMODYNAMICS

13

Uses and Needs of Thermodynamics in the Oil Industry

JAMES C. W. KUO

Thermodynamics is something that nobody working in science and engineering can do without. In fact, we all comprehend the truth that the whole universe is driving toward a physical and chemical limit defined by thermodynamics, and this truth can also be applied to many branches of social science, such as economics and philosophy. Whether you like it or not, we are all living following the laws of thermodynamics. One can never get away from it or contradict it. No wonder that some people say that "thermodynamics is a religion." If there is a law of nature, then thermodynamics is playing the most important role in this law.

Another fact about thermodynamics is that its four basic laws[1] were developed years ago. Therefore, unless there is a fifth law coming soon, all on-going work in thermodynamics consists of practical applications of these four laws. The objective of this chapter is to communicate the uses and needs of applied thermodynamics from the perspective of a scientist working in the oil industry.

Since the uses and needs of thermodynamics in the oil industry are very

[1] In addition to the three well-known basic laws of thermodynamics, there is a so-called "Zeroth Law of Thermodynamics" that states that if body "A" is in thermal equilibrium with both bodies "B" and "C" then "B" and "C" must also be in equilibrium.

widespread, it is impossible to cover the entire area here. Therefore, the chapter is restricted to the uses and needs of thermodynamics related to "mixtures."

In the oil industry, we very often deal with mixtures of chemical compounds. In a sense, this chapter is concerned with these mixtures and thermodynamics. Although the volume focuses on problems related to complex mixtures in the chemical reaction system, this chapter includes some thermodynamic aspects of nonreactive systems.

When dealing with mixtures of compounds, one needs the "lumping of mixtures" and the thermophysical data bases for the "lumps." It is important to clarify one point about the thermophysical data bases. Of course, the data bases include those properties associated with the four basic laws of thermodynamics, but they also include other properties, particularly those used in characterizing products (for instance, viscosity and octane number). Lumping, thermophysical data bases, and thermodynamics constitute the essence of this chapter.

DRIVING FORCES BEHIND THE INCREASING INDUSTRIAL USES

Uses of thermodynamics and thermophysical data bases in the oil industry have always been very substantial. The present and future trends in the industry, however, indicate an increasing need for these uses. There are large driving forces today and in the future that demand large changes in the industry. As is well known, changes demand increasing research and development efforts, and these demand increasing needs in the use of thermodynamics and thermophysical data bases. Some of these driving forces are explained here.

The industry is increasing its efforts to enhance recovery of huge amounts of oil from the aged oil fields of North America. These efforts result in an increasing need for applications of phase equilibrium, particularly in the area of supercritical extraction.

Similarly, there are substantial increases in our concern for environmental protection today that have a great impact on the oil industry. Two examples can be given. The U.S. government imposes a sulfur limit of 500 ppmw in diesel fuels in order to minimize the SO_x pollution. This raises a question of whether this sulfur limit violates the chemical equilibrium limits of our hydrodesulfurization reactions. Another example is the increasing use of oxygenates as gasoline octane boosters or pollution depressants. In this area, we often encounter the problems of blending and phase separation.

The industry is also heavily involved in the research and development of various new technologies and increasing the optimization of refinery operation to improve efficiency. These trends provide substantial impetus to the use of thermodynamics and thermophysical data bases. Of course, engineers and scientists in the oil industry always demand detailed characterizations of various streams and fundamental quantifications of various processes. However, this desire was and still is limited by existing analytical and computational capabilities. Fortunately, in the last 20 years or so we have seen tremendous progress in these capabilities. This progress enables us to deal with more complex and heavier compounds and mixtures today and in the future.

LUMPING ACCORDING TO OBJECTIVES

As mentioned earlier, we often need to establish "lumping" in dealing with complex mixtures. It is essential to establish a proper but general guideline for "lumping." My guideline for "lumping" is rather simple, i.e.,

"LUMPING ACCORDING TO OBJECTIVES."

There are many objectives of lumping. The following list is by no means complete.

- Descriptions of phase equilibrium and related properties
  - —Phase equilibria
  - —Dew and bubble points
  - —Boiling point distributions
  - —ASTM D86 distillations
  - —Reid vapor pressures and initial vapor pressures
- Descriptions of chemical equilibrium
  - —Chemical equilibria without phase separation
  - —Chemical equilibria with phase separation
  - —Gibbs enthalpies of formation
- General characterizations of various streams (enthalpy and stoichiometric balances)
  - —Enthalpies or heat capacities
  - —Enthalpies of vaporization
  - —Enthalpies of formation
  - —Temperatures from given enthalpies
  - —Heats of combustion
  - —Weights or masses
  - —Liquid volumes
  - —Atomic stoichiometries
- Specific characterizations of product streams
  - —Octane or cetane numbers
  - —Viscosities

—PONA (paraffins, olefins, naphthenes, aromatics) distributions
—Liquid specific gravities
—Viscosity indexes
—Freezing points
—Flash points
—Smoke points

- Other process calculations
  —Heats of reaction
  —Adiabatic temperatures from chemical reactions with and without phase separation
  —Gas compression work

Obviously, a lumping scheme can be applied to many objectives. For instance, all the properties related to the phase equilibrium category can have one consistent lumping scheme. Same or similar lumping schemes may also be used for general characterization of various streams and other process calculations. However, special care must be given to the schemes for describing chemical equilibria and special characterizations of product streams.

A detailed discussion on lumping for describing chemical equilibrium will be given later. In this case, a complication arises from the fact that one has to account for entropies of mixing when dealing with chemical equilibria.

Caution must be exercised in the construction of a lumping scheme for characterization of specific products. For example, one must separate the *n*-paraffin from *i*-paraffins of the same carbon number for describing octane and cetane numbers and viscosities. In fact, one may even have to separate single-branched *i*-paraffins from double-branched *i*-paraffins in order to describe those properties satisfactorily.

In my long experience in dealing with lumpings, I have found that it is a luxury if one can find a universal lumping scheme that will fit all objectives. Consequently, algorithms for delumping and lumping are often needed to tie together various lumping schemes.

THERMOPHYSICAL DATA BASES OF PURE COMPOUNDS

An important aspect of applied thermodynamics is to have a good thermophysical data base. In fact, the development and construction of thermophysical data bases might very well be the most important task in thermodynamics today. Although the volume is concerned with "complex mixtures," we need to start with some discussion of the thermophysical data bases of pure compounds because the properties of a mixture depend a priori on the properties of its constituent pure compounds.

We face two major problems when we construct the thermophysical data bases for pure compounds, i.e.,

- Data on certain properties, such as viscosities, cetane numbers, freeze points, flash points, and smoke points, are very scarce.
- Data on many compounds, such as heavier hydrocarbons (particularly polycyclic hydrocarbons) and heteroatom compounds, are essentially nonexistent.

In dealing with properties, one may not need the data for each individual compound if a good "property model" can be constructed to predict the properties of the stream based on the composition of the stream. This situation may occur in smoke point prediction but is generally not applicable to other property predictions.

We need to develop further techniques for predicting various properties for various compounds. Among the various property estimation methods, I consider the Group Contribution Methods (GCMs) to be the most fundamental, advanced, accurate, and comprehensive. Basically, they first partition a known molecular structure into groups and then estimate the properties based on group additivity and interactions among groups. In other words, they utilize the detailed molecular structure of the compound which is fundamentally an excellent approach.

Among the many GCMs, I consider the Benson Method (BGCM) the best because it has excellent fundamental chemical and physical understanding in defining group classes and contains many detailed secondary corrections. It also has a very extensive data base, and continuing efforts in group identification and data base improvements can be expected in the future.

The BGCM was invented in 1958 (Benson and Buss 1958) and has been under extensive development since then. It still has many aspects that need further development. The following list is by no means complete

- Data for many groups and corrections are nonexistent.
- Properties that can be estimated are limited to
  - —Ideal gas enthalpies
  - —Ideal gas enthalpies of formation at 25°C
  - —Intrinsic ideal gas absolute entropies at 25°C
  - —Liquid enthalpies of formation at 25°C (limited to hydrocarbons only)
- New groups and corrections may be needed.

The "intrinsic entropies" indicated above may need some explanation. These are the entropies estimated by adding the group entropy values and the group correction values. These values do not include the entropy contributions resulting from the symmetry of the molecule. Consequently, in order to obtain the real entropies one must add additional contributions relating to the "number of optical isomers" and the "internal and external symmetry numbers" of the molecule (see the Appendix for details) according to the following equation:

$$S = S_{ins} + R(\ln N_{opt} - \ln N_{int} - \ln N_{ext}). \tag{13–1}$$

N_{opt} can be obtained from a consideration of stereochemistry and N_{int} can be obtained by a simple algorithm of counting the number of terminal rotors. Nevertheless, finding N_{ext} is a complicated matter. It generally requires a rigorous knowledge of "point group theory" which is usually not available to most engineers and scientists. We urgently need simpler algorithms to make this task easier.

It is clear now that the use of BGCM for estimating ideal gas absolute entropy (or Gibbs enthalpy of formation) is much more complicated than its use for estimating ideal gas enthalpy and enthalpy of formation. Unfortunately, the ideal gas Gibbs enthalpy of formation is a very important property needed for evaluating chemical equilibria, one of the most useful calculations in thermodynamics.

I would like to emphasize the need of extending the BGCM to estimate properties other than those listed above. These properties can include

- Normal boiling points (NBP)
- Enthalpies of evaporation at a temperature other than NBP
- Normal freezing points
- Critical temperatures and pressures
- Viscosities.

Of course, one can always use different GCMs to estimate the above properties; however, the process can be very tedious, time-consuming, and confusing. It will be much better if one can build up a library (i.e., a data base) of contributions of BGCM and use it to estimate all properties.

Of course, more work is needed to find out whether the BGCM is good for all (or most of the) properties. However, because the BGCM is probably the most powerful GCM (it contains more groups and corrections than any other GCM that I know of), there is a very good chance that our goal can be achieved.

Another sensible task to be done in relation to the above task is to build up a central library of the BGCM contributions data base. I am not aware that this has been done before.

SOME GENERAL ASPECTS OF LUMPING AND THERMOPHYSICAL DATA BASES

We are now ready to discuss the problem of how to treat the hydrocarbon or other chemical mixtures such as petroleum fractions. In these cases, it becomes either impossible to measure or impractical to use the compositional distribution within each mixture. We need to "lump," and there are basically two ways to do this. I will call one way the "old approach" and the other the "new approach."

The old approach was introduced in the earlier days when analytical techniques were very rudimentary. In this approach, hydrocarbon mixtures were characterized according to some easily measured properties, such as boiling range, specific gravity, and viscosity (see Hougen et al. 1959). Many other properties, such as hydrogen content, heat capacity, heat of vaporization, and heat of combustion, could be estimated only with a high degree of inaccuracy. Although this approach had the advantages of simplicity and convenience, it was totally devoid of the fundamental understanding of the stoichiometry and molecular types and structure of the compounds that existed in the mixture. The old approach will not be discussed any further in this chapter.

With advances in analytical techniques and computer technology and increased goals of scientists and engineers to model feedstocks, processes, and products, newer and improved lumping approaches have been adopted in the last 30 years or more. These newer approaches use "isomer[2] lumps" so that the atomic stoichiometry is maintained. They also take into account the molecular types and, to some extent, the molecular structures of the compounds that existed in the mixture. In other words, they adopted a componential approach by treating mixtures as "lumped" compounds with intrinsic properties similar to those of an actual compound.

Although one can always treat all isomers as individual compounds, very often one does not want or cannot afford to do so. For example, optical isomers, which always come in pairs (of mirror images in molecular symmetry)[3], are usually "lumped" into one compound because they have identical thermophysical properties except for the ability to rotate a polarized light. The only thermophysical property that differs between the optical

[2] There are many definitions of "isomer." The one adopted here is the "Composition Isomer" (i.e., isomers with the same molecular formula), except that any functional groups, such as –OH,–CHO, –COOH, etc., and any ring structures are considered to be inseparable entities.

[3] Throughout this chapter, an "optical isomer" and an "isomer of mirror image" are synonymous.

isomers and the "isomer lump" is the Gibbs enthalpy of formation. The difference in the Gibbs enthalpies of formation is

$$RT \ln 2. \tag{13–2}$$

Lumping is also useful when there are so many isomers that it is impractical to treat them as individual compounds.

In making isomer lumping, the first task is usually the identification of each isomer and its molecular structure.

One of the newer approaches involves expanding the thermophysical data base to include many pure compounds, selecting mixture lumps based on isomers, and representing each isomer lump as a proper pure component. With this approach, the integrity of the stoichiometry and, to some degree, the molecular structure of the isomers, can be maintained. For example, during the development of Mobil's MTG (Methanol-to-Gasoline) process, our analytical personnel were able to analyze the products of the MTG reactions using gas chromatography resulting in the identification of hundreds of compounds. Based on the analytical results, we assembled a compositional group containing about 40 pure components and lumps which are represented by a typical compound of the isomer lump. To perform process calculations, we assembled a thermophysical data base containing all the pure and representative compounds. This approach is definitely more elegant and fundamental than the older approach.

There is a difficulty in estimating the Gibbs enthalpy of formation for the compound representing the lump. In addition to the Gibbs enthalpy of formation of the pure compound, one must add a contribution from the entropy of mixing. When the isomer contribution within the lump is known, the entropy of mixing is given as

$$R \textstyle\sum_i x_i \ln x_i, \tag{13–3}$$

where x_i, $i = 1, 2, \ldots, N$, are the molar fractions of the isomers in the lump. However, most often the isomer distribution of the lump is unknown. In this case, one may assume that all isomers in the lump have the same thermophysical properties and Eq. (13–3) becomes

$$-R \ln N, \tag{13–4}$$

where N is the total number of isomers in a lump. In this highly simplified case, one needs to know the total number of isomers. This assumption may be the best one can hope for if the isomer distribution in the lump is unknown.

For our purposes, we shall call the above approach the Lumping Approach-I (LA-I).

ISOMER LUMPING WITH A KNOWN ISOMER DISTRIBUTION

The above approach works quite well for most of the process and product calculations. However, one shortcoming is that many thermophysical properties of the lump may be quite unreliable or inaccurate because of the large spread of the values of the corresponding properties of the isomers in the lump. Nevertheless, because the isomer distribution of the lump is known, the accurate value of the property of the lump can always be estimated from the value of the corresponding property of the isomers according to some mixing rules (or shall we say "lumping rules").

All enthalpy-related properties, except for those related to entropy, may be approximated by a linear combination of the corresponding properties of isomers using the known isomer compositions as weighting factors. For entropy of the lump, an amount equal to the entropy of mixing must be subtracted from the entropy of the lump obtained by the rule of linear combination. This entropy of mixing is given as Eq. (13–3). A similar correction resulting from the entropy of mixing must be added to the Gibbs enthalpy of formation of the lump obtained by the rule of linear combination.

A word of caution must be given to those using the above approach. Some thermophysical properties of the isomers in an isomer group can vary widely according to their molecular structures or other factors. Let us take the normal freezing point (NFP) as an extreme example. The difference between the maximum and minimum NFP within an isomer group can be very large. The maximum and minimum NFPs for octanes are 374 K (for 2,2,3,3-tetramethylbutane) and 147 K (for 2,3-dimethylhexane), respectively, which gives a difference of 227 K. Another well-known example is the NFP of durene (1,2,4,5-tetramethylbenzene), the most notorious product of our MTG process. The NFP of durene is 352 K whereas that of *n*-butyl benzene is only 185 K, a difference of 167 K. Durene is an ideal gasoline component in every way except for its unacceptably high NFP. It crystallizes in gasoline when its concentration is high enough. In our New Zealand MTG plant, a HGT (Heavy-Gasoline-Treatment) process was installed to lower the durene concentration in the MTG gasoline to an acceptable level. An obvious way to deal with this disparity in isomer properties is to choose the isomer or isomers that have distinct values of the properties as a separate lump. Of course, this is necessary only if these properties are the objectives of the lumping.

From the above observation on NFPs, I would like to emphasize that a very strong relationship exists between isomer molecular structure and NFP. The isomers whose molecular structures have the highest degree of symmetry within the isomer group almost always possess the highest NFP within the group. Similar relationships also exist for many other thermophysical properties. In fact, properties of symmetrical isomers often differ very much from those of asymmetrical isomers. In a way, the symmetrical isomers are so different from the rest of the isomers that they stand out just like tall trees in a forest. Obviously, one cannot ignore these differences and it probably makes a lot of sense to treat symmetric isomers as a separate lump.

The mixing rules for obtaining properties, such as critical temperatures and pressures, that are related to the use of corresponding states are very complicated in general. They usually require knowledge of the binary interaction parameters to account for the effect due to the interaction of any two components in the mixture. This interaction parameter is unity when there is no interaction between the two components and deviates from unity when there is interaction. For a mixture containing a large number of components, the total number of binary interaction parameters is astronomical (with 100 components in a mixture, the number of parameters is 9,900). Obviously, it is impractical to obtain all the parameters in these cases. Fortunately, we are dealing only with isomers in each mixture and the mixing rules may be simplified. However, I do not know if any studies have been done in this area for isomer mixtures. Can one assume all binary interaction parameters to be unity? Or, if they are not unity, can one estimate them according to the disparity between two molecular structure?

For some properties, such as viscosity, very complicated mixing rules may be needed. Reid et al. (1987) give an excellent review of these rules. Still, other properties, such as octane number, have complicated but proprietary mixing rules that are being developed separately by many industrial outfits.

To distinguish the above approach from the earlier LA-I, we shall call it the Lumping Approach-II (LA-II).

In LA-II, some thermophysical properties of the isomers in the lump are very often not known. In these cases, the thermophysical properties can be estimated using various proper estimation techniques (see Reid et al. 1987).

ISOMER LUMPING WITH AN UNKNOWN ISOMER DISTRIBUTION

In a majority of cases, one does not have the luxury of knowing the isomer distribution within a lump. The question now is how to approximate or estimate the isomer distribution for the lump. We need to make a distinction on whether the streams or system under consideration are related to any isomerization reactions.

When the streams or the system under consideration are not related to any isomerization reactions, we have two options:

- Approximating the lump with a representative isomer
- Requesting help from our analytical personnel.

In the first option, one will take whatever information one has on the types of isomers in the lump and choose one isomer whose properties represent mid-ranged values among all possible isomers. Of course, one needs only to consider those properties that are related to the objectives of the lumping. This option may be unsatisfactory from a rigorous point of view but is the best option for achieving short-term objectives. This option is the same as LA-I.

The second option is usually very time consuming and expensive. It is the best option if one has the luxuries of time and money.

If the streams or the systems under consideration are related to a set of isomerization reactions, then we may assume the isomer distribution of each isomer lump is at chemical equilibrium with respect to the set of isomerization reactions. It is very important to select the proper set of isomerization reactions based on the reaction mechanisms of the catalytic system used. This selection is particularly important when using shape-selective catalysts. Shape-selectivity may restrict the isomers to those with certain molecular structures. Once the set of isomerization reactions is fixed, the isomers in the lump are defined as the possible isomers derived from the isomerization reactions. As an example, if a catalytic system only promotes the moving of the double bond in olefins, then the isomers within each lump are restricted to olefins with various double bond positions but the same carbon-atom skeleton.

Next, one needs the Gibbs enthalpies of formation of all the isomers in the lump. If they are unknown, then one needs to enumerate the isomers in the lump and to estimate the Gibbs enthalpy of formation of each isomer based on its molecular structure. BGCM is the one method most often used for this purpose (Benson 1976). However, in order to obtain a correct Gibbs enthalpy of formation of any isomer, this and all other estimation methods require an adjustment of the estimated enthalpies by an amount due to an effect resulting from molecular symmetry and the existence of any optical isomer (see the section on Thermophysical Data Bases of Pure Compounds).

Once the isomer distribution of the lump is established, one can go back to the procedure given in the preceding section to estimate various thermophysical properties of the lump.

The need to deal with a subset of isomers instead of a set of all isomers has one serious consequence. One may need data bases of more than one isomer lump within the set of all isomers. In fact, instead of having permanent data bases for all possible subsets of isomers, one may want to maintain computer software that can establish a data base for any subset of isomers whenever needed.

We shall call the above approach of assuming isomer equilibrium within each lump as Lumping Approach-III (LA-III). Alberty and his colleagues at MIT have pioneered this type of work on many isomer systems (Alberty 1987; Alberty and Reif 1988). Mobil's effort on the kinetic modeling of the MOGD (Mobil Olefin-to-Gasoline and Distillate) process is also based on the same concept.

LA-III is obviously not proper for the streams before any chemical reactions take place because the assumption of an isomer equilibrium would be invalid.

It is clear now that knowledge of the Gibbs enthalpies of formation of all isomers in the lump is crucial in constructing a thermophysical data base of the lump. Unfortunately, estimation of these enthalpies using BGCM is quite troublesome (see the section on Thermophysical Data Bases of Pure Compounds).

LUMPING AND KINETIC MODELING

This section emphasizes the importance of constructing a kinetic lumping scheme without violating fundamental thermodynamic principles. In my opinion, a good kinetic lumping scheme must satisfy the following points in the compositional space:

- Initial compositions
- Chemical equilibrium composition.

If it becomes impossible to satisfy these compositions, then the errors of lumping at these compositional points should be within an acceptable limit. When a lumping scheme satisfies the above criteria, we have completed half of the job. The next step would be to minimize the lumping errors of the first derivatives at the initial points and the equilibrium point. In fact, in the neighborhood of the equilibrium point, the kinetic system becomes a mono-molecular reaction system; the lumping of such a system has been thoroughly studied by Wei and Kuo (Kuo and Wei 1969; Wei and Kuo 1969). One can intuitively perceive also that the lumping errors will become smaller when higher derivatives at the initial points and the equilibrium point are satisfied. In this sense, an exact lumping scheme is obtained when all the lumped derivatives at the initial and the equilibrium points contain no error.

Based on the above reasoning, determination of the thermophysical roperties of the lumped species of any lumping scheme becomes a very rucial issue. This is particularly true for the determination of the Gibbs nthalpies of formation of all lumps because these are required for estimation f the lumped equilibrium compositions.

ROM ISOMER MOLECULAR STRUCTURE ELUCIDATION TO HERMOPHYSICAL DATA OF "ISOMER LUMPS"

t is now possible to summarize the steps required to establish a thermo-hysical data base for isomer lumps whose isomer distributions are restricted y some isomerization reactions. One must

1. Identify the set of isomerization reactions that will restrict the types of isomers in each lump.
2. Select lumps according to our lumping objective or objectives and according to the restrictions imposed by Step 1.
3. Enumerate all possible isomers within the restrictions imposed by Steps 1 and 2, and identify their structures.
4. Obtain for each isomer its group and correction contributions of various GCMs (hopefully we need only one GCM, such as BGCM).
5. Obtain for each isomer its total symmetry number and the number of its optical isomers.
6. Establish ideal gas enthalpy correlations, enthalpies of formation (at 25°C), and absolute entropies (at 25°C) of all isomers in the lump.
7. Estimate Gibbs enthalpies of formation at a selected isomerization temperature for all isomers in the lump and then calculate the isomer distribution in the lump.
8. Estimate for each isomer all other required properties using proper GCMs.
9. Obtain all binary interaction parameters for all isomer-pairs in the lump (or set them to be unity).
10. Estimate all the required properties using proper mixing rules.

All these steps are necessary although approximations or simplifications may be adopted for some steps. Many of these steps need further development. So far I have identified the following:

- Establishment of algorithms for "isomer enumeration" (Steps 1, 2, and 3).
- Expansion of data bases of GCMs, particularly that of BGCM (Step 4).
- Establishment of simple algorithms or algorithms for determining the "external symmetric numbers" of any molecule (Step 5).

- Extension of use of BGCM to yet more thermophysical properties (Step 8)
- Establishment of an algorithm or algorithms for determining the binar interaction parameters of isomer-pairs which are needed for setting mixin rules for the equation of state (Step 9).

The area of "isomer enumeration" is already under active research an development, such as the work within the DENDRAL project under th leadership of J. Lederberg (Lindsay et al. 1980), and many others (Dias 1984 Kleywegt et al. 1987). Most of the work involves use of Expert Systems. However, as far as I know, none of the work involves such restrictions a those imposed by Steps 1 and 2. Nevertheless, I do not believe that suc restrictions will create major difficulties in using Expert Systems, but some body has to revise them so that their algorithms can be tailored for our us Furthermore, the final Expert Systems must be "user friendly," which ma be a difficult task.

It is well known that the number of isomers within certain types c compounds (such as alkanes or alkenes) increases geometrically with respec to increasing carbon number. For example, there are 46 million isomers c C_{20} alkenes. This means that we must learn to handle the cases of a larg number of isomers in the lumps. Professor Alberty is a leader in this area He and his colleagues at MIT have reported ideal gas heat capacities enthalpies of formation, and Gibbs enthalpies of formation of "isome lumps" of many hydrocarbon and hetero-atom hydrocarbon system (Alberty 1987; Alberty and Reif 1988). However, it is not clear to me how the handled entropy corrections related to the "total symmetry number" of th isomers.

There is another question of maintaining an "internal consistency rela tionship" among the three ideal gas enthalpies. For any pure compound, th ideal gas enthalpy of formation and Gibbs enthalpy of formation at al temperatures are defined if the following enthalpies are known:

- Ideal gas enthalpies at all temperatures
- Ideal gas enthalpy of formation at one temperature, usually at 25°C
- Ideal gas Gibbs enthalpy of formation at one temperature, usually at 25°C

The above relationship is the "internal consistency relationship among th three enthalpies." To maintain the integrity of the data base, the enthalpie of the isomer lumps must also maintain this "internal consistency relation

[4] There is no unique definition of an Expert System. Here, I define it as a system (fo example computer software) that does what an expert can do, such as executing a algorithm.

ship." Is it necessary to maintain such integrity in the data base? If not, what are the penalties? And if yes, what is the best way to do it?

For lumps with a large number of isomers, many of the steps given above that deal with individual isomer must be handled by Expert Systems. In an ideal situation, one can tie all the steps together once the Expert Systems for those steps become operational. This type of setup will be very useful if one has to evaluate many types of isomer lumps.

FIVE NONLUMPING BUT INTERESTING AREAS CONCERNING COMPLEX MIXTURES

There are five interesting areas in thermodynamics that apply to mixtures but have no relationship to lumping. However, they are all in need of more study, and are listed below.

- Mixing rules for the equations of state
- Mixing rules for some essential properties (such as critical temperatures and pressures)
- Supercritical extraction
- Liquid–liquid phase equilibria
- Phase envelopes that generate retrograde condensation.

These areas are discussed separately in the following.

The problem of mixing rules for the equation of state of mixtures is similar to what I have discussed for the isomer lumps in the section on Isomer Lumping with a Known Isomer Distribution. We are facing the same problem of establishing a data base of a very large number of binary interaction parameters for large number of pure compounds and isomer lumps. The problem here is more critical than that of the isomer lumps because we are now dealing with compounds and isomer lumps that may have drastically different molecular types. Again, a similar question can be asked. Can an algorithm or algorithms be established so that the binary interaction parameters can be estimated based on differences in molecular structures of components and lumps?

Mixing rules for essential properties, such as acentric factors, cetane numbers, etc., are nonexistent and should be developed. These rules are expected to be more complicated and more difficult to develop than those for the isomers and the equations of state. But their development is essential for the establishment of thermophysical data bases.

Applications of supercritical extraction are greatly in need today in the oil industry, particularly in the tertiary recovery of crude oils from existing reservoirs. More studies are needed to identify potential solvents, and to quantify their effects.

The oil industry today encounters a lot more liquid–liquid phase separation problems than before. This results from increasing use and interest in oxygenated fuels due to increasing environmental protection regulations. The liquid–liquid phase separation can occur either during fuel blending or during synthesis of oxygenates. I believe that the UNIFAC method (Fredenslund et al. 1975), a group contribution method, is probably the most versatile method for estimating data needed for phase separation calculations. There is a large demand for more extensive data base for UNIFAC groups. Furthermore, we also need a simpler algorithm to identify the possible existence of liquid–liquid phase separation.

The vapor–liquid equilibria of mixtures can behave very strangely in the neighborhood of the true critical point of the mixture. This phenomenon is called "retrograde condensation" (see Hougen et al. 1959; and Reid et al. 1987). The shape of the pressure–temperature phase envelope of a mixture in the neighborhood of the true critical point may be in such a form that multiple dew points and/or bubble points are possible. The existence of multiple dew and bubble points can cause difficulties in calculating the phase equilibria in the same neighborhood. Unfortunately I do not know if there is a simple algorithm that can easily identify the possible existence of the "retrograde condensation" phenomenon.

CONCLUSIONS

The oil industry faces great changes today and in the near future. The changes result from demands for higher oil recovery, increasing environmental protection, and improved refinery operations. All these forces dictate more research and development and increasing use of thermodynamics. We thoroughly discussed these uses as well as the needs of thermodynamics in the oil industry.

"Lumping" is a necessity in the oil industry and it is most important that one must have "lumping according to objectives." These objectives of lumping are carefully defined. However, before lumping, one must develop thermophysical data bases of pure components. Methods of estimating these thermophysical data are essential for data base construction. Among various estimation methods, we recommended those based on group contribution techniques, particularly the BGCM. Various interesting areas in GCMs were discussed.

The importance of "isomer lumping" is emphasized as well as the construction of data bases of thermophysical properties for "isomer lumps." In a sense, how to construct "lumps" and their thermophysical data base are the most important issues confronting many of today's industrial researchers. Depending on the objective or objectives of the lumping and how much information one has on the lumped mixtures, we have discussed three Lumping Approaches. A strong relationship between some thermophysical properties and isomer molecular structures, particularly those with high degree of molecular symmetry, is emphasized. Particular attention was given to lumping of isomers whose distribution can be defined by a set of isomerization reactions. With great progress in both analytical and computational capability today and in the near future, one should be able to construct a thermophysical data base of any isomer lump starting with isomer enumeration and molecular structure elucidation using Expert Systems. Detailed steps for such a calculation were summarized.

We have also discussed four nonlumping but interesting areas concerning complex mixtures. These areas are mixing rules for the equations of state, supercritical extraction, liquid–liquid phase equilibria, and phase envelopes that generate retrograde condensation. Specific problems in these areas that require further research and development were given.

NOMENCLATURE

- N total number of isomers in a lump
- N_{ext} "external symmetry number" of a molecule
- N_{int} "internal symmetry number" of a molecule
- N_{tsym} "total symmetry number" of a molecule
- R ideal gas law constant
- S real entropy [see Eq. (A–1)]
- S_{ins} intrinsic entropy, the entropy estimated by adding the group entropy values of the structural parts of the molecule
- S_{opt} entropy correction resulting from the existence of optical isomers of a molecule [see Eq. (A–3)]
- S_{sym} entropy correction related to the symmetry of a molecule [see Eq. (A–2)]
- S_{tsym} entropy correction related to the total symmetry number of a molecule [see Eq. (A–4)]
- T absolute temperature
- x_i isomer molar fractions in a lump, $i = 1, 2, \ldots, N$

APPENDIX: ENTROPY CORRECTIONS DUE TO NUMBERS OF OPTICAL ISOMERS AND TOTAL SYMMETRY NUMBER OF A MOLECULE

Entropy Corrections

Readers are referred to Benson (1976), Cotton (1971), Davies et al. (1988), Domalski and Hearing (1988), Herman and Lievin (1977), Levine (1976),

Orchin and Jaffe (1970a,b,c), and Schonland (1965) for the derivation of the relations given in this Appendix.

The entropy estimated by adding the group values for the entropies of the structural parts of the molecule is called "intrinsic entropy," S_{ins}. To obtain the real entropy from it, one must add to S_{ins} an entropy corrections, S_{sym}, related to the symmetry of the molecule

$$S = S_{ins} + S_{sym} \tag{A-1}$$

in which S is the real entropy of the molecule. The S_{sym} is in turn divided into two parts

$$S_{sym} = S_{opt} - S_{tsym} \tag{A-2}$$

in which S_{opt} is the entropy correction resulting from the existence of the optical isomers of the molecule and S_{tsym} is the entropy correction related to the total symmetry number of the molecule. These two entropy corrections are related to the number of optical isomers, N_{opt}, and the total symmetry number, N_{tsym}, of the molecules by the following equations

$$S_{opt} = R \ln N_{opt} \tag{A-3}$$

$$S_{tsym} = R \ln N_{tsym} \tag{A-4}$$

where R is the ideal gas constant. Note that the above treatment of optical isomers is equivalent to lump all optical isomers into a single component. The S_{opt} is simply the entropy of mixing.

We purposely leave out of this text the methods for determining both the number of optical isomers and the total symmetry number of a molecule. For readers who are interested in such details, please see references in the first paragraph of this Appendix.

Total Symmetry Number of a Molecule

The total symmetry number of a molecule, N_{tsym}, is composed of two parts: the internal symmetry number, N_{int}, and the external symmetry number, N_{ext}, governed by the following equation

$$N_{tsym} = N_{int} \times N_{ext}. \tag{A-5}$$

Internal Symmetry Number of a Molecule

The internal symmetry number of a molecule depends completely on the type and number of terminal rotors that the molecule possesses. A "terminal otor" is defined as a group of atoms attached to the rest of the molecule by a single bond.[5] The internal symmetry arises when a terminal rotor has more than one indistinguishable conformation as it makes one complete otation about an axis colinear with the single bond linking it to the rest of he molecule. The axis is called an n-fold axis of symmetry if there are n such conformations. N_{int} for that rotor is then equal to $n(>1)$.

External Symmetry Number of a Molecule

The external symmetry number is defined as the total number of independent permutations of identical atoms (or pseudo-atoms, i.e., terminal otors) in a molecule that can be arrived at by rigid rotations of the entire molecule (or pseudo molecule, i.e., the molecule with the pseudo-atoms).

Number of Optical Isomers

One of the ways in which a molecule may exhibit optical isomerism occurs when its mirror image cannot be superimposed on itself as a result of simple molecular rotations. Such mirror images are called "enantiomers" are are aid to be chiral. The enantiomers have identical physical properties except or their interaction with transmitted polarized light.

REFERENCES

Alberty, R. A. (1987) Kinetics of the polymerization of alkenes on zeolites. *J. Chem. Phys.* **87**:660–667.

Alberty, R. A. and A. K. Reif (1988) Standard chemical thermodynamic properties of polycyclic hydrocarbons and their isomer groups, I. Benzene series. *J. Phys. Chem. Ref. Data* **17**:241–253.

Benson, S. W. (1976) *Thermochemical Kinetics*, 2nd edit. John Wiley, New York.

Benson, S. W. and J. H. Buss (1958) Additivity rules for the estimation of molecular properties—thermodynamic properties. *J. Chem. Phys.* **29**:546–572.

Cotton, F. A. (1971) *Chemical Applications of Group Theory*. Wiley-Interscience, New York.

Davies, C. A., A. N. Syverud, and E. C. Steiner (1988) Definition of symmetry numbers and optical isomers used in CHETAH calculations and a simple method for choosing these numbers. (personal communication).

This definition is valid except for compounds of $\mathbf{CY_4}$ or $\mathbf{NY_3}$ structure or when the "rest of the molecule" consists of a single atom or a linear group of atoms, all on the axis of rotation. In these cases, the concept of one group rotating relative to the other becomes meaningless and the symmetry is considered as external symmetry.

Dias, J. R. (1984) A periodic table for polycyclic aromatic hydrocarbons. 4. Isomer enumeration of polycyclic conjugated hydrocarbons. *J. Chem. Inf. Comput Sci* **24**:124–135.

Domalski, E. S. and E. D. Hearing (1988) Estimation of the thermodynamic properties of hydrocarbons at 298.15 K. *J. Phys. Chem. Ref. Data* **17**:637–678.

Fredenslund, A., R. L. Jones, and J. M. Prausnitz (1975) Group-contribution esfimation of activity coefficients in nonideal liquid mixtures. *A.I.Ch.E.J* **21**:86–99.

Herman, M. and J. Lievin (1977) Group theory. *J. Chem. Educ.* **54**:596–598.

Hougen, O. A., K. M. Watson, and R. A. Ragatz (1959) *Chemical Process Principles Part II—Thermodynamics*, 2nd edit. John Wiley, New York.

Kleywegt, G. J., H. J. Luinge, and H. A. Van't Klooster (1987) Artificial intelligence used for the interpretation of combined spectral data. Part II. PEGASUS: a PROLOG program for the generation of acyclic molecular structures. *Chemomet Intell. Lab. Syst.* **2**:291–302.

Kuo, J. C. W. and J. Wei (1969) A lumping analysis in monomolecular reaction system. Analysis of approximately lumpable system. *I.E.C. Fund.* **8**:124–133.

Levine, I. N. (1976) Molecular symmetry. In *Quantum Chemistry*, 2nd ed., Chap 12. Allyn & Bacon, Needham Heights, Massachusetts.

Lindsay, R. K., B. G. Buchanan, E. A. Feigenbaum, and J. Lederberg (1980) *Applications of Artificial Intelligence for Organic Chemistry.* McGraw-Hill, New York.

Orchin, M. and H. H. Jaffe (1970a) Symmetry, point groups, and character tables—I Symmetry operations and their importance for chemical problems. *J. Chem. Educ* **47**:246–252.

Orchin, M. and H. H. Jaffe (1970b) Symmetry, point groups, and character tables—II Classification of molecules into point groups. *J. Chem. Educ.* **47**:372–377.

Orchin, M. and H. H. Jaffe (1970c) Symmetry, point groups, and character tables—III Character tables and their significance, *J. Chem. Educ.* **47**:510–516.

Reid, R. C., J. M. Prausnitz, and B. E. Poling (1987) *The Properties of Gases and Liquids*, 4th edit. McGraw-Hill, New York.

Schonland, D. S. (1965) *Molecular Symmetry.* D. Van Nostrand, London.

Wei, J. and J. C. W. Kuo (1969) A lumping analysis in monomolecular reaction system. Analysis of the exactly lumpable system. *I.E.C. Fund.* **8**:114–123.

14

Chemical Equilibrium in Complex Organic Systems with Various Choices of Independent Variables

ROBERT A. ALBERTY

In considering chemical equilibrium in complex organic systems, it is necessary to find ways to simplify the problem so that there can be some understanding of what is happening. The first simplification is to use isomer groups as species (Alberty 1983; Smith 1959; Smith and Missen 1974, 1982), since the standard Gibbs energy of formation $\Delta_f G_n^\circ$ of an isomer group with carbon number n can be calculated using

$$\Delta_f G_n^\circ = -RT \ln \sum_{i=1}^{N_I} \exp(-\Delta_f G_i^\circ/RT), \qquad (14\text{–}1)$$

where N_I is the number of isomers in the isomer group and $\Delta_f G_i^\circ$ is the standard Gibbs energy of formation of the ith isomer in the group. The equilibrium mole fraction r_i of isomer i in its isomer group is given by

$$r_i = \exp[(\Delta_f G_n^\circ - \Delta_f G_i^\circ)/RT]. \qquad (14\text{–}2)$$

In these equations, and the rest of the chapter, ideal gases are assumed. If some of the isomers in the isomer group are excluded by the catalyst, they can be left out of the summation in Eq. (12–1) (Alberty 1987; Quann et al. 1988; Tabak et al. 1986). We have prepared thermodynamic tables on the first several isomer groups of 16 homologous series and have found that, after the first one or two members of a series, all of the isomer group thermodynamic properties are linear in carbon number (Alberty and Reif 1988). For example,

$$\Delta_f G_n^\circ = A + Bn, \tag{14–3}$$

where A and B are functions of temperature only for a given homologous series. The parameters A and B contain all of the information on the thermodynamics of the whole homologous series. In the case of the benzenoid polycyclic aromatic hydrocarbons, the first member of the series also follows Eq. (14–3).

The concept of an isomer group can be extended to whole homologous series of the usual type with an increment of CH_2 if the equilibrium partial pressure of ethylene is specified (Alberty 1984, 1985). If the equilibrium partial pressure of ethylene is specified, the successive isomer groups in a homologous series become pseudoisomers; that is, their equilibrium ratios are functions only of temperature. The standard Gibbs energy of formation of an isomer group at a specified value of $P_{C_2H_4}$ can be calculated using a formation reaction involving C_2H_4; this adjusted Gibbs energy of formation is represented by $\Delta_f G_N^*$ where N is the ordinal number of the isomer group in the homologous series. For the homologous series of benzenoid polycyclic aromatic hydrocarbons, the increment between isomer groups is C_4H_2. Therefore, $P_{C_2H_2}$ and P_{H_2} are specified as independent variables, rather than $P_{C_2H_4}$. These adjusted Gibbs energies can be used to calculate the Gibbs energy of formation $\Delta_f G_{HSG}^*$ of the whole homologous series under the ambient conditions by using an equation analogous to Eq. (14–1).

$$\Delta_f G_{HSG}^* = -Rt \ln \sum_{N=1}^{\infty} \exp(-\Delta_f G_N^*/RT). \tag{14–4}$$

In this case the summation has an infinite number of terms and can diverge. For partial pressures of ethylene for ordinary homologous series, or partial pressures of acetylene and hydrogen for polycyclic aromatic hydrocarbons, where the summation converges, the equilibrium mole fraction r_N of an isomer group within a homologous series can be calculated using an analog of Eq. (14–2).

$$y_N = \exp[(\Delta_f G_{HSG}^* - \Delta_f G_N^*)/RT]. \tag{14–5}$$

If the standard Gibbs energies for the isomer groups in a homologous series follow Eq. (14–3), the adjusted Gibbs energies will follow

$$\Delta_f G_N^* = a + bN. \tag{14–6}$$

Alberty and Oppenheim (1986) showed that when the partial pressure of ethylene is fixed, the equilibrium distribution in a homologous series following Eq. (14–3) is given by a simple analytic function that depends only on B.

When the equilibrium partial pressure of molecular hydrogen is specified, the calculation of the equilibrium composition in a mixture of hydrocarbons reduces to an element polymerization problem (Alberty 1989b, 1990b), which has been described by Smith and Missen (1982). When the successive isomer groups in a homologous series follow Eq. (14–3), the equilibrium distribution within the homologous series at specified P_{H_2} is given by simple analytic function that depends only on the value of A. The equilibrium distribution within the alkenes is, of course, independent of P_{H_2}, but the distribution function for the alkenes can be derived by the same method as the other homologous series. Thus A and B are really thermodynamic properties in their own right, although they are a little different from the usual thermodynamic properties.

Strictly speaking, the adjustment of standard Gibbs energies of formation to specified partial pressures of H_2, C_2H_4, and C_2H_2 really results from the definition of new thermodynamic potentials by use of Legendre transforms (Alberty and Oppenheim 1988). These new potentials have their values at equilibrium when partial pressures of certain reactants are taken as independent variables. In practice, it is convenient to simply adjust standard Gibbs energies of formation by use of formation reactions involving these reactants. However, the derivation of a new thermodynamic potential G_{12} for the benzene series of polycyclic aromatic hydrocarbons at specified $P_{C_2H_2}$ and P_{H_2} is given in the Appendix. Isomer groups and homologous series groups have their counterparts in isothermal–isobaric (Alberty 1988) and semigrand ensembles (Alberty and Oppenheim 1989) of statistical mechanics.

The various types of calculations described above can be viewed as equilibrium calculations with different choices of the degrees of freedom for a system. The number of degrees of freedom for a gaseous reaction system at constant T and P depends on the number of components. The number of components depends on the number of element balances and other constraints. The fact that other constraints can come in through the mechanism of reaction has been known for a long time. In 1921 Jouguet pointed out the advantages of calculating the equilibrium composition using the reaction paths actually available for chemical change. If the number of independent reactions in the mechanism is greater than the number of reactions required to allow all possible composition changes permitted by the element balances, the equilibrium composition calculated using the mechanism will be the same as given in the preceding section. However, if the number of independent steps in the mechanism is smaller, the mechanism itself involves a constraint or constraints, and the equilibrium composition will be different (Alberty 1989a; Smith and Missen 1982). In order to illustrate this and the use of various choices for degrees of freedom, a single system is considered here from six points of view. To keep the system as simple as possible, we will ignore inert species and systems made up of completely independent subsystems.

The general problem of calculating the equilibrium composition of a complex system has been solved since there are computer programs for calculating the equilibrium amounts given (1) the system matrix **A** (isomer groups can be considered to be species), (2) the initial amounts vector $\mathbf{n}_0$, and (3) the Gibbs energies of formation of the species corrected to the pressure of the system (Smith and Missen 1982). We have used EQUCALC, written in APL by Krambeck (1978), and given in Chapter 3, which we have found to be very dependable and convenient to use. In EQUCALC the system matrix **A** is represented by AS, the initial amount vector $\mathbf{n}_0$ is represented by NO, and the standard Gibbs energies of formation and total pressure are used to calculate LNK, which is the vector $-\boldsymbol{\mu}^\circ(T, P)/RT$.

Linear algebra is very useful in equilibrium calculations on complex systems, and the following relations are described by Smith and Missen (1982). The number C of components is equal to the number of species (or isomer groups) N minus the number of independent reactions R.

$$C = N - R. \tag{14–7}$$

Under circumstances to be examined later, the number of components is equal to the rank of the formula matrix.

$$C = \text{rank } \mathbf{A}. \tag{14–8}$$

A set of reactions (mechanism) can be represented by a stoichiometric number matrix $\mathbf{v}$ that has as many rows as species and as many columns as reactions. The number of independent reactions R is equal to the rank of the $\mathbf{v}$ matrix.

$$R = \text{rank } \mathbf{v}. \tag{14–9}$$

Thus under circumstances to be examined later

$$N = \text{rank } \mathbf{A} + \text{rank } \mathbf{v}. \tag{14–10}$$

The relation between **A** and $\mathbf{v}$ is

$$\mathbf{A}\mathbf{v} = 0 \quad \text{or} \quad \mathbf{v}^T\mathbf{A}^T = 0. \tag{14–11}$$

Thus, **v** is the null space of the **A** matrix, and we can calculate **v** from **A**. Since $\mathbf{A}^T$ is the null space of the $\mathbf{v}^T$ matrix, we can also start with a mechanism, as described in connection with Cases 2 and 5, and calculate the system matrix **A**.

THE CHEMICAL SYSTEM CONSIDERED

As an example, consider chemical equilibrium in a system consisting of acetylene, molecular hydrogen, and the successive isomer groups in the benzene series of polycyclic aromatic hydrocarbons (Alberty and Reif 1988; Dias 1987). The isomer groups in this series have the formulas $C_{2+4N}H_{4+2N}$, where N is the ordinal number of the isomer group in the series, so that the increment between isomer groups is C_4H_2. The formulas for the successive isomer groups can also be written as $C_nH_{n/2+3}$, where N is the carbon number 6, 10, 14, ... In calculations here using EQUCALC, the first 10 isomer groups are included, but we are really interested in the whole series. In the six cases considered the temperature is 1,000 K and the pressure on this system is 1 bar. The system contains initially 0.125 mol C_2H_2 and 0.125 mol C_6H_6 so that there is one mole of carbon atoms and one mole of hydrogen atoms. The standard Gibbs energies of formation at 1,000 K for the successive isomer groups of the benzene series obeys Eq. (14–3) with $A =$ 46.753 kJ mol^{-1} and $B = 37.222$ kJ mol^{-1} (Alberty and Reif 1988). (These parameters are not known this accurately, but these are the values obtained by linear regression of the data on the first six isomer groups and are the values used in the calculations presented here.) The standard Gibbs energy of formation of acetylene at 1,000 K is 169.607 kJ mol^{-1} (Chase et al. 1985/1987).

Case 1: Calculation of the Equilibrium Composition at $T, P, n_H/n_C$

Since there are two components and one phase, the phase rule tells us that three intensive variables have to be specified to fix the intensive state of the system. The equilibrium composition is independent of the starting material so long as there are equal amounts of carbon atoms and hydrogen atoms; therefore, the degrees of freedom have been listed as $T, P, n_H/n_C$. The isomer groups of the benzene series of polycyclic aromatic hydrocarbons are pseudospecies and so the system matrix **A** is

$$\mathbf{A} = \begin{matrix} 2 & 0 & 6 & 10 & 14 & 18 & 22 & 26 & 30 & 34 & 38 & 42 \\ 2 & 2 & 6 & 8 & 10 & 12 & 14 & 16 & 18 & 20 & 22 & 24 \end{matrix} \qquad (14\text{–}12)$$

The initial amount vector is $\mathbf{n}_0 = 0.125\ 0\ 0.125\ 0\ 0\ 0\ 0\ 0\ 0\ 0\ 0\ 0$. The equilibrium amounts calculated using EQUCALC are given in Table 14–1. This is a bad example in the sense that the equilibrium mole fractions within the benzene series increase with carbon number, and substantially different results would be obtained by adding more isomer groups. But, it is a good example because it shows that if the whole homologous series can indeed be formed, the system can never reach equilibrium. We can see that most of the initial acetylene and benzene is used to form higher isomer groups in the homologous series.

If we want an independent set of reactions for this system, we can use the following 10 reactions.

$$\begin{aligned} 3C_2H_2 &= C_6H_6, \\ 5C_2H_2 &= C_{10}H_8 + H_2, \\ &\vdots \end{aligned} \tag{14–13}$$

Case 2: Calculation of the Equilibrium Composition at T, P, n_H/n_C, n_{ar}/n_C Assuming a Particular Mechanism

Now suppose we have an idea as to what reactions are actually occurring in this system. Assume that the mechanism for chemical change is

$$\begin{aligned} C_6H_6 + 2C_2H_2 &= C_{10}H_8 + H_2, \\ C_{10}H_8 + 2C_2H_2 &= C_{14}H_{10} + H_2, \\ &\vdots \end{aligned} \tag{14–14}$$

Note that this mechanism has nine independent steps. Each of these steps has a mechanism involving free radicals, but these steps give the driving force for polymerization. Sometimes, such a set of reactions is called a reaction scheme, but the term mechanism is often used in this more general way and is not restricted to a set of elementary reactions. The **A** matrix that corresponds with this mechanism can in principle be obtained by writing down the **ν** matrix and using Eq. (14–11). This can be done by hand using only the two reactions actually given in Eq. (14–14) to obtain the first part of the following **A** matrix. Alternatively, we can see from the mechanism that aromatic character is conserved by this mechanism and simply put this constraint in like an element balance.

$$\mathbf{A} = \begin{matrix} 2 & 0 & 6 & 10 & 14 & 18 & 22 & 26 & 30 & 34 & 38 & 42 \\ 2 & 2 & 6 & 8 & 10 & 12 & 14 & 16 & 18 & 20 & 22 & 24 \\ 0 & 0 & 1 & 1 & 1 & 1 & 1 & 1 & 1 & 1 & 1 & 1 \end{matrix} \tag{14–15}$$

Table 14–1 Equilibrium Amounts (n/mol) of Species at 1,000 K and P = 1 bar

	Case 1	Case 2	Case 3	Case 4	Case 5	Case 6
C_2H_2	2.60×10^{-6}	6.69×10^{-7}	6.69×10^{-7}	—	—	absent
H_2	0.201	0.062	—	—	—	—
C_6H_6	9.49×10^{-4}	0.083	0.083	0.083	0.064	0.044
$C_{10}H_8$	1.19×10^{-3}	0.028	0.028	0.028	0.031	0.020
$C_{14}H_{10}$	1.50×10^{-3}	0.009	0.009	0.009	0.015	0.009
$C_{18}H_{12}$	1.87×10^{-3}	0.003	0.003	0.003	0.007	0.004
$C_{22}H_{14}$	2.36×10^{-3}	0.001	0.001	0.001	0.004	0.002
$C_{26}H_{16}$	2.96×10^{-3}	0	0	0	0.002	0.001
$C_{30}H_{18}$	3.71×10^{-3}	0	0	0	0.001	0
$C_{34}H_{20}$	4.65×10^{-3}	0	0	0	0	0
$C_{38}H_{22}$	5.84×10^{-3}	0	0	0	0	0
$C_{42}H_{24}$	7.33×10^{-3}	0	0	0	0	0
Initial amounts						
C_2H_2	0.125	0.125	0.125	—	—	0
C_6H_6	0.125	0.125	0.125	0.125	0.125	0.125
Equilibrium pressures						
$P_{C_2H_2}$/bar	1.11×10^{-5}	3.57×10^{-6}	3.57×10^{-6}	3.57×10^{-6}	3.57×10^{-6}	0
P_{H_2}/bar	0.862	0.333	0.333	0.333	0.333	0.998
P_H/bar	0	0	0	0	1.57×10^{-9}	0
P_{hc}/bar	0.138	0.666	0.666	0.666	0.666	0.002

The third row is the aromatic constraint. The initial amount vector is still $\mathbf{n}_0 = 0\ 0.125\ 0\ 0.125\ 0\ 0\ 0\ 0\ 0\ 0\ 0\ 0\ 0$. Now the initial amount vector is important. The matrix product of $\mathbf{A}$ and $\mathbf{n}_0$ shows that the system contains 1 mol of carbon, 1 mol of hydrogen atoms, and 0.125 mol of aromatics. Of course, it is not the initial amounts but the ratios n_H/n_C and n_{ar}/n_C that are important. The amount of aromatics in the system is represented by n_{ar}. LNK is the same as for Case 1.

According to Eq. (14–8), the number of components is now three, and so there are four degrees of freedom. These can be chosen to be $T, P, n_H/n_C, n_{ar}/n_C$. The equilibrium amounts calculated using EQUCALC are given in Table 14–1, and the equilibrium mole fractions within the benzene series are 0.666, 0.222, 0.074, 0.025, 0.008, 0.003, 0.001, 0, 0, 0. This equilibrium composition is quite different from that in Case 1 because benzene can only be converted to other aromatics and the amount of acetylene available is small. This is the equilibrium composition that will be obtained by a numerical integration of the rate equations for the mechanism given by Eq. (14–14).

The mechanism given by (14–14) is not the only mechanism that will give this equilibrium composition. Another mechanism that will is

$$\begin{aligned} C_6H_6 + 2C_2H_2 &= C_{10}H_8, \\ C_6H_6 + 4C_2H_2 &= C_{14}H_{10}, \\ &\vdots \end{aligned} \tag{14–16}$$

This mechanism is less credible from a kinetic standpoint, but the null space for the $\mathbf{v}$ matrix for this mechanism is also given by Eq. (14–15).

Case 3: Calculation of the Equilibrium Composition at $T, P, P_{H_2}, n_{ar}/n_C$

In this case we replace the intensive variable n_H/n_C with P_{H_2} and continue to assume that the reaction occurs only via mechanism, Eq. (14–14). We consider the system to be in contact with a reservoir of molecular hydrogen at a specified partial pressure through a semipermeable membrane. This removes hydrogen from the $\mathbf{A}$ matrix because it is supplied to the extent needed from the reservoir. The system formula matrix is now

$$\mathbf{A} = \begin{matrix} 2 & 6 & 10 & 14 & 18 & 22 & 26 & 30 & 34 & 38 & 42 \\ 0 & 1 & 1 & 1 & 1 & 1 & 1 & 1 & 1 & 1 & 1 \end{matrix} \tag{14–17}$$

The initial amount vector $\mathbf{n}_0 = 0.125\ 0.125\ 0\ 0\ 0\ 0\ 0\ 0\ 0\ 0\ 0$ and LNK do not contain terms for hydrogen. The matrix product of $\mathbf{A}$ and $\mathbf{n}_0$ shows that the system contains 1 mol C and 0.125 mol of aromatics. The mechanism for this case can be written as

$$\begin{aligned} C_6 + 2C_2 &= C_{10}, \\ C_{10} + 2C_2 &= C_{14}, \\ &\vdots \end{aligned} \tag{14–18}$$

since molecular hydrogen is supplied to the extend needed. The equilibrium composition depends on the total pressure P_{hc} of these species. The equilibrium composition can be calculated using EQUCALC by correcting the Gibbs energies of formation of these 11 species to the desired P_{H_2} and P_{hc} by use of the formation reactions

$$2C(\text{graphite}) + H_2(P_{H_2}) = C_2H_2(P_{hc}), \tag{14–19}$$

$$(2 + 4N)C(\text{graphite}) + (2 + N)H_2(P_{H_2}) = C_{2+4N}H_{4+2N}P_{hc}), \quad N = 1, 2, \ldots. \tag{14–20}$$

The adjusted Gibbs energies of formation are given by

$$\Delta_f G^*_{C_2H_2} = \Delta_f G^\circ_{C_2H_2} + RT \ln P_{hc} - RT \ln P_{H_2}, \tag{14–21}$$

$$\Delta_f G^*_N = \Delta_f G^\circ_N + RT \ln P_{hc} - (2 + N)RT \ln P_{H_2}. \tag{14–22}$$

This method was used by Krambeck in the late 1970s (Ramage et al. 1980). Since hydrogen is an element and the standard state pressure is arbitrary, we simply adjust $\Delta_f G^\circ$ to whatever hydrogen partial pressure wanted. As mentioned above, this can also be viewed as making a Legendre transform to a new thermodynamic potential.

Calculations can be made for any partial pressure of hydrogen, but, if we use $P_{H_2} = 0.333$ bar and $P_{hc} = 0.667$ bar, we obtain the same equilibrium partial pressures of reactants as in the preceding case. The equilibrium amounts calculated with EQUCALC are given in Table 14–1, and the equilibrium mole fractions of the isomer groups within the benzene series are the same as in Case 2. The contribution of acetylene to the mole fraction is almost negligible, but its presence is important.

Case 4: Calculation of the Equilibrium Composition at $T, P, P_{C_2H_2}, P_{H_2}$

For the system following mechanism Eq. (14–14), another choice of four independent variables is $T, P, P_{C_2H_2}, P_{H_2}$. This happens to be very useful for the study of flames. For example, in a near-sooting benzene flame, Bittner and Howard (1981) have determined these intensive variables, as well as the mole fractions of benzene, naphthalene, anthracene–phenanthrene, and pyrene, as a function of height in the flame. To calculate the equilibrium distribution of isomer groups of the benzene series polycyclic aromatic hydrocarbons at each level in the flame, we can imagine that they are in contact with reservoirs of acetylene and hydrogen at the observed partial pressures through a semipermeable membrane. Under these conditions the successive isomer groups are pseudoisomers. EQUCALC can be used to calculate the equilibrium amounts of the successive isomer groups by omitting C_2H_2 and H_2 from the system formula matrix and using

$$\mathbf{A} = 1\ 1\ 1\ 1\ 1\ 1\ 1\ 1\ 1\ 1. \tag{14–23}$$

The initial amount matrix is 0.125 0 0 0 0 0 0 0 0 0, and the terms for acetylene and hydrogen are omitted from LNK. The system contains 0.125 mol of the pseudoisomers. The Gibbs energies of formation of the successive isomer groups are adjusted to the actual $P_{C_2H_2}$ and P_{H_2} at a given level of the flame using the formation reactions

$$(1 + 2N)C_2H_2(P_{C_2H_2}) = C_{2+4N}H_{4+2N}(P_{hc}) + (N - 1)H_2(P_{h_2}), \quad N = 1, 2, \ldots, \tag{14–24}$$

where N is the ordinal number of the isomer group in the benzene series. The Gibbs energy of formation of the Nth isomer group under the ambient conditions is

$$\Delta_f G_N^* = \Delta_f G_N^\circ + RT \ln P_{hc} + (N - 1)RT \ln P_{H_2} - (1 + 2N)(\Delta_f G_{C_2H_2}^\circ + RT \ln P_{C_2H_2}). \tag{14–25}$$

Substituting $n = 2 + 4N$ for the carbon number in Eq. (14–3) yields

$$\Delta_f G_N^\circ = A + 2B + 4BN \tag{14–26}$$

and substituting this in Eq. (14–25) yields Eq. (14–6) with

$$a = A + 2B + RT \ln P_{hc} - RT \ln P_{H_2} - (\Delta_f G_{C_2H_2}^\circ + RT \ln P_{C_2H_2}), \tag{14–27}$$

$$b = 4B + RT \ln P_{H_2} - 2(\Delta_f G_{C_2H_2}^\circ + RT \ln P_{C_2H_2}). \tag{14–28}$$

At 1,000 K, $B = 37.222$ kJ mol^{-1} and $\Delta_f G_{C_2H_2}^\circ = 169.607$ kJ mol^{-1}. In order to compare the results with Case 2, we take $P_{C_2H_2} = 3.566 \times 10^{-6}$ bar and $P_{H_2} = 0.3333$ bar. The use of the adjusted Gibbs energies $\Delta_f G_N^*$ in

EQUCALC yields the equilibrium amounts given in Table 14–1, and the equilibrium mole fractions within the polycyclic aromatic hydrocarbons are the same as in Case 2.

This result can be obtained without a general equilibrium program by using the following analytic function derived by Alberty and Oppenheim (1986)

$$y_N = (1 - q)q^{N-1}, \quad N = 1, 2, 3, \ldots, \tag{14–29}$$

where

$$q = \exp(-b/RT). \tag{14–30}$$

This is a geometric distribution (Ross 1988). It is important to note that y_N depends only on the value of b. Since $b = 9.14$ kJ mol^{-1}, $q = 0.333$. Thus,

$$y_N = 0.667\ (0.333)^{N-1}. \tag{14–31}$$

This yields the equilibrium mole fractions within the benzene series that were first obtained in Case 2. The equilibrium mole fractions obtained using Eq. (14–31) can be converted to equilibrium pressures by multiplying by (1 − 0.3333) bar or to equilibrium amounts by multiplying by 0.125. The mean ordinal number in the homologous series at equilibrium is 1.50. The standard deviation and other characteristics of the distribution are readily calculated. Under conditions where b is negative, no equilibrium is reached. The fact that the equilibrium distribution is independent of a can be understood a little better by noting that a contains P_{hc}, which cannot affect the equilibrium distribution.

This method for adjusting the Gibbs energies of formation of the isomer groups of polycyclic aromatic hydrocarbons is justified by the fact that it is equivalent to using a Legendre transform. The definition of a new thermodynamic potential G_{12} for use at specified partial pressures of acetylene and molecular hydrogen is given in the Appendix.

Case 5: Calculation of the Equilibrium Composition of the Benzene Series at T, P, $P_{C_2H_2}$, P_{H_2}, P_H

Another mechanism that has been suggested for the formation of benzenoid polycyclic aromatic hydrocarbons in flames (Frenklach et al. 1985) is

$$\begin{aligned} 2H + C_6H_6 + 2C_2H_2 &= C_{10}H_8 + 2H_2, \\ 2H + C_{10}H_8 + 2C_2H_2 &= C_{14}H_{10} + 2H_2 \\ &\vdots \end{aligned} \tag{14–32}$$

This mechanism involves two constraints (Alberty 1989b), and the **A** matrix showing these constraints is

$$\mathbf{A} = \begin{matrix} 2 & 0 & 0 & 6 & 10 & 14 & 18 & 22 & 26 & 30 & 34 & 38 & 42 \\ 2 & 2 & 1 & 6 & 8 & 10 & 12 & 14 & 16 & 18 & 20 & 22 & 24 \\ 0 & 1 & 1 & 0 & 0 & 0 & 0 & 0 & 0 & 0 & 0 & 0 & 0 \\ 0 & 0 & 0 & 1 & 1 & 1 & 1 & 1 & 1 & 1 & 1 & 1 & 1 \end{matrix} \tag{14–33}$$

where the third column is for hydrogen atoms, the third row is the "hydrogen" constraint, and the fourth row is the "aromatic" constraint. Specifying $P_{C_2H_2}$ eliminates the first column and first row, specifying P_{H_2} eliminates the second column and the second row, and specifying P_H eliminates the third column and the third row, so that the **A** matrix for this calculation is given by Eq. (14–23).

The formation reaction for this calculation is

$$(2N)H(P_H) + (1 + 2N)C_2H_2(P_{C_2H_2}) = C_{2+4N}H_{4+2N}(P_{hc}) + (2N - 1)H_2(P_{H_2}), \tag{14–34}$$

where N is the ordinal number of the isomer group in the benzene series. The Gibbs energy of formation of the Nth isomer group under the ambient conditions is

$$\begin{aligned}\Delta_f G_N^* = {} & \Delta_f G_N^\circ + RT \ln P_{hc} + (2N - 1)RT \ln P_{H_2} \\ & - 2N(\Delta_f G_H^\circ + RT \ln P_H) - (1 + 2N)(\Delta_f G_{C_2H_2}^\circ + RT \ln P_{C_2H_2}).\end{aligned} \tag{14–35}$$

Substituting Eq. (14–26) in Eq. (14–35) yields Eq. (14–6) with

$$a = A + 2B + RT \ln P_{hc} - RT \ln P_{H_2} - (\Delta_f G_{C_2H_2}^\circ + RT \ln P_{C_2H_2}) \tag{14–36}$$

$$\begin{aligned}b = {} & 4B + 2RT \ln P_{H_2} - 2(\Delta_f G_H^\circ + RT \ln P_H) \\ & - 2(\Delta_f G_{C_2H_2}^\circ + RT \ln P_{C_2H_2}),\end{aligned} \tag{14–37}$$

where $\Delta_f G_H^\circ = 165.485 \text{ kJ mol}^{-1}$ at 1,000 K. In order to compare the results with Case 4, we take $P_{C_2H_2} = 3.566 \times 10^{-6}$ bar and $P_{H_2} = 0.3333$ bar. If the partial pressure of hydrogen atoms is taken as their equilibrium pressure, the equilibrium composition is the same as for Case 4. However, if the partial pressure of hydrogen atoms is taken as 20% greater than the equilibrium value, the equilibrium amounts are given in Table 14–1. It is clear that the superabundance of hydrogen atoms increases the equilibrium amounts of higher polycyclic aromatic hydrocarbons. According to mechanism Eq. (14–32) the only way that two hydrogen atoms can combine to form a hydrogen molecule is via the polycyclic aromatic hydrocarbons. Thus, EQUCALC can be used to calculate what is usually considered to be a nonequilibrium composition.

Again this result can be obtained by using the geometric distribution given in Eq. (14–29). The value of b calculated from Eq. (14–37) is 6.102 kJ mol^{-1} and so

$$y_N = (0.520)(0.480)^{N-1}, \qquad (14\text{–}38)$$

which yields the amounts given in Table 14–1.

Case 6: Calculation of the Equilibrium Composition of the Benzene Series at T, P, P_{H_2}

This case is different from Case 3 in that acetylene is absent. The isomer groups in the benzene series are imagined to be in contact with a reservoir of molecular hydrogen through a semipermeable membrane. The system formula matrix is

$$\mathbf{A} = 6\ \ 10\ \ 14\ \ 18\ \ 22\ \ 26\ \ 30\ \ 34\ \ 38\ \ 42. \qquad (14\text{–}39)$$

The mechanism is unspecified but must have at least nine independent steps. In using EQUCALC, the Gibbs energies are adjusted by use of Eq. (14–20); that is, Eq. (14–22) is used. The initial amount vector is $\mathbf{n}_0 = 0.125\ 0\ 0\ 0\ 0\ 0\ 0\ 0\ 0\ 0$. The formation reaction for an isomer group from graphite and molecular hydrogen is given in Eq. (14–20), but it is more convenient in this case to use carbon number. The Gibbs energies of formation are adjusted by use of the formation reaction

$$n\text{C(graphite)} + (3/2 + n/4)\text{H}_2(P_{\text{H}_2}) = \text{C}_n\text{H}_{n/2+3}(P_{\text{hc}}) \quad n = 6, 10, 14, \ldots \qquad (14\text{–}40)$$

which yields

$$\Delta_f G_n^* = A + Bn + RT \ln P_{\text{hc}} - (3/2 + n/4)RT \ln P_{\text{H}_2} \qquad (14\text{–}41)$$

and

$$a = A + RT \ln P_{\text{hc}} - (3/2)RT \ln P_{\text{H}_2}, \qquad (14\text{–}42)$$

$$b = B + (1/4)RT \ln P_{\text{H}_2}. \qquad (14\text{–}43)$$

When the equilibrium partial pressure of molecular hydrogen is arbitrarily taken to be 0.998 bar and the equilibrium partial pressure of the hydrocarbons is taken arbitrarily to be 0.002 bar, the equilibrium amounts calculated using EQUCALC are given in Table 14–1. The equilibrium mole fractions within the hydrocarbons are 0.548, 0.248, 0.112, 0.051, 0.023, 0.010, 0.005, 0.002, 0.001, 0.000.

In this case, the equilibrium distribution is also given by a geometric distribution, this time involving A in Eq. (14–3). It can be shown that the equilibrium distribution (Alberty 1990b) is given by the geometric distribution, Eq. (14–29), but in this case q is the root of

$$q^{1.5}\exp(-a/RT) + q - 1 = 0. \qquad (14\text{–}44)$$

This polynomial has a single root between 0 and 1, and it is readily determined by use of a computer. Helzer's CENTSECT (1983) is very convenient to use. Thus, the distribution depends only on the value of a and A. At $P_{\text{hc}} = 0.002$ bar and $P_{\text{H}_2} = 0.998$ bar, $a = -4.8934\ \text{kJ mol}^{-1}$. This yields $q = 0.4522$ so that Eq. (14–29) becomes

$$y_N = 0.5478(0.4522)^{N-1}. \qquad (14\text{–}45)$$

This equation gives the same mole fractions within the hydrocarbons as were obtained with EQUCALC. In using EQUCALC to calculate the distribution, the sign of B in Eq. (14–3) can be changed without changing the solution. This is a rather dramatic demonstration of the fact that you do not need the Gibbs energies of formation of the successive isomer groups to solve this problem, just A in Eq. (14–3).

DEGREES OF FREEDOM FOR EQUILIBRIUM CALCULATIONS

We have been proceeding on the basis of a rather informal consideration of degrees of freedom, and so now we want to be very explicit about that. The number of degrees of freedom F is given by the phase rule

$$F = C - p + 2, \qquad (14\text{–}46)$$

where p is the number of phases. The numbers of degrees of freedom for the six cases and the quantities leading to it are summarized in Table 14–2. The columns N, R, C, and F summarize the calculations of the numbers of degrees of freedom F using Eq. (14–38), which agrees with the numbers of intensive variables selected in the five cases. However, this raises a question about the interpretation of rank $\mathbf{A}$, which is referred to as C' and is given in the next

Table 14–2 Numbers of Species, Independent Reactions, Components, and Degrees of Freedom for the Six Cases

Case	N	$R = \text{rank } \mathbf{v}$	$C = N - R$	F	$C' = \text{rank } \mathbf{A}$	F'
1. T, P, n_H/n_c	12	10	2	3	2	3
2. T, P, n_H/n_C, n_{ar}/n_C	12	9	3	4	3	4
3. T, P, P_{H_2}, n_{ar}/n_C	12	9	3	4	2	3
4. T, P, P_{H_2}, $P_{C_2H_2}$	12	9	3	4	1	2
5. T, P, P_{H_2}, $P_{C_2H_2}$, P_H	13	9	4	5	1	2
6. T, P, P_{H_2}	11	9	2	3	1	2

column. The form of the system matrix **A** is determined after the choice of specific variables to hold constant for the equilibrium calculation. Thus, $C = C'$ for the first two cases where the choice of independent variables does not affect the **A** matrix. In Case 3, the partial pressure of molecular hydrogen is specified and so the hydrogen row and the hydrogen column are removed from Eq. (14–15), and the Gibbs energies of formation of the remaining species are adjusted to the desired P_{H_2}. Thus, $C' = C - 1$. When this smaller number of components is used in the phase rule, three degrees of freedom F' are obtained. These are the degrees of freedom in addition to P_{H_2}, which was specified before writing the **A** matrix. In Case 4, $P_{C_2H_2}$ and P_{H_2} were specified prior to writing the **A** matrix, and the number of components C' under these conditions is 1; $C' = C - 2$. The corresponding number of degrees of freedom F' is 2, which is the number of degrees of freedom after setting $P_{C_2H_2}$ and P_{H_2}. In Case 5, three partial pressures were specified before writing the **A** matrix, and the number of components C' is 1; $C' = C - 3$. The corresponding number of degrees of freedom F' is 2. Case 6 is similar to Case 3.

CONCLUSION

Cases 2, 3, and 4 give three views of the same equilibrium of the benzene series of polycyclic aromatic hydrocarbons with acetylene and molecular hydrogen when mechanism Eq. (14–14) is followed. In Case 2, the equilibrium partial pressures of acetylene and molecular hydrogen are determined by T, P, and the initial composition of the system. In Case 3, the equilibrium partial pressure of molecular hydrogen is specified to be the same as in Case 2, but other values of P_{H_2} could be used. In Case 4, the equilibrium partial pressures of acetylene and molecular hydrogen are specified to be the same as in Case

2, but again other values could have been used. Case 4 is easy to study because the parameter for a geometric distribution is readily calculated. Case 5 involves a new mechanism, and the calculation of a nonequilibrium composition in the sense that the hydrogen atoms and hydrogen molecules do not have to be in equilibrium. Case 6 is different in that no C_2H_2 is present, but again the equilibrium composition can be calculated using the geometric distribution.

These cases have illustrated the profound effect that mechanism can have on equilibrium composition. They have also illustrated the usefulness of the geometric distribution in calculating equilibrium compositions when the standard Gibbs energy of formation is a linear function of carbon number.

In a flame there are other series of polycyclic aromatic hydrocarbons; there is, in principle, an infinite number of series of this type. When $P_{C_2H_2}$ and P_{H_2} are fixed, all of these isomer groups become pseudoisomers, but the analytic function used here only applies to a single series.

These calculations show that a subsystem of a complicated equilibrium system can be dissected out of the more complicated system by choosing independent degrees of freedom and making certain other changes. The equilibrium distribution of the species of interest can be considered to be a function of the partial pressures of the species from which they can be formed. This relationship can be investigated outside of the whole complex system by using a general equilibrium program or, in certain cases, by using analytic functions.

ACKNOWLEDGMENTS

This research was supported by a grant from Basic Energy Sciences of the Department of Energy (Grant DE-FG02-85ER13454). I am indebted to Irwin Oppenheim for helpful discussions.

NOMENCLATURE

A	system matrix (C × N)
A	Function of T only for a specified homologous series, kJ mol^{-1}
a	function of T, P, and specified partial pressures of reactants for a particular homologous series, kJ mol^{-1}
B	function of T only for a particular homologous series, kJ mol^{-1}
b	function of T, P, and specified partial pressures of reactants for a particular homologous series, kJ mol^{-1}
$C = N - R$	number of components in a system

C'	rank $\mathbf{A}$ = number of components in system after specifying the partial pressures of one or more reactants
F	number of degrees of freedom
F'	number of degrees of freedom after specifying the partial pressures of one or more reactants
G_{12}	transformed Gibbs energy at specified chemical potentials of acetylene and molecular hydrogen, kJ mol^{-1}
$\Delta_f G_i^\circ$	standard Gibbs energy of formation of species i, kJ mol^{-1}
$\Delta_f G_n^\circ$	$\Delta_f G_N^\circ$, standard Gibbs energy of formation of the isomer group with n carbon atoms or ordinal number N, kJ mol^{-1}
$\Delta_f G_n^*$	$\Delta_f G_N^*$, Gibbs energy of formation of the isomer group with n carbon atoms or ordinal number N at specified partial pressures of one or more reactants, kJ mol^{-1}
$\Delta_f G_{\mathrm{HSG}}^*$	Gibbs energy of formation of a whole homologous series at specified partial pressures of one or more reactants, kJ mol^{-1}
N	number of species (or isomer groups) in a system
N	ordinal number of an isomer group in a homologous series
N_{I}	number of isomers in an isomer group
n	amount, mol
q	parameter in the geometric distribution function
P	pressure, bar
R	gas constant, J K^{-1} mol^{-1}
R	number of independent reactions
r_i	equilibrium mole fraction of isomer i in its isomer group at equilibrium
T	thermodynamic temperature, K
y_N	equilibrium mole fraction of the isomer group with ordinal number N in a homologous series at a specified partial pressures of one or more reactants

Greek Letters

$\boldsymbol{\nu}$	stoichiometric number matrix ($N \times R$)
μ_n	μ_N, chemical potential of the isomer group with carbon number n or ordinal number N, kJ mol^{-1}
$\tilde{\mu}_N$	μ_{HSG}, chemical potential of an isomer group with ordinal number N with respect to the chemical potentials of acetylene and molecular hydrogen, kJ mol^{-1}
μ_N^*	$\Delta_f G_N^*$, chemical potential or Gibbs energy of formation of an isomer group at specified partial pressures of acetylene and molecular hydrogen, kJ mol^{-1}
μ_{HSG}^*	$\Delta_f G_{\mathrm{HSG}}^*$, Gibbs energy of formation of a homologous series group at specified partial pressures of acetylene and molecular hydrogen, kJ mol^{-1}

APPENDIX: USE OF A LEGENDRE TRANSFORM TO DERIVE A THERMODYNAMIC POTENTIAL G_{12} FOR THE BENZENE SERIES OF POLYCYCLIC AROMATIC HYDROCARBONS THAT IS AT A MINIMUM FOR SPECIFIED $P_{C_2H_2}$ AND P_{H_2}

Consider a system containing the whole benzene series of polycyclic aromatic hydrocarbons

$$(C_6H_6, C_{10}H_8, C_{14}H_{10}, \ldots)$$

at equilibrium with C_2H_2 and H_2 that are provided through a semipermeable membrane at $P_{C_2H_2}$ and P_{H_2}. The temperature and total pressure are also fixed. The possible chemical reactions in this system are

$$\phi_N + 2C_2H_2 = \phi_{N+1} + H_2, \tag{A–1}$$

where N is the ordinal number of an isomer group in the benzene series. The differential of the Gibbs energy at constant temperature and pressure is given by

$$\begin{aligned}(dG)_{T,P} &= \mu_{C_2H_2}\, dn_{C_2H_2} + \mu_{H_2}\, dn_{H_2} + \sum_{N=1}^{\infty} \mu_N\, dn_N \\ &= (2\mu_{C_2H_2} - \mu_{H_2})\, dn_e + \sum_{N=1}^{\infty} \mu_n\, dn_N,\end{aligned} \tag{A–2}$$

where $dn_e = (dn_{C_2H_2})/2 = -dn_{H_2}$ is the negative differential extent of polymerization. It is possible to write the second form of Eq. (A–2) in this way because of the stoichiometry of reaction (A–1).

At equilibrium,

$$\mu_n + 2\mu_{C_2H_2} = \mu_{N+1} + \mu_{H_2} \tag{A–3}$$

and at fixed T, P, $\mu_{C_2H_2}$, μ_{H_2}, the values of μ_N and μ_{N+1} differ by a constant. It is convenient to define the quantity $\tilde{\mu}_n$ by

$$\begin{aligned}\tilde{\mu}_N &\equiv \mu_N - N(2\mu_{C_2H_2} - \mu_{H_2}) \\ &= \mu_n^\circ - N(2\mu_{C_2H_2} - \mu_{H_2}) + RT \ln P_N \\ &\equiv \mu_N^* + RT \ln P_n,\end{aligned} \tag{A–4}$$

where P_N is the partial pressure of ϕ_N and μ_N^* is defined by this equation.

'he chemical potential $\tilde{\mu}_N$ is independent of N. For example, eliminating μ_N nd μ_{N+1} from Eq. (A–3) by use of Eq. (A–4) yields

$$\tilde{\mu}_n + N(2\mu_{C_2H_2} - \mu_{H_2}) + 2\mu_{C_2H_2} = \tilde{\mu}_{N+1} + (N+1)(2\mu_{C_2H_2} - \mu_{H_2}) + \mu_{H_2}$$

$$\tilde{\mu}_N + (N+1)2\mu_{C_2H_2} - N\mu_{H_2} = \tilde{\mu}_{N+1} + (N+1)2\mu_{C_2H_2} - N_{H_2},$$

$$\tilde{\mu}_N = \tilde{\mu}_{N+1}. \tag{A–5}$$

Equation (A–2) can now be rewritten

$$(dG)_{T,P} = (2\mu_{C_2H_2} - \mu_{H_2})\, d\tilde{n}_e + \sum_{N=1}^{\infty} \tilde{\mu}_N\, dn_N, \tag{A–6}$$

'here

$$d\tilde{n}_e = dn_e + \sum_{N=1}^{\infty} N\, dn_N. \tag{A–7}$$

The desired new thermodynamic potential G_{12} is obtained by making the ›llowing Legendre transform of the Gibbs energy

$$G_{12} \equiv G - \tilde{n}_e(2\mu_{C_2H_2} - \mu_{H_2}). \tag{A–8}$$

'hus,

$$(dG_{12})_{T,P} = \sum_{N=1}^{\infty} \tilde{\mu}_N\, dn_N - \tilde{n}_e(2d\mu_{C_2H_2} - d\mu_{H_2}). \tag{A–9}$$

ince $\mu_{C_2H_2}$ and μ_{H_2} are constant by virtue of the availability of C_2H_2 and I_2 through a semipermeable membrane,

$$(dG_{12})_{T,P,\mu_{C_2H_2},\mu_{H_2}} = \sum_{N=1}^{\infty} \tilde{\mu}_N\, dn_N. \tag{A–10}$$

Thus,

$$(dG_{12}/dn_N)_{T,P,\mu_{C_2H_2},\mu_{H_2},n_{i=N}} = \tilde{\mu}_N. \tag{A–11}$$

Since $\tilde{\mu}_N$ is independent of N, Eq. (A–10) can be written

$$(dG_{12})_{T,P,\mu_{C_2H_2},\mu_{H_2}} = \mu_{HSG}\, dn_{HSG}, \tag{A–12}$$

where HSG refers to the homologous series group (in this case the benzen series of polycyclic aromatic hydrocarbons), and

$$\mu_{\mathrm{HSG}} = \tilde{\mu}_N \tag{A–13}$$

and

$$dn_{\mathrm{HSG}} = \sum_{N=1}^{\infty} dn_N \tag{A–14}$$

Thus, for the system at specified T, P, $\mu_{C_2H_2}$, and μ_{H_2}, there is one pseudo species and one term in the fundamental equation for G_{12}. The standar chemical potential for the homologous series group is given by

$$\mu^*_{\mathrm{HSG}} = -RT \ln \sum_{N=1}^{\infty} \exp(-\mu^*_N/RT) \tag{A–15}$$

and the equilibrium mole fraction y_N for the Nth isomer group in th homologous series is given by

$$y_N = \exp[(\mu^*_{\mathrm{HSG}} - \mu^*_N)/RT]. \tag{A–16}$$

In Eqs. (A–15) and (A–16), μ^*_{HSG} can be replaced by $\Delta_f G^*_{\mathrm{HSG}}$ and μ^*_N can b replaced by $\Delta_f G^*_N$, as they are in the body of the chapter. Actually, a specified T, P, $P_{C_2H_2}$, P_{H_2}, all polycyclic aromatic hydrocarbons are pseudo isomers and so the series do not have to be treated separately.

REFERENCES

Alberty, R. A. (1983) Chemical thermodynamic properties of isomer groups. *I.E.C Fund.* **22**:318.

Alberty, R. A. (1985) Chemical equilibrium in complex organic systems. *J. Phys Chem.* **89**:880.

Alberty, R. A. (1987) Thermodynamics of the catalytic polymerization of alkenes i the gas phase. *Chem. Engin. Sci.* **42**:2325.

Alberty, R. A. (1988) The effect of the catalyst on the thermodynamic properties an partition functions of a group of isomers. *J. Chem. Educ.* **65**:409.

Alberty, R. A. (1989a) Thermodynamics of the formation of benzene series polycycli aromatic hydrocarbons in a benzene flame. *J. Phys. Chem.* **93**:3299.

Alberty, R. A. (1989b) Equilibrium distribution of alkanes at a fixed partial pressur of molecular hydrogen. (Note). *J. Chem. Phys.* **91**:7999.

Alberty, R. A. (1990a) Analytic expressions for equilibrium distributions of isomer groups in homologous series of alkanes, alkenes, and alkynes at a specified partial pressure of molecular hydrogen. *J. Chem. Phys.* **92**:5467.

Alberty, R. A. (1990b) Equilibrium distributions of isomer groups in homologous series of hydrocarbons at a specified partial pressure of molecular hydrogen. *J. Chem. Phys.* **93**:5979.

Alberty, R. A. and C. A. Gehrig (1984) Standard chemical thermodynamic properties of alkane isomer groups. *J. Phys. Chem. Ref. Data* **13**:1173.

Alberty, R. A. and I. Oppenheim (1986) Analytic expressions for the equilibrium distributions of isomer groups in homologous series. *J. Chem. Phys.* **84**:917.

Alberty, R. A. and I. Oppenheim (1988) Fundamental equation for systems in chemical equilibrium. *J. Chem. Phys.* **89**:3689.

Alberty, R. A. and I. Oppenheim (1989) Use of semigrand ensembles in chemical equilibrium calculations on complex, organic systems. *J. Chem. Phys.* **91**:1824.

Alberty, R. A. and A. K. Reif (1988) Standard chemical thermodynamic properties of polycyclic aromatic hydrocarbons I. Benzene series. *J. Phys. Chem. Ref. Data* **17**:241.

Bittner, J. D. and J. B. Howard (1981) Pre-particle chemistry in soot formation. In *Particulate Carbon Formation During Combustion* (Siegla, D. C. and G. W. Smith, Eds.), Plenum, New York.

Chase, M., M. W. Chase, C. A. Davies, J. R. Downey, D. J. Frurie, R. A. McDonald, and A. N. Syrerad (1985/1987) *JANAF Thermochemical Tables*, 3rd ed., *J. Phys. Chem. Ref. Data*, **14** (Suppl. 1 and 2), 1985; Addenda, Nov., 1987.

Dias, J. R. (1987) *Handbook of Polycyclic Hydrocarbons, Part A.* Elsevier, New York.

Frenklach, M., D. W. Clary, W. C. Gardiner, Jr., and S. E. Stein (1985) Detailed kinetic modeling of soot formation in shock-tube pyrrolysis of acetylene. In *Twentieth Symposium (International) on Combustion*, The Combustion Institute, Pittsburgh, p. 887.

Helzer, G. (1983) *Applied Linear Algebra with APL.* Little, Brown, Boston.

Krambeck, F. J. (1978) APL in an industrial environment. In 71st Annual Meeting of A.I.Ch.E., Miami Beach, FL, Nov. 16, 1978.

Quann, R. J., L. A. Green, S. A. Tabak, and F. J. Krambeck (1988) Chemistry of olefin oligomerization over ZSM-5 catalyst. *Ind. Engin. Chem. Res.* **27**:565.

Ramage, M. P., K. R. Graziani, and F. J. Krambeck (1980) Development of Mobil's kinetic reforming model. *J. Chem. Engin. Sci.*, **35**:41.

Ross, S. (1988) *A First Course in Probability*. Macmillan, New York.

Smith, B. D. (1959) Simplified calculation of chemical equilibrium in hydrocarbon systems containing isomers. *A.I.Ch.E.J.* **5**:26.

Smith, W. R. and R. W. Missen (1974) The effect of isomerization on chemical equilibrium. *Can. J. Chem. Engin.* **52**:280.

Smith, W. R. and R. W. Missen (1982) *Chemical Reactions Equilibrium Analysis.* John Wiley, New York.

Tabak, S. A., F. J. Krambeck, and W. E. Garwood (1986) Conversion of propylene and butylene over ZSM-5 catalyst. *A.I.Ch.E.J.* **32**:1526.

15

Chemical and Phase Equilibrium Calculations in Complex Mixtures

W. R. SMITH

The computation of the equilibrium composition of complex chemical mixtures has a long history, beginning in the 1940s during World War II and accelerating in the 1960s with the simultaneous advent of the space program and the development of digital computers. The arrival of microcomputers has enabled performing calculations for relatively large and complex systems with ease. Current interest in equilibrium calculations centers in two main directions. One of these is in new areas of application of chemical equilibrium computations. These include aspects of geochemistry, kinetics, and fluorescence spectroscopy. The other direction concerns fundamental aspects of the chemical equilibrium computations themselves.

In the second section of this chapter, we present the two main mathematical formulations of the chemical equilibrium problem, and discuss the relationships between them. The next section briefly discusses the structure of numerical algorithms for the solution of the problem. The fourth section discusses sensitivity analysis with respect to the problem parameters. The fifth and sixth sections discuss respectively the special situations of stoichiometric restrictions and equilibrium constraints. The seventh section discusses some currently available microcomputer algorithms for chemical equilibria, and the eighth section concludes with a discussion of some recent applications of chemical equilibria.

FORMULATION OF THE PROBLEM

The chemical equilibrium problem for complex mixtures (*complex* implies that the number of species may be up to the order of 100 and the number of elements the order of 10) can be formulated as a nonlinear optimization problem that may be expressed in one of the following two forms

P1:
$$\min_{\mathbf{n}\in\Omega} G(\mathbf{n}) = \mathbf{n}^T\boldsymbol{\mu} \tag{15–1}$$

where Ω is the constraint set defined by

$$\mathbf{An} = \mathbf{b} \tag{15–2}$$

$$n_i \geq 0 \tag{15–3}$$

and $\mathbf{n}\in E^N$, $\boldsymbol{\mu}\in E^N$, $b\in E^M$. N is the number of chemical species and M is the number of elements in the system. The rank of the *system formula matrix* $\mathbf{A}$ is denoted by C, where it is assumed that $C \leq N$. The columns of $\mathbf{A}$ are *formula vectors* $\mathbf{a}_i$, giving the chemical formula of each species.

P2:
$$\min G(\xi) = \mathbf{n}^T\boldsymbol{\mu}$$

where

$$\mathbf{n} = \mathbf{n}^0 + \mathbf{N}\xi \tag{15–4}$$

and

$$n_i \geq 0. \tag{15–5}$$

In this formulation, $\xi\in E^R$, $\mathbf{N}\in E^{NR}$, where R is the number of *stoichiometric equations.* n^0 is an initial system composition defined for $\xi = 0$. The columns of the *stoichiometric matrix* $\mathbf{N}$ are stoichiometric vectors $\mathbf{v}_j$, and we assume that it is of full rank R, where $R \leq N$. (We note in passing that the correct uppercase form of $\mathbf{v}$ is $\mathbf{N}$.)

P1 is called the *nonstoichiometric* formulation and P2 is called the *stoichiometric* formulation (Smith and Missen 1982). Formulation P1 is readily transformed to formulation P2. To achieve this, the linear constraints of Eq. (15–2) are effectively removed by computing a matrix $\mathbf{N}$ that appears in Eq. (15–4) of formulation P2 such that its columns form a basis for the null-space of the system formula matrix $\mathbf{A}$. The two matrices are related by

$$\mathbf{AN} = \mathbf{0}. \tag{15–6}$$

Equation (15–6) may be regarded as a shorthand way of writing sets of stoichiometric equations among the species of the system (Smith and Missen 1979). We prefer this term to the term *chemical reactions*, since Eq. (15–6) in general need have no mechanistic implications in a kinetic sense.

It should be emphasized that there need be no direct link between problems P1 and P2 for a chemical system comprised of a given set of chemical species. However, if one problem statement is given at the outset, a form of the other statement can be determined that is consistent with the original problem. By this, we mean that if **A** is given for problem P1, we can determine a stoichiometric matrix **N** for formulation P2 such that

$$\text{rank}(\mathbf{N}) + \text{rank}(\mathbf{A}) = N \tag{15–7}$$

and alternatively, if **N** is given for problem P2, we can compute a formula matrix **A*** for formulation P1 which may be used in place of **A** in Eq. (15–7). Although in the usual situation **A** may be used for **A*** in Eq. (15–7), we consider circumstances later when this is not the case.

Both problems P1 and P2 can be recast in other alternative forms. For example, the necessary conditions for the optimum in each case can be written as a set of nonlinear equations. In the special case when $C = 1$, a polynomial equation of degree N results (Norval et al. 1989). Also, the dual form of problem P1 is a geometric programming problem (Duffin et al. 1967). Finally, we remark that, although we assume that temperature T and pressure P are fixed [and hence G in Eq. (15–1) is the Gibbs free energy], the formulations remain unchanged in the other three cases of fixed (T, V), (S, P), and (T, U).

In order to solve either P1 or P2 for the equilibrium composition n, explicit expressions for the chemical potential μ must be given, which we assume are available in the form

$$\mu_i = \mu_i^* + RT \ln a_i, \tag{15–8}$$

where R is the gas constant, T the absolute temperature, and a_i the *activity* of species i. The simplest form of a_i is that for an *ideal solution*, for which the activity of each species is identical to its concentration variable in the phase in which it is present. For example, if Raoult's Law ideality holds, then a_i is the mole fraction, x_i.

Each individual chemical species in formulation P1 is characterized by its formula vector, $\mathbf{a}_i$, and its *standard chemical potential*, μ_i^*. It is important to emphasize that, by characterizing each species in this way, we are able to subsume all phase equilibrium problems. Typically, phase equilibria are customarily studied using the concept of fugacity rather than chemical potential, and the generalization of the former to reacting systems is awkward at best.

NUMERICAL ALGORITHMS

The details of numerical algorithms for solving the chemical equilibrium problem depend on the formulation used, but in general algorithms implement either a method for nonlinear optimization or for solution of sets of nonlinear equations. A review of general strategies for constructing equilibrium algorithms is given by Smith and Missen (1988). Although some researchers have attempted to apply general optimization or nonlinear equation solving computer codes calculating chemical equilibria (e.g., Ruda and Thompson 1985), this simplistic approach is fraught with difficulties due to the special structural features of the chemical equilibrium problem. These must be taken into account in any successful algorithm. The most important of these are numerical problems due to scaling and due to the behavior of μ_i as n_i approaches zero.

The scaling problem relates to the fact that the values of the individual equilibrium mole numbers often differ from each other by many orders of magnitude. This can cause computer under- and overflows unless special precautions are taken. As n_i approaches zero, the chemical potential approaches negative infinity, which can also cause computer overflows. Finally, as the amount of a phase approaches zero, the mole fraction of each species in it behaves like 0/0.

THE PROBLEM PARAMETERS

The systematic study of the dependence of the solution to a problem on its underlying parameters is called *sensitivity analysis*. The parameters of problem formulation P1 are $T, P, \boldsymbol{\mu}^*, \mathbf{A}, \mathbf{b}$. For formulation P2 they are $T, P, \boldsymbol{\mu}^*, \mathbf{N}, \mathbf{n}^0$.

Provided the dependence of the problem solution on its underlying parameters is sufficiently smooth (as is generally the case here), the first-order sensitivity matrix with entries $\partial n_i/\partial p_j$, where p_j is a problem parameter, is useful in studying this dependence. The calculation of these coefficients is straightforward, and was first discussed by Smith (1969). As one example of their use, they enable equilibria at conditions other than fixed T and P to be computed efficiently. Although T and P are the usual thermodynamic variables that are fixed in a chemical equilibrium computation, any two variables may be chosen from the set $\{P, V, T, S, U, H, A\}$ (in fact, any two functionally independent combinations may be chosen from this set). The sensitivity coefficients $\partial n_i/\partial T$ and $\partial n_i/\partial P$ for a (T, P) calculation may be employed to solve the problem of interest. For example, to compute the equilibrium composition at fixed (H, P), one may augment formulations P1 or P2 with the additional equation

$$\sum_{i=1}^{N} n_i(T)\bar{h}_i[T, \mathbf{n}(T)] = H^0, \tag{15–9}$$

where $\bar{h}_i$ is the partial molar enthalpy of species i and H^0 is the given fixed value of the enthalpy, H. In this way, an equilibrium problem using variables other than (T, P) can be solved if one has available an efficient algorithm for solving the (T, P) problem. Equation (15–9) would be solved as a single nonlinear equation in T, using values of $\mathbf{n}$ obtained from a subprogram that computes equilibrium at fixed (T, P).

Another use of the sensitivity coefficients is to assess the precision of the computed equilibrium solution, given estimates of the precision of the underlying parameters. For example, if the thermodynamic data $\mu_j{}^*$ have estimated precision $\delta\mu_j^2$, then the precision of the solution is given approximately by

$$\delta n_i^2 = \sum_{j=1}^{N} \left(\frac{\partial n_i}{\partial \mu_j^*}\right)^2 \delta\mu_j^2. \tag{15–10}$$

CHEMICAL STOICHIOMETRY AND STOICHIOMETRIC RESTRICTIONS

In general, chemical stoichiometry may be defined as the constraints placed on the composition of a closed chemical system by the necessity of conserving the total mass of each component of the system. These constraints may be expressed by means of the system formula matrix $\mathbf{A}$ and element abundance vector $\mathbf{b}$ in the form of element-abundance equations via Eq. (15–2), or by means of the stoichiometric matrix $\mathbf{N}$ and an initial composition $\mathbf{n}^0$ via Eq. (15–4).

The term *stoichiometric restrictions* refers to whether or not Eq. (15–7) holds for the system. If it holds, then we say that no stoichiometric restrictions are present. Normally, we are given $\mathbf{A}$ in problem formulation P1, and we then may determine $\mathbf{N}$ satisfying Eq. (15–7). This may be used in the solution of the problem according to formulation P2, if desired. However, if $\mathbf{N}$ is specified at the outset in formulation P2 and Eq. (15–7) does not hold for the given system formula matrix, then we have

$$\text{rank}(\mathbf{N}) < N - \text{rank}(\mathbf{A}) \tag{15–11}$$

nd we say that stoichiometric restrictions are present in the system, the umber of which is given by

$$r = N - \text{rank}(\mathbf{A}) - \text{rank}(\mathbf{N}). \tag{15–12}$$

uch restrictions may arise from a postulated kinetic mechanism (e.g., lberty 1989; Missen and Smith 1990).

Chemical equilibrium algorithms based on formulation P1 essentially only andle the case of unrestricted systems. However, stoichiometricly restricted roblems posed via formulation P2 can be readily handled in the context of uch algorithms by computing from **N** a pseudo-formula matrix **A*** that atisfies

$$\text{rank}(\mathbf{N}) = N - \text{rank}(\mathbf{A}^*). \tag{15–13}$$

ormulation P1 may then be used in conjunction with this pseudo-formula iatrix (Cheluget et al. 1987).

QUILIBRIUM CONSTRAINTS

Vhen equilibrium constraints are present, the chemical potentials of one or iore species are fixed at equilibrium at preassigned values. Examples of such roblems include aqueous solution equilibria in which the pH is specified in dvance, and gaseous equilibria in which the partial pressure of one or more onstituents is specified. Such problems are discussed more fully by Norval t al. (1990a), but we give a brief description here.

We consider the general case of constrained equilibrium to be formula- ons P1 or P2 augmented by the additional equations

$$\mathbf{f}(\boldsymbol{\mu}) = 0, \tag{15–14}$$

here $\mathbf{f}: E^N \rightarrow E^L$, where L is the number of constraints. The simplest case is hen **f** is linear, the general case of which is

$$\sum_{i=1}^{N} D_{ji}\mu_i = \mathrm{d}_j, \quad j = 1, 2, \ldots, L \tag{15–15}$$

here D and d are given constants. Such problems may be solved by erforming an equilibrium computation in the usual way without constraints, ut using modified data parameters. One modification is to the set of andard chemical potential data $\boldsymbol{\mu}^*$, and the other is to the element-abundance equations (see Norval et al. 1990b for details). The solution of the equilibrium problem with this modified data and no constraints yields the solution to the original problem.

One effect of the second modification is to reduce the number of unknown Lagrange multipliers in formulation P1 by the number of constraints. This suggests that it is possible to underspecify the overall system composition. Thus, instead of the M values of elemental abundances normally required for the vector **b**, only M-L need be specified. The remaining abundances are determined by the equilibrium calculation itself. For example, Norval et al. (1990) consider a problem involving the removal of SO_2 from the stack gas resulting from the regeneration of spent H_2SO_4 in air. Due to pollution considerations, it is desired to restrict the partial pressure of SO_2 in the stack gas to 10^{-5} atm (this represents the constraint). The problem is to determine the amount of H_2SO_4 to be added to a given feedstock to achieve this (this represents the undetermined system composition). By treating the appropriate constrained equilibrium problem, the solution is readily obtained.

COMPUTER ALGORITHMS

A number of microcomputer-based software packages have recently become available for computing chemical equilibria. These include Chemreact (Infochem Computer Services 1990), EQCAL (Biosoft 1990), EQS (Technical Database Services 1990a), and UMRPC/Solgasmix (Morris 1990). The EQS package was developed by the present author, and a brief description is given here.

EQS runs on MS-DOS machines ranging upwards in power from a PC-level machine. It handles up to 99 species and 10 elements, although this is not an intrinsic limitation. Calculations may be performed at constant (T, P) or constant (T, V). In the liquid phase, either Raoult's or Henry's Law ideality may be assumed. The equilibrium composition of all species, even those present in minute amounts, are accurately computed. (The scaling problem discussed in the third section is explicitly addressed.)

EQS is menu driven and all input data are entered by means of input screens, which are easily edited. Context-sensitive help is available for all input fields. Data may be saved to disk for future use, and problem output may be directed to any combination of screen, printer, or disk.

Thermodynamic data may be entered either in the form of species-specific data (formulation P1), or via sets of stoichiometric equations with their associated free energy data (formulation P2). Problems involving stoichiometric restrictions are treated directly by EQS, as is the case for problems with equilibrium constraints. Finally, EQS is linked to a thermodynamic database of approximately 1,000 compounds, based on data from a number of sources, including DIPPR (Technical Database Services, 1990b) and JANAF (American Chemical Society 1986).

As an indication of the speed of EQS, calculations for a 22-species three-element system take about 20 seconds on a standard 8088-based PC without a numeric coprocessor. A coprocessor increases the speed by about factor of 6.

NEW APPLICATIONS

Chemical equilibrium calculations are important in many areas, and new applications are continually evolving. Applications to geochemical problems have been recently discussed by Cheluget et al. (1987).

A recent application is to the elucidation of kinetic mechanisms (Norval et al. 1989). The most recent development in this area is a technique that identifies, from a specified experimental system composition, subsets of species in equilibrium with each other (Norval et al. 1990b). The technique systematically generates and tests all possible sets of equilibria involving reactions among $C + 1$ or fewer species.

Another recent application is to the analysis of complex protein and ligand systems (Royer et al. 1990). Using experimental spectroscopic data, consistent sets of free energy data may be computed for the underlying species. Nonlinear regression analysis is used to adjust iteratively the species free energy parameters to fit the observed data. The underlying equilibrium compositions required on each iteration are rapidly computed using a variant of the EQS algorithm (Technical Database Services 1990a).

REFERENCES

Alberty, R. A. (1989) Thermodynamics of the formation of benzene series polycyclic aromatic hydrocarbons in a benzene flame. *J. Phys. Chem.* **93**:3299–3304.

American Chemical Society (1986) *JANAF Thermochemical Tables*, 3rd Edit.

Biosoft (1990) 22 Hills Road, Cambridge CB2 1JP, United Kingdom, suppliers of "EQCAL" software.

Cheluget, E. L., R. W. Missen, and W. R. Smith (1987) Computer calculation of ionic equilibria using species- or reaction-related thermodynamic data. *J. Phys. Chem.* **91**:2428–2432.

Duffin, R. J., E. L. Peterson, and C. Zener (1967) *Geometric Programming.* John Wiley, New York.

Infochem Computer Services Ltd. (1990) South Bank Technopark, 90 London Road, London SE1 6LN, United Kingdom, suppliers of "Chemreact" software.

Missen, R. W. and W. R. Smith (1990) The permanganate–peroxide reaction: illustration of a stoichiometric restriction. *J. Chem. Educ.* **67**:876–877.

Morris, A. E. (1990) Department of Metallurgical Engineering, University of Missouri at Rolla, Rolla, MO. 65401, UMR PC/Solgasmix.

Norval, G. W., M. J. Phillips, R. W. Missen, and W. R. Smith (1989) Application of equilibrium sensitivity analysis to aromatization processes. *Ind. Engin. Chem Res.* **28**:1884–1887.

Norval, G. W., M. J. Phillips, R. W. Missen, and W. R. Smith (1989) On chemical equilibrium of pseudo-one-element systems: application to oligomerization. *Can J. Chem. Engin.* **67**:652–657.

Norval, G. W., M. J. Phillips, R. W. Missen, and W. R. Smith (1990a) Identification of partial equilibria. Paper presented at the 73rd CIC-CSChE Conference, Halifax Nova Scotia, July 15–20.

Norval, G. W., M. J. Phillips, R. W. Missen, and W. R. Smith (1990b) On chemical equilibrium in constrained systems. *Can. J. Chem. Engin.*

Royer, C. A., W. R. Smith, and J. M. Beecham (1990) Analysis of binding in macromolecular complexes: a generalized numerical approach. Paper presented at the International Biophysical Society Conference, Vancouver, British Columbia July 28–31.

Ruda, M. M., and D. W. Thompson (1985) Comparison of primal and dual solution methods for the chemical equilibrium problem using GRG and sensitivity analysis. *Can. J. Chem. Engin.* **63**:113–121.

Smith, W. R. (1969) The effects of changes in problem parameters on chemical equilibrium calculations. *Can. J. Chem. Engin.* **47**:95–97.

Smith, W. R., and R. W. Missen (1979) What is chemical stoichiometry? *Chem. Engin Educ.* Winter, pp. 26–32.

Smith, W. R. and R. W. Missen (1982) *Chemical Reaction Equilibrium Analysis Theory and Algorithms*, p. 118. Wiley-Interscience, New York.

Smith, W. R. and R. W. Missen (1988) Strategies for solving the chemical equilibrium problem and an efficient microcomputer-based algorithm. *Can. J. Chem. Engin* **66**:591–598.

Technical Database Systems (1990a) 10 Columbus Circle, New York, New York 10019, suppliers of "EQS" software.

Technical Database Systems (1990b) 10 Columbus Circle, New York, New York 10019, suppliers of DIPPR Database Software.

Appendix
Panel Discussion

Fred Krambeck moderated the panel discussion and the following commentary is transcribed from this free-flowing exchange of ideas, and captures the essence of the meeting.

Krambeck: In discussions following a previous paper, we used the expression "more is less" to describe how considerations of reactions at a more detailed, fundamental level actually can greatly *reduce* the number of parameters needed to describe a complex system. Sandler's paper shows an entirely different way in which the same expression applies. In going from a discrete distribution to a continuous one he is able to reduce the number of components from 40 to 2.

In my opinion, we do have to come to grips with the computational problems of very detailed modeling approaches. For example, what do you do with these detailed kinetics you have developed for a process when you apply them to a three-dimensional simulation of a reactor? It seems likely that some kind of approximation is necessary. The orthogonal collocation approach may provide the necessary approximation. Some of these discrete variables such as carbon number can be treated as continuous to apply analytical techniques to reduce the complexity. This is something we all can think about.

Sandler: Some other types of bases are also possible. For example, lumping which is based on free energy minimization is another approach. A variety of work is currently under way.

Quadrature methods give you a basis to choose components that you then use in an existing code, such as a reservoir simulation model. One uses the quadrature method only once to identify the components, and after that everything is done with the components identified in this manner.

In phase equilibrium studies the use of quadrature approach is obvious. In these calculations, the mathematics operator for the phase splitting is simple, unlike that for catalytic reactions problem.

Froment: I disagree with Fred's suggestion to wait till we get to the reactor design, then probably you'll have to simplify our fundamental single event approach and we may have to get into a continuous reaction mixture approach. When I go back to my experience in thermal cracking, which is very similar to catalytic cracking, the main problem is the kinetic model. Insertion of kinetic models into momentum, mass, and energy balances and solving the resulting reactor models is trivial.

We do the detailed modeling of industrial thermal crackers with the most complicated kinetics. The reactor model includes cracking coils, and we also model the fire box, generating temperature distribution and heat flux in a multidimensional space which affects the rates of cracking reactions. Such detailed modeling with the complex reaction schemes is not a problem at all. It is just a problem of computation time. There is no reason to sacrifice the accuracy of the detailed kinetic model which you have derived with such great pain for the reactor design. No lumping is required as far as I can see.

Krambeck: Let me give you an example where it may be necessary to lump. We have been looking at three-dimensional simulations of a trickle-bed reactor to investigate the impact of flow maldistribution. And in these simulations, we follow gas and liquid distribution through the reactor. We find that, to get accurate answers, we need 10,000 nodes in the reactor. Using a Convex mini-supercomputer or the Cray computer (actually Cray is a problem due to its limited memory), we find that in going past five or six components computations become prohibitive. If we talk about thousands of species with thousands of nodes for this problem, there is no computer that can solve it today.

Froment: Why do you insist on 10,000 nodes for the reactor but only five components for the reactions? That sounds unreasonable.

Krambeck: One implies the other. In this particular case you are interested in the flow maldistribution effects, and you are trying to focus mainly on the fluid flow. The reaction comes into the picture only because it affects vaporization which changes the fluid mechanics, which feed back to the reactor performance. And that is the only reason you need reactions. You need that many nodes to get reasonable accuracy on the fluid flow problem. Only a limited number of reactant species can be included, only because that is all we can fit on the current generation computers.

Bischoff: It clearly shows the lumping is problem specific. If the goal is to look at the effects of flow maldistribution in a reactor, then lumping is necessary. Reactions are affecting only the heat of reaction and vaporization for this reactor problem. Whereas when the primary interest is reactions and not reactors, then detailed kinetic modeling is appropriate.

Jacob: When you are looking at some other situations, for example control of a chemical reactor, don't you have to simplify and lump your detailed kinetic model? How do you tackle this problem for ethylene cracking furnaces?

Froment: We are now talking about an entirely different problem. The main focus of our effort is steady-state simulation of ethylene cracking furnaces. For control problems, we will look only at a subset of key reactions, but this is not the case for steady-state simulations of these systems.

Jaffe: The flow maldistribution problem is a steady-state problem though.

Froment: When you are looking at flow maldistribution, which is a reactor specification issue, I agree that reaction lumping may be necessary. I would accept some sacrifice in the composition specifications then.

Luss: We have here two kinds of audiences with two different missions. We have industrial practitioners and academic researchers, and our objectives are distinctly different. Problems associated with detailed industrial reactor design and optimization are out of the academic research interest. The academic goal is to advance the basic knowledge and science of reaction kinetics, to generate new knowledge, and to teach new concepts. The information we learned in this Workshop about detailed kinetic models which incorporate a lot of fundamental chemistry is very valuable. We need to teach our students how to develop these detailed models and when and how to simplify them to make these problems more manageable. The objective is to define when these simplifications are meaningful from the kinetics perspective and not to use reactor design consideration to simplify the kinetics.

Bischoff: It seems to me that until recently except for one or two cases, most of the work on the continuous lumping has really been concerned with parallel reactions. J. Baily tried it years ago for series reactions. He did not integrate all the way, he looked at only the finite integration, and even then it was very complex. I would like to ask the advocates of continuous lumping, What is the future for this approach in series parallel reactions?

Astarita: My feeling is that our current research is focusing on these issues. I think once we have a continuous description, it is very easy to chop it into parts when we need it. One way to deal with this issue may be to use a combination of discrete and continuous approaches.

Allen: We can describe each lump as a continuum. I am familiar with the aerosol literature where this approach is used. The method of moments could be used to characterize the continuum in each lump. Dimitris Liguras briefly looked at this approach in his Ph.D. thesis. Modeling how the moments would change as a function of the extent of reaction is a very difficult problem though.

Luss: It is not necessary to use the continuum approach. We should develop general concepts on how to treat different reaction networks, and how they may be simplified in specific cases. More research is needed about how to simplify complex reaction networks and the impact of the simplifications on the accuracy of the model.

Krambeck: I suggest we should consider adding stoichiometry and thermodynamics to the continuum approximation.

Froment: I have some reservations on how one could account for interaction between species and bimolecular reactions and so on with the continuum approach. I don't see much future in it.

Klein: From our discussions so far it is clear that the level of detail in a given model is necessarily closely tied to the level of detail we desire in the solution. However, let me raise an issue here: Do we need two distinct models to represent two different levels of complexity? In the reactor model we needed only five reaction components for reaction, as the primary focus was fluid mechanics. David Allen's problem is catalyst design and chemistry, and he is concerned about the product selectivity description of which needs roughly 200 components. Do we need to develop two different models to solve these two different problems? From what we read in Sandler's chapter, it seems that his techniques may be used to simplify the complex model only when we need such simplifications. It is easy to create less information from more and it is harder to develop more information from the lumped models. Therefore, I am a strong proponent of the detailed models.

Franklach: Lumping is not just one particular approach. There are many ways of lumping, or what is called in combustion modeling, reduction of a reaction model. In the field of combustion modeling, the need for detail reaction models has been appreciated long ago. Now, when a great deal of such models have been developed, it becomes more and more apparent that these models need to be reduced substantially in size to be useful for simulations of realistic reactive flows. Several reduction methods have been proposed recently. What is important, in my opinion, is to recognize that model reduction or lumping constitutes an integral part of model development. As there can be no all-purpose model, there can be no all-purpose lumping. One should choose a lumping method, and hence a model, that suits the needs of a given application. Thus, there can exist several models for the same process or phenomenon. Obviously, detail models offer a "foundation" on which other, lumped models can be developed. At the same time, there are situations when an exactly lumped model can be developed for a highly complex, nonlinear system without the precise knowledge of the individual parts. One example of such a case is what I call chemical lumping, when a replicating-type reaction network of an infinite size can be exactly lumped into a small number of differential equations for the moments of the distribution function based on similarity in chemical structure. In this case, the lumped model and its rate parameters are obtained from "the first principles."

Krambeck: There is one additional feature with these detailed models though. For example, it provides a means of correlating molecular properties in terms of their component substructures. Now we can describe the kinetics in terms of these molecular substructures equivalent to the use of Benson's group contribution approach widely used in thermodynamics. Mathematical description of molecular substructure to represent kinetics appears to be one key area for research.

Allen: The carbon center approach definitely shows promise in this regard.

Klein: From the thermodynamics talks we see that the Gibbs free energy is a function of carbon number. To the extent you have a reaction family, and the rate of reaction depends on enthalpy of transition state, you can relate kinetics to molecular structure. As you saw in my chapter, enthalpy of transition state does correlate with the molecular structure. Therefore, the rate constants are a function of structures and the underlying chemistry. But there is more than one reaction family. You can come up with these detailed molecular features and then you can avoid the traditional lumping.

Wei: I am not sufficiently motivated by the continuum approach. In reaction engineering, generally, we get into a new approach only when there is a problem with the old one. "What is it that we are trying to fix by the continuum approach?"

Krambeck: From the chapters and discussions so far, we can see two reasons: The continuum approach is an approximation of a large number of discrete lumps, such as Sandler's approach. This is just a mathematical simplification of the underlying complexity. The continuum approach also represents a new learning model, to try and understand some general principle, to get an insight into this area.

Astarita: I agree, the continuum approach can be viewed as a mathematical simplification. When the components become too numerous, integrals are a lot easier to handle than thousands of species with multitude of reactions.

Wei: To summarize then, the proponents of the continuum approach agree with Fred in that there are advantages. First, it is a first simulation of the true discrete model. Second, you may learn something valuable. I still don't see the merits of the second point. I say, God bless you. Come back in a few years with additional results with the continuum approach, and share them with us.

Jaffe: We were attracted to the continuum approach because historically data was always represented as continuous functions, for example, boiling point distributions for hydrocarbon streams, which is a continuous function. Only recently with the breakthroughs in the analytical laboratory is continuity now broken down into discrete chemical compounds that make up that mixture. Today, therefore, it makes sense to treat our problems as discrete species and not as continuous functions.

Froment: I am not sure we can take Sandler's approach from thermodynamics and apply it to problems in kinetics. Because, essentially Sandler has used the methodology for describing parallel events. In kinetics we have series events, parallel events, interactive events, and so on. I don't believe we can be that optimistic for extending it to kinetics.

Astarita: Conceptually, I don't see the difference between thermodynamics and kinetics. Let's consider the stage separation device in Sandler's case. When he wants to move to the multistage problem, it is obvious that he has to change his pseudocomponents, at each stage. As the separation proceeds the distribution of light and heavy components shifts. The reason he may not want to change pseudocomponents is perhaps that he wants to construct something that can be fed directly to an existing computer code that doesn't allow components to change their nature from one stage to another. In kinetics, we start with a certain known composition and as reaction proceeds with time, the more reactive species disappear faster and as a result the composition distribution shifts, as well as the reaction rate distribution. Therefore, in principle, even the kinetics could be treated in the same fashion. In my opinion, multistage separation is then similar to consecutive kinetics.

Luss: A major advantage of the continuous approach is that it often enables us to develop a model which is not sensitive to the exact form of the distribution. Thus, we can carry out design without having complete kinetic information. Many industrial kinetic models use the boiling point as the continuous index. However, rate constants need not be a function of the boiling point alone. A better approach may be to carry out a separation of the feed by a chromatographic column that simulates the adsorption characeristics of the catalyst under reacting conditions. The resulting distribution will be better related to that based on the rate constants.

Sapre: For hydrodesulfurization, we know that kinetic rates are not just a function of boiling point alone. Certain specific components such as dibenzothiophenes and substituted dibenzothiophenes react very slowly compared to the rest. You can see that these peaks remain in the GC trace as all other sulfur peaks are reacting away. A detailed discrete modeling approach in this case would definitely pay off.

Allen: What should we call this new type of molecular approach? I would appreciate some suggestions from the panel.

Froment: As a member representing the new elegant kinetic schemes, I propose “Functional Group Kinetics” (FGK).

Krambeck: The key element in the new approach is correlating kinetic parameters as a function of substructures or functional groups. Therefore, it sounds like a reasonable name.

Allen: For this method to be successful, though, you need more than just the groups used in thermodynamics group contribution methods. It needs to include heuristics rules which are based on the overall structure of the molecule, and not just the functional groups alone. Certain functional groups may react differently on a particular molecular structure; therefore it is more of a structural modeling approach.

Jaffe: Let me introduce the Mobil nomenclature. We call it “Structure-Oriented Lumping” (SOL).

Krambeck: Dave Allen raised a technical issue which we need to appreciate. In order to calculate the reaction rates, the molecule is more than just the sum of its parts. Maybe what it means is that we haven't yet determined the right structural subunits that describe the molecule. While putting together a molecule somehow, either by Monte Carlo methods or other methods, we need to consider what are the properties that we really need to describe its kinetics. We are describing these pseudocomponents by a series of numbers that carry all the kinetic information. For example, how many β bonds are in molecules such that we can properly represent the carbenium ion chemistry and corresponding reaction rates? A few essential features of the molecule are perhaps more important than the accurate description of a molecule.

Sandler: There is a whole literature on graph theory, structural indices, and molecular connectivity to characterize molecules that is used in thermodynamics, environmental engineering, and to estimate pharmacological and toxicological effects that perhaps kineticists could use.

Klein: This may be simplifying it too much. Because, not only the reactivity of that functional group depends on the rest of the molecule in which it is sitting, but also the environment of the mixture itself. You are describing what I'll call the intrinsic chemistry. We also have to consider interactions, for example, interactions of carbenium ions, free radicals, competitive adsorption, and so on. All these effects make the actual chemistry and kinetics different from its intrinsic behavior.

Krambeck: It may be true. But, I think we should be able to extract essential features of molecules and its surroundings that go into its description and kinetics calculations, and there may be an extra baggage of molecular description that we may not have to worry about. The Benson Group approach and other graph theoretic approaches used in thermodynamics may be just what we need.

Jacob: Could we discuss some issues that Jim Kuo raised in his talk on use of thermodynamics and our needs in the industry? For example, symmetry numbers, group contribution methods, and others—"What sort of work is going on in the universities? Could you give us some feedback?"

Frenklach: Benson's group approach will soon be replaced by quantum mechanical calculations. In my opinion, these quantum mechanical calculations are much easier to use than Benson's rule. The future is in quantum chemistry.

Sandler: You are right about the future, but I believe it is a bit premature to use quantum chemistry methods routinely in the context of petroleum processing. We spent a large amount of computer time to calculate energies and atomic charges for about 90 different molecules by ab initio molecular orbital calculations. Our goal was to develop some generalizations of energies and properties. At present we are limited to about 10 heavy atoms. It may be several years before we can do the things we are talking about here in a reasonably rigorous and accurate manner.

Franklach: I thought the most recent handbooks include examples that give results on polycyclics, starting with benzene and include up to 6-ring aromatics. Perhaps you don't want ab initio methods but some semiempirical methods that may be manageable computationally.

Alberty: When you talk about quantum mechanical calculations you are considering only energy. For chemical equilibrium, entropy is equally important and it doesn't come out of quantum mechanical calculations. You also need statistical mechanics.

Klein: Lots of this theoretical chemistry, which is not so exotic any more, was built upon the problems of what Mobil and other oil companies have long been interested in, for example, chemistry of aromatics. For such calculations, the semiempirical methods are just fine. Semiempirical methods are some correlations of data bases and provide the key information for thermodynamics. However, these methods may not be sufficient for kinetics calculations.

Krambeck: We are looking at systems with millions of isomers. And then ab initio or even some semiempirical approximate quantum calculations are impractical.

Frenklach: For homologous series though, we may not have to compute for each isomer. We can develop simple correlations based on some detailed calculations of low carbon number molecules and then simply extrapolate. Most of these systems asymptote to linear dependences quickly as molecular size increases.

Alberty: If you look at olefin oligomerization, for example, in the ZSM-5 zeolite catalyst, you are not interested in the whole isomer group, but you need only a smaller subgroup. For example, you need just the straight chains to reflect the specificity of the catalyst. So, you are not just increasing the carbon number, but looking at molecule by molecule to determine the thermodynamic properties only of the linear ones that are relevant to this specific problems.

Wei: I have heard some good ideas in the last two days. The focus of this panel discussion—**Where do we go from here?**—asks us to look into the next generation of models for the refining and petrochemicals industry. We talked about a lot of different things so far, but it seems to me we need to focus on some deficiencies of the current generation of models and fix them in the future. For example, we need to add product properties such as viscosity of hydrocarbon streams from first principles. Product qualities is something process engineers are most interested in. Therefore, we need to develop relationships between molecular structure and physical properties. Modern analytical tools have been advancing in this area. We need to incorporate these results in our kinetic models. Perhaps in that sense we have to delump our last generation models. And therefore there was a suggestion before that we should coin a new name for these models. Just a moment ago I checked with Dwight Prater on the origin of the word "lumping" in Mobil and he had an interesting story on this subject.

Prater: I am a physicist by training and I know giving big names to things is colorful, for example, quantum electron dynamics. Big names rarely describe the true meaning to a guy on the street. Lumping, on the other hand, is very precise. I proposed its use to describe our models in the mid-1950's. If you look in the dictionary it is very descriptive. It means putting things together. It describes so well what you are doing that you have a difficult time to get rid of it. Your objection to it seems to be it is not a colorful jargon word. But believe me, a guy on the street understands it perfectly. Therefore, we should not change it. Whatever you are doing is still lumping. In Mobil, management wanted to call it something else when we first used it. I refused to change it then and I object to changing it now.

Wei: To continue with what I was saying before, including product qualities in our models is something I would like to see. The other question relates to the fact that there are frequent changes in catalysts, operating conditions, and different feedstocks. There was not much discussion about how these lumps or pseudocomponents would change with changing catalysts or significantly different operating conditions such as pressure and temperature. They may lead to different mechanisms and kinetic regimes, and so the lumps may look different. How would these variables affect the detailed modeling approach? This is a generic issue and transcends all model types. I was astonished to learn that even the Cray is not much help once the number of components increases for a good reactor model. I do believe that we will have to do some simplifications, although the level of simplifications today is quite different from that of the early days of my involvement in petroleum refining. In the mid-1950's in Mobil, to use a two-dimensional slide rule for the Thiele Modulus versus effectiveness factor was a step forward relative to a regular slide rule. In those days, Dwight Prater introduced the use of exponential integrals for kinetic analysis and it was a major breakthrough. We had to do a lot of things in those days to make things simple. There was a small pile of problems that could be dealt with using linear algebra and exact integrals, and there were a vast majority of problems which were untouched by rigorous methods. We had a set of rules, good ones, to solve these difficult problems. A lot of systematic analysis went into it. I don't know if we have done a systematic study of the spectrum of models discussed here to determine what makes sense and try to simplify, if possible. We need to take up this challenge, and pass on only the critical information to the next generation.

Right now, I am doing what I consider to be a fascinating study for the MIT provost. We are looking at the greenhouse effect and global warming. I learned several new concepts in this Workshop and I assure you some of the modeling issues discussed here will shape our thinking to solve these problems. That makes this Workshop quite valuable. I want to apply all I have learned here. We have a basic problem—explaining the greenhouse effect to people in political science and economics is very difficult. Explaining delumping to them is next to impossible. The first thing you explain to them is that the greenhouse effect is related to CO_2 emissions. And then the most straightforward things to do in terms of kinetics is to tell them we have 5 billion tons per year of carbon emissions and it is increasing at a rate of 2 per cent per annum. This is our one-lump model of the greenhouse effect. This is a vastly simplified model. It doesn't give you any details. It doesn't tell you anything about the dynamics. The next model is where you delump into number of countries—160 countries in the world, and that gives you 160 lumps. The problem is that most countries don't even have emissions data. The next question is, How do you delump further to assign values to different sectors such as transportation, commerce, industry, agriculture, and so on? Every country has different energy resources and utlization. Extent of use of petroleum, nuclear energy, etc., is also a function of national policy. The biggest driving force in this problem is population growth and GNP per capita. So you can see that this is really a complex kinetic model. We will spend the next few years just defining some key variables. A few years down the road, I'll share the results of our study with you. I assure you, one way or another it will affect all of you.

rambeck: The detailed models discussed here provide the natural framework to integrate product quality information and also the catalyst effects. Looking at these complex mixtures, the message is to keep the components separate as long as possible and don't lump it if you don't have to. The only way to unscramble an egg is not to scramble it in the first place. You really can't delump something once you have lumped it.

The next issue is "What could we do in industry? How should we proceed in academia to move this area forward?"

starita: One thing we clearly lack in academia is data. A lot of information that you have in your files could be of great help to us in academia to develop theories, reasonable assumptions to validate our models. For example, data on olefin oligomerization would be of great help for my own work. But perhaps what we should discuss is a mechanism of developing an effective dialogue between industry and academia.

llen: I was intrigued by the second order plot you (Chapter by F. J. Krambeck) showed for FCC pilot plant data. And you mentioned that, if data doesn't follow this trend it must be wrong.

I presume it reflects your collective wisdom of many years (working in this area. We on the other hand are looking at basi chemistry and trying to develop models to represent industri processes. What is lacking is a mechanism whereby you ir dustrial folks could transfer the essence of your knowledge t academia.

Froment: We have to make the industrial engineers talk more abo their work.

Bischoff: To follow up on Jim Wei's suggestion on including properti in the kinetic models. I think, here again, we need lots of dat from industry. We in academia very rarely look at produ properties, except perhaps some work done by thermodyn micists. To make serious progress then you have to give u your data bases.

Luss: As we are coming to the end of the Workshop, on behalf of all (us, I would like to thank Mobil Research and Developme Corporation for its hospitality. The last two days were ver refreshing, we learned several new concepts, and realize that lot of work still needs to be done.

Index

THE NATIONAL ACADEMIES

Advisers to the Nation on Science, Engineering, and Medicine

The **National Academy of Sciences** is a private, nonprofit, self-perpetuating society of distinguished scholars engaged in scientific and engineering research, dedicated to the furtherance of science and technology and to their use for the general welfare. Upon the authority of the charter granted to it by the Congress in 1863, the Academy has a mandate that requires it to advise the federal government on scientific and technical matters. Dr. Ralph J. Cicerone is president of the National Academy of Sciences.

The **National Academy of Engineering** was established in 1964, under the charter of the National Academy of Sciences, as a parallel organization of outstanding engineers. It is autonomous in its administration and in the selection of its members, sharing with the National Academy of Sciences the responsibility for advising the federal government. The National Academy of Engineering also sponsors engineering programs aimed at meeting national needs, encourages education and research, and recognizes the superior achievements of engineers. Dr. Charles M. Vest is president of the National Academy of Engineering.

The **Institute of Medicine** was established in 1970 by the National Academy of Sciences to secure the services of eminent members of appropriate professions in the examination of policy matters pertaining to the health of the public. The Institute acts under the responsibility given to the National Academy of Sciences by its congressional charter to be an adviser to the federal government and, upon its own initiative, to identify issues of medical care, research, and education. Dr. Harvey V. Fineberg is president of the Institute of Medicine.

The **National Research Council** was organized by the National Academy of Sciences in 1916 to associate the broad community of science and technology with the Academy's purposes of furthering knowledge and advising the federal government. Functioning in accordance with general policies determined by the Academy, the Council has become the principal operating agency of both the National Academy of Sciences and the National Academy of Engineering in providing services to the government, the public, and the scientific and engineering communities. The Council is administered jointly by both Academies and the Institute of Medicine. Dr. Ralph J. Cicerone and Dr. Charles M. Vest are chair and vice chair, respectively, of the National Research Council.

www.national-academies.org